Springer Series on

Wave Phenomena 15

Edited by L.M. Brekhovskikh

Springer Series on
Wave Phenomena

Editors: L.M. Brekhovskikh L.B. Felsen H.A. Haus
Managing Editor: H.K.V. Lotsch

Volume 1 **Mechanics of Continua and Wave Dynamics**
By L. Brekhovskikh, V. Goncharov

Volume 2 **Rayleigh-Wave Theory and Application**
Editors: E.A. Ash, E.G.S. Paige

Volume 3 **Electromagnetic Surface Excitations**
Editors: R.F. Wallis, G.I. Stegeman

Volume 4 **Asymptotic Methods in Short-Wave Diffraction Theory**
By V.M. Babič, V.S. Buldyrev

Volume 5 **Acoustics of Layered Media I** Plane and Quasi-Plane Waves
By L.M. Brekhovskikh, O.A. Godin

Volume 6 **Geometrical Optics of Inhomogeneous Media**
By Yu.A. Kravtsov, Yu.I. Orlov

Volume 7 **Recent Developments in Surface Acoustic Waves**
Editors: D.F. Parker, G.A. Maugin

Volume 8 **Fundamentals of Ocean Acoustics** 2nd Edition
By L.M. Brekhovskikh, Yu. Lysanov

Volume 9 **Nonlinear Optics in Solids**
Editor: O. Keller

Volume 10 **Acoustics of Layered Media II** Point Source and Bounded Beams
By L.M. Brekhovskikh, O.A. Godin

Volume 11 **Resonance Acoustic Spectroscopy**
By N.D. Veksler

Volume 12 **Scalar Wave Theory** Green's Functions and Applications
By J. DeSanto

Volume 13 **Modern Problems in Radar Target Imaging**
Editors: W.-M. Boerner, F. Molinet, H. Überall

Volume 14 **Random Media and Boundaries**
By K. Furutsu

Volume 15 **Caustics, Catastrophes and Wave Fields**
By Yu.A. Kravtsov, Yu.I. Orlov

Yu.A. Kravtsov   Yu.I. Orlov

Caustics, Catastrophes and Wave Fields

With 59 Figures

Springer-Verlag
Berlin Heidelberg New York London Paris
Tokyo Hong Kong Barcelona Budapest

Yu.A. Kravtsov

General Physics Institute,
Division GROT, Russian Academy of Sciences
Vavilov Street 38, 117942 Moscow, Russia

Yu.I. Orlov

Formerly at the
Power Engineering Institute, Moscow

Translated by

Dr. M.G. Edelev

Butyrskaya 95-31, 125015 Moscow, Russia

ISBN 3–540–56587–6 Springer-Verlag Berlin Heidelberg New York
ISBN 0–387–56587–6 Springer-Verlag New York Berlin Heidelberg

Typeset by Macmillan India Ltd, Bangalore-25

54/3140/SPS – 5 4 3 2 1 0 – Printed on acid-free paper

Preface

This monograph is envisaged as a natural continuation of our previous work (Vol. 6 of the Springer Series on Wave Phenomena) devoted to the fundamentals of geometrical optics. Our intention is to present, in a concise form that is convenient for practical use, the key generalizations of geometrical optics related to caustics and caustic fields, such as the complex method of geometrical optics, and the local and uniform asymptotic descriptions of caustic fields.

Available monographs on the asymptotic methods of wave theory (for example those by L.B. Felsen and N. Marcuvitz, V.P. Maslov, V.P. Maslov and M.V. Fedoryuk, L.M. Brekhovskikh, and V.M. Babich and V.S. Buldyrev) do not meet all the demands of applications in optics, acoustics and radar techniques. This book aims to fill that gap. From a modern stand point, regarding caustics as catastrophes, we describe the principal asymptotic methods used to compute caustic fields and illustrate this discussion with specific application-oriented problems.

Our general aim is to demonstrate the potential of a method that solves wave problems with the aid of rays that form a geometric framework spanning the diffraction field. This principle has proved to be advantageous for problems concerning the geometrical theory of diffraction and for the description of caustic fields. These objectives determined the presentation of material and govern the structure of the book. The introductory chapter (Chap. 1) focuses on two interrelated facets of the caustic problem: geometric and wave aspects. The general geometric description of rays, caustics and associated fields is given in Chap. 2.

Chapter 3 presents a geometric classification of typical caustics based on catastrophe theory. The geometric principles underlying the theory have been useful vehicles in constructing the standard integrals that describe the fields near typical caustics. The properties of such integrals are outlined in Chap. 4.

In addition to local expansions, standard integrals may be involved in the construction of uniform asymptotic field expansions in the presence of caustics. This last problem has been extensively studied by the authors of this book and forms the subject matter of Chap.5

A rather general approach to the caustic problem, as suggested by V.P. Maslov, is delineated in Chap. 6. We present a simplified version of Maslov's method, oriented toward physical applications. Another treatment, developed by Yu.I. Orlov and called the interference integral technique, is outlined in Chap. 7.

Chapter 8 is devoted to the theory of broken caustics associated with edge catastrophes. This type of caustic occurs in problems with jump-wise variation of parameters.

Chapter 9 touches upon modifications of the method of standard integrals, specifically as applied to structurally unstable caustics, contour integrals, and combinations of functions satisfying ordinary differential equations.

Finally, Chap. 10 summarizes available evidence on particular types of caustic, such as caustics in anisotropic media, space-time caustics, random caustics, complex caustics, and caustics with anomalous phase shifts.

Some results given in this book have never been published before. These relate above all to the general proof of the fact (established earlier for particular cases only) that the standard integral equations in amplitude and phase variables reduce, in the final analysis, to equations of geometrical optics. A second innovation is the concept of a longitudinal caustic scale, which specifies a distance along the caustic within which the diffraction of the wave field must be taken into consideration. Finally, we have formulated the improved sufficient applicability conditions for uniform and local descriptions of wave fields in the presence of caustics.

Unfortunately, Yu.I. Orlov was not able to participate in the preparation of this manuscript. His death in 1982 deprived the scientific community of a talented and ingenious researcher. The book would not exist in its present form without the aid of N.S. Orlova, who passed to me relevant work by her late husband, some of which served as a basis for Chaps. 3–5, 7 and 8. I am very grateful to Mrs. Orlova for her generous assistance.

I am greatly indebted also to A.A. Asatryan, E.D. Gazazyan, and K.S. Tropkin for their critical reading of separate chapters of the manuscript. I thank M.P. Firsova for her help in the preparation of the text and B.Ya. Zeldovich and S.S. Moiseyev for their critical remarks and suggestions made in reviewing the manuscript. I acknowledge the help of Dr. M.I. Charnotskii and Prof. A.I. Saichev in completing the list of references. The assistance of B.S. Agrovsky, A.S. Gurvich and S.M. Flatté, who kindly provided us with photographs of caustics in turbulent media, is greatly appreciated. Finally, my special thanks go to M.G. Edelev for his literary advice, painstaking reading and translation of the manuscript into English.

Moscow, November 1993 *Yu.A. Kravtsov*

Contents

1 Introduction

Until recently caustics have been treated predominantly on an elementary level like geometrical objects – the envelopes of families of rays. However, physical measurements treat caustics as wave objects, namely, diffuse regions with enhanced amplitude of the wave field.

In this chapter we first discuss the geometric and wave aspects of the caustic problem separately to pass over to the synthetic description later.

1.1 Caustic Fields in Physical Problems

Consistent interest in caustics traced throughout all stages of the history of physics beginning from the ancient times stems from the fact that caustics are where marked concentration of the field occurs. Indeed, the light intensity on a caustic can be sufficient to burn a paper, for example, thus explaining the origin of the name. Thc caustic concentration of a field is clearcut at the focal plane of a simple lens where a caustic surface degenerates into a point. Under certain conditions a rather high concentration of the field can be observed on simple (non-singular) caustics, say in a cup of coffee illuminated by slant sunlight.

Whereas for light waves or waves on a smooth water surface caustics can be observed with a naked eye, for most other wave fields, say acoustic, electromagnetic, and seismic, caustics can be recorded only with physical devices.

Description of wave fields in the presence of caustics is essential for many physical disciplines among which instrumental optics (lenses, prisms, mirrors, etc.) is far from the prime consumer. The problem of caustic fields is topical in radio engineering, optics and acoustics of natural media. Focusing and caustic phenomena have to be taken into account, for example, in considering the propagation of radio waves through solar plasma and terrestrial atmosphere. By way of example, Fig. 1.1 shows a family of rays in the ionosphere representing radio waves emitted by a point terrestrial source. This figure, borrowed from [1.1], indicates both the regions of shadow and the regions of focusing of the electromagnetic (EM) field distinctive of an increased density of rays.

A rather complicated focusing pattern occurs when sound waves propagate in a deep ocean under the conditions of an underwater sound duct. An example of such a ray pattern, borrowed from [1.2], is presented in Fig. 1.2.

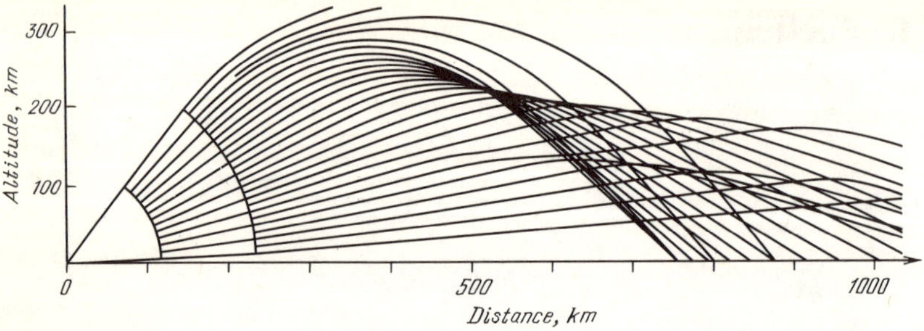

Fig. 1.1. Ray pattern for a terrestrial point source of radio waves emitted in the ionosphere [1.1]

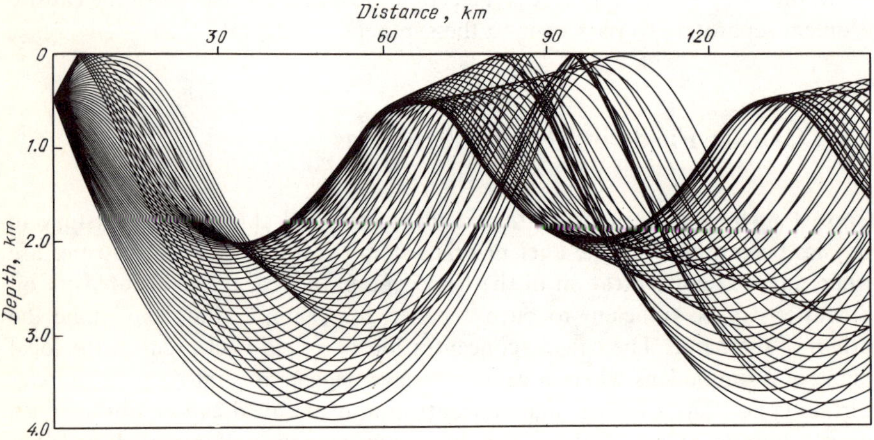

Fig. 1.2. Example of a ray pattern in an ocean acoustic channel [1.2]

Light propagating in a turbulent atmosphere and radio waves passing through a random inhomogeneous plasma also produce rather intricate caustic patterns. Figure 1.3a shows a distribution of intensity over the cross section of a laser beam passing through a cell with turbulent liquid. This figure, vividly illustrates the caustic spots due to random focusing [1.3]. Computer simulations, Fig. 1.3b, of such focusings have been demonstrated by *Martin* and *Flatte* [1.4].

Evaluation of caustic fields is of prime significance in the theory of scattering of light and particles (suffice it to mention the rainbow phenomenon in optics and nuclear physics), in optical instrumentation, in microwave and radiowave antenna engineering, in the problems of eigenmodes and waves in cavities and waveguides for optical range and microwaves, for astronomy (theory of gravitation lenses), for nonlinear optics, and for problems of wave hydrodynamics.

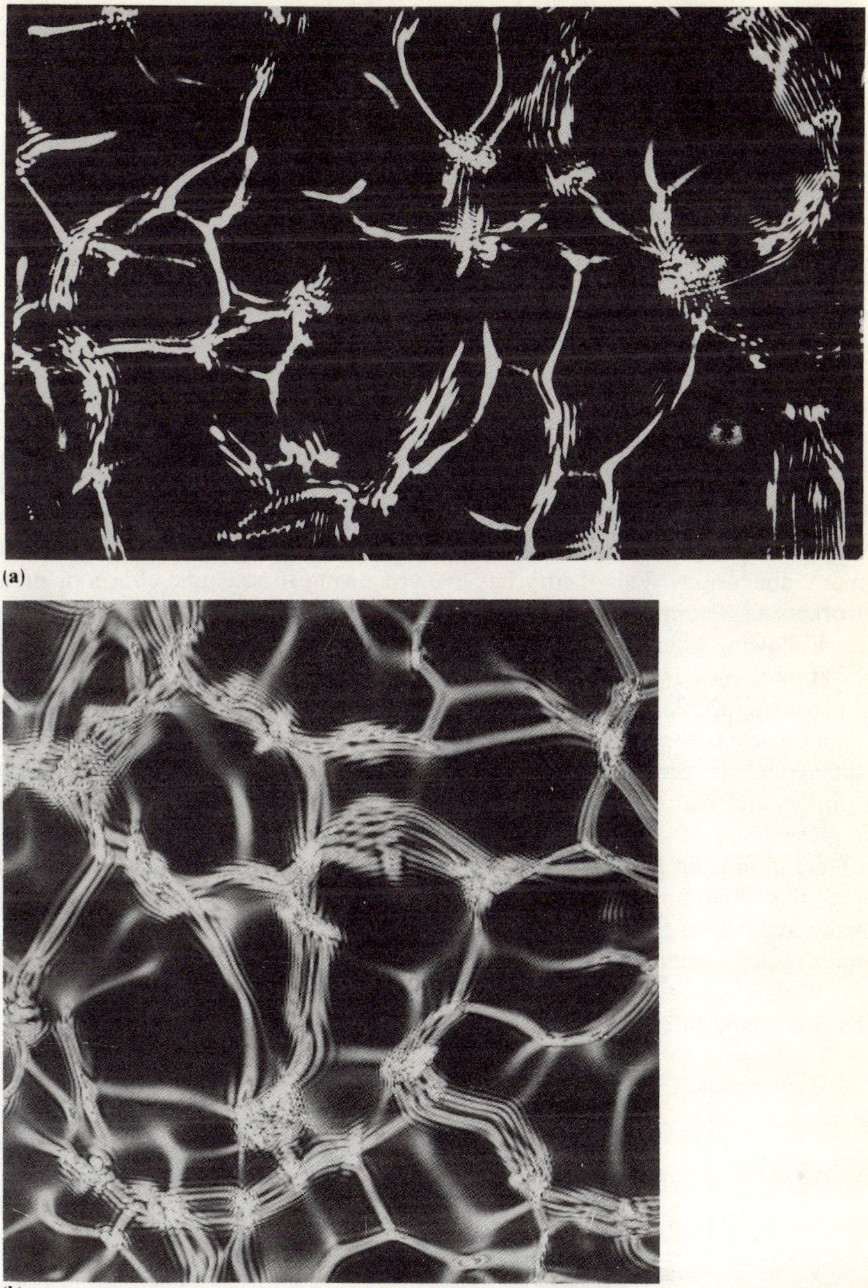

Fig. 1.3. (a) Background intensity distribution over a cross section of a light beam passed through a cell with a turbulent liquid [1.3]. Light spots corrupted by interference represent random caustics and foci formed in the light beam [1.3]. (b) Results of a computer simulation performed by *Martin* and *Flatté* [1.4]

A common feature of all these theories and applications has to do with a concentration of the field on caustics and the formation of complex interference patterns in regions of multipath propagation. Analysis of the typical features of caustic fields will be the major objective of this treatise.

Caustic fields may be evaluated from two aspects, geometric and field, each of which has been considerably advanced in recent years. In fact this upsurge of interest in the field has stimulated the appearance of this book.

1.2 The Geometrical Aspect of the Caustic Problem

An important step in the geometrical approach to the problem has been the treatment of caustics as singularities of differentiable mappings induced by families of rays. The theory of such singularities is a new branch of mathematics that has established the general properties of mappings of different manifolds onto spaces of different dimensionality. The theory of differentiable mappings is essentially a grand generalization of the ordinary extrema analysis on the case of many dimensions. This theory has evolved owing to scientific efforts of such workers as *Whitney* [1.5], *Arnold* [1.6], and *Thom* [1.7].

Following René Thom the theory of singularities in mappings has come to be known as catastrophe theory. The grounds for this term may be as follows: Intersecting the locus of singularities corresponding to a given system brings about a "catastrophe", i.e., a qualitative and jumpwise variation of the state of the system. In geometrical optics, the jump occurs in moving across caustic surfaces and manifests itself as a change in the number of rays coming into a given point of space. For a simple caustic shown in Fig. 1.4, the point of observation is hit by two rays – this is the lit region – or does not receive any rays, this is the caustic shadow. The field varies accordingly – it decays exponentially away from the caustic into the shadow region, while exhibiting intense interference oscillations in the lit region.

With reference to Fig. 1.5 representing a caustic cusp, every point within the cusp is hit by three rays, whereas outside of the cusp by one ray only. In the

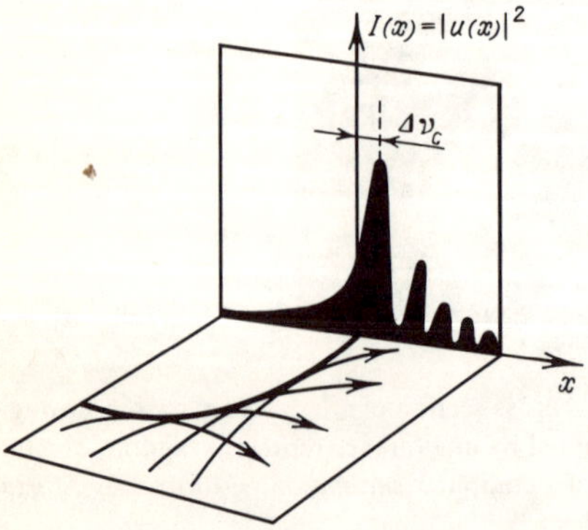

Fig. 1.4. Distribution of intensity in the vicinity of a nonsingular caustic

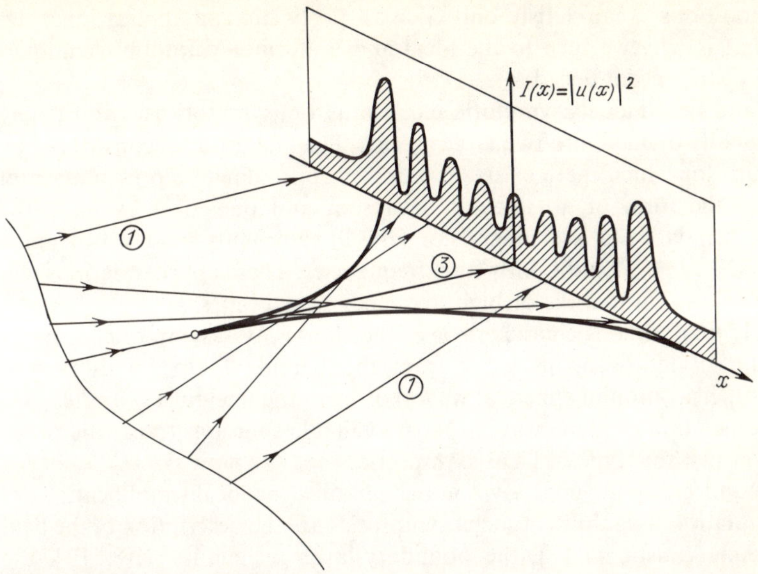

Fig. 1.5. Intensity distribution in the range of three-ray interference (encircled figures represent the number of rays in a particular area)

three-ray region the field exhibits an intricate interference structure which disappears in the one-ray region.

Despite the notorious disadvantages of the term "catastrophe theory" associated with its unjustifiably wide interpretation (e.g., see [1.8, 9]) we shall use it in this text, treating catastrophe theory as another name of the theory of singularities of differentiable mappings.

An important role of the theory of singularities for the topics under discussion was that it provided a basis for categorization of a wide class of typical or, what is the same, structurally stable singularities. This subject has been reflected in several monographs [1.10–13], and in the exposition of caustics we shall follow primarily the developments of the books [1.10, 13], research papers [1.9, 14] and our own investigations [1.15, 16] which addressed the caustic problems.

1.3 The Wave Aspect of the Caustic Problem

The field treatment of the caustic problem will be fundamental to this book. Therefore at the vary outset we note that the discussion will hinge predominantly around the approximate, asymptotic descriptions of the field in the presence of caustics. Available exact solutions – valid as a rule under rather specific conditions – can satisfy only a small proportion of applicational demands. That is why we turn to the local and uniform asymptotic techniques developed in the recent decades.

Many and varied local asymptotic expansions, or asymptotics for short, may be conditionally divided into two large groups. The first group is constituted by the problems for which there exists an exact or approximate representation of the field in the form of an integral or a sum and one must evaluate the asymptotic behavior of these integrals or sums at sufficiently small wavelengths or, what is the same, at sufficiently high frequencies. These approximations will be referrred to as short-wave or high frequency asymptotics.

Precisely this origin is characteristic of the short-wave asymptotic formulas obtained for a homogeneous space from the Kirchhoff type of diffraction integrals (superposition of spherical waves) or from the integral Fourier expansions (superposition of plane waves). Most available solutions to caustic problems rely on just this type of local asymptotics.

The second group includes asymptotics obtained by local simplifications of the wave equation. Examples of such asymptotics are the description of the field near a simple caustic [1.17], the boundary-layer technique [1.18, 19], the evaluation of the fields concentrating near reference rays [1.20], and some other techniques listed in [1.20, 21]. These approaches are significant in that they are applicable both in homogeneous and inhomogeneous media.

A still higher applicational potential relates to the asymptotic techniques that allow field description to be made not only near the caustic but also at a significant separation from it. These are more sophisticated formalisms than the local asymptotic treatments, yet they offer new potentialities for the researcher. The point here is not only in higher distance allowances, but also in the capability of establishing the applicability limits of the local methods and of evaluating the relation between the global and local asymptotics.

By the time of this writing, several uniform asymptotic techniques have been developed. One of them has been initially developed by *Kravtsov* [1.22] for simple caustics. It is based upon the Airy function, which is simplest among the standard functions. *Ludwig* [1.23] has extended this method with the aid of generalized Airy functions to the cuspoid type of caustics, followed by extensions to arbitrary types of caustic [1.24, 15]. A suitable solution for arbitrary caustics evolves from the typical diffraction integrals of catastrophe theory.

Use of special functions adapted to describe caustic fields led to an alternative name for the method of uniform asymptotics, namely, the method of standard functions or integrals. A version of this method, under the name of uniform theory of diffraction, has been invoked to solve problems of the geometrical theory of diffraction [1.25].

Another approach, suggested by *Maslov* [1.26] in 1965, assumes the construction of a quasi-classical approximation in the momentum-coordinate representation, i.e., in the form of a Fourier integral in a part of the variables. Such an asymptotic is local in nature, but having matched different asymptotic representations in the domain where they overlap results in a uniform asymptotic of the field. This radical approach, known as Maslov's method of canonical operator, has rapidly become popular in various fields of physics. Simplified treatments of the method may be found in a number of publications [1.27–30].

A third method to tackle the problem, set forth by *Orlov* [1.31] in 1972, is based on representing the field in the form of an integral superposition of the ray fields or caustic fields in a more complex case. In the literature this approach is known as Orlov's method of interference integral, the method of virtual rays (suggested by *Vainstein* [1.32]), the method of oscillatory integrals [1.29], and the spectral approach [1.33]. A discussion on the potentialities of the method of interference integral may be found in a recent review [1.34].

There are other promising uniformly asymptotic techniques, such as the summation of Gaussian beams [1.35] which is well amenable to numerical computations. This by no means implies that the other mentioned methods are hard to implement on a computer. We shall touch upon this topic later in more detail and with examples of successful agreement of the uniform asymptotic techniques with the methods of numerical analysis.

In this text we would like to reflect the advantages of a new direction in wave theory, which is underlaid by the principle of "spanning a wave-field cloth on a ray framework". All the aforementioned uniform asymptotic techniques provide examples of application of this principle. In conjunction with catastrophe theory they open up new possibilities for effective description of fields in inhomogeneous media. This topic has been subjected to analysis in a few excellent review papers [1.36–40] which, however, almost exclusively address the theoretical aspects of the caustic problem. We wish to place more weight onto applications by means of numerous examples for caustic fields taken from various branches of physics [1.41, 42]. In doing so we keep the approach of our previous volume [1.43] which served as a point of departure for the present treatise.

We hope that the systematic exposition of the theory of caustic fields will facilitate evaluation of ever new relationships between wave fields and the spanning ray systems and, in this way, the development of new efficient techniques of field computation so much needed by applied physics.

2 Rays and Caustics

In this chapter we will briefly recapitulate the fundamentals of the ray optical method which requires an approximate description of the wave field on the basis of ray representations. Assigning a mathematical ray with the Fresnel volume converts it into a physical object. The concept of Fresnel's volume yields a versatile criterion of applicability of the ray theory to be derived. It defines caustic zones as regions where rays cease to be physically distinctive. In the shadow region, ordinary geometrical optics gives way to its complex counterpart.

2.1 Equations of Geometrical Optics

2.1.1 The Scalar Problem

Let the scalar monochromatic field with time factor $\exp(-\mathrm{i}\omega t)$ be described by the Helmholtz equation

$$\Delta u + k^2 n^2 u = 0 \ , \tag{2.1.1}$$

where $n = n(r)$ is the refractive index of the inhomogeneous medium. In the geometrical optics approximation the wave field is commonly represented as an asymptotic series in negative powers of wavenumber $k = \omega/c$

$$u(\mathbf{r}) = \left(U_0(\mathbf{r}) + \frac{U_1(\mathbf{r})}{\mathrm{i}k} + \frac{U_2(\mathbf{r})}{(\mathrm{i}k)^2} + \dots \right) \mathrm{e}^{\mathrm{i}k\psi(\mathbf{r})} \ . \tag{2.1.2}$$

Substituting (2.1.2) into (2.1.1) and equating the coefficients of equal powers of k yields the equations for ψ, U_0, U_1, etc. These are the equation of eikonal

$$(\nabla\psi)^2 = n^2(\mathbf{r}) \ , \tag{2.1.3}$$

and the transfer equations in the amplitudes U_m

$$\begin{aligned} &2\nabla U_0 \cdot \nabla\psi + U_0 \Delta\psi = 0 \ , \\ &2\nabla U_1 \cdot \nabla\psi + U_1 \Delta\psi = -\Delta U_0 \ , \\ &\dots \end{aligned} \tag{2.1.4}$$

This derivation of equations of geometrical optics was devised by Debye in 1911 and is delineated in depth in the classical work of *Born* and *Wolf* [2.1] and in our previous text [2.2]. The same set of equations given by (2.1.3, 4) may be obtained with the aid of Rytov's procedure [2.3] which unlike (2.1.2) relies on the expansion in a small dimensionless parameter $\mu = 1/kL$, where L is the characteristic scale of the problem (see also [2.2]).

The solution of (2.1.3, 4) may be written with the aid of rays that occur as the characteristics of the eikonal equation (2.1.3) or, what is the same, as the bicharacteristics of the Helmholtz equation (2.1.1). It is convenient to represent the equations of rays in the canonical Hamiltonian form

$$\frac{\mathrm{d}\mathbf{r}}{\mathrm{d}\tau} = p, \qquad \frac{\mathrm{d}\mathbf{p}}{\mathrm{d}\tau} = \frac{1}{2}\nabla n^2 , \tag{2.1.5}$$

where $\mathbf{p} = \nabla\psi$ is the "momentum" of the ray, which indicates the direction of propagation $\mathbf{l} = \mathbf{p}/p$, and τ is a parameter along the ray coupled with its length σ by the relation $\mathrm{d}\tau = \mathrm{d}\sigma/n$.

Under typical conditions, the initial field u^0 is specified on a certain surface Q endowed with curvilinear ray coordinates ξ and η as illustrated in Fig. 2.1 as follows:

$$u^0 = u^0(\xi, \eta) = U_0^0(\xi, \eta)\exp[ik\psi^0(\xi, \eta)] . \tag{2.1.6}$$

Having given an equation for this surface $\mathbf{r} = \mathbf{r}^0(\xi, \eta)$ we obtain for the ray trajectory $\mathbf{r} = \mathbf{R}(\xi, \eta, \tau)$ an initial condition

$$\mathbf{R}(\xi, \eta, 0) = \mathbf{r}^0(\xi, \eta) . \tag{2.1.7}$$

For two components of the initial vector $\mathbf{p}^0$ lying in Q we can obtain equations by differentiating the initial eikonal $\psi^0(\xi, \eta)$ with respect to ξ and η,

$$\mathbf{p}^0 \frac{\partial\mathbf{r}^0}{\partial\xi} = \frac{\partial\psi^0}{\partial\xi}, \qquad \mathbf{p}^0 \frac{\partial\mathbf{r}^0}{\partial\eta} = \frac{\partial\psi^0}{\partial\eta} . \tag{2.1.8}$$

The third equation for the component p_n normal to Q will be the eikonal equation (2.1.3).

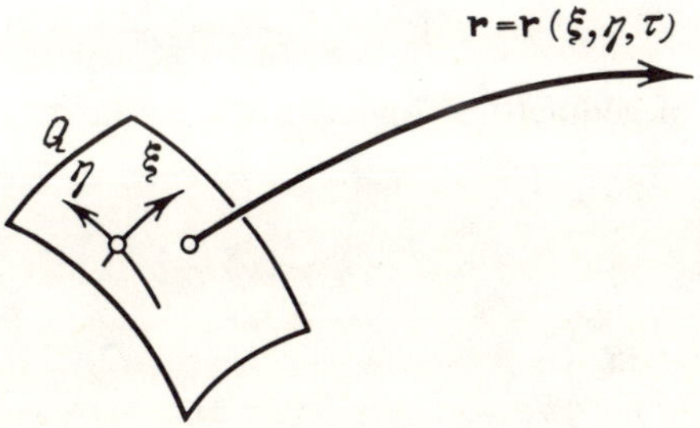

Fig. 2.1. Ray path bearing ray coordinates

Solution of the ray equations (2.1.5) subject to the initial conditions (2.1.7, 8) defines in the configuration space (x, y, z) a family (congruence) of rays

$$\mathbf{r} = \mathbf{R}(\xi, \eta, \tau) \; , \tag{2.1.9}$$

where the parameters ξ and η label the rays leaving Q, while the parameter τ (or σ) indicates the position of a point on a fixed ray. In the six-dimensional phase space $\{\mathbf{r}, \mathbf{p}\} = \{x, y, z; p_x, p_y, p_z\}$, the solution (2.1.9) along with the solution

$$\mathbf{p} = \mathbf{P}(\xi, \eta, \tau) \tag{2.1.10}$$

defines a three-dimensional surface called a Lagrangian manifold.

Given a family of rays (2.1.9), the eikonal ψ may be represented as an integral along the ray leading to the given point of space,

$$\psi = \psi^0(\xi, \eta) + \int_0^\tau n^2[\mathbf{R}(\xi, \eta, \tau)]\,\mathrm{d}\tau \; , \tag{2.1.11}$$

and the zero-approximation amplitude U_0 may be represented in terms of the generalized divergence of rays $\mathscr{J}$, viz.,

$$U_0 = U_0^0(\xi, \eta)/\mathscr{J}^{1/2} \; . \tag{2.1.12a}$$

Here,

$$\mathscr{J} = \frac{n\,\mathrm{d}a}{n^0\mathrm{d}a^0} = \frac{D(\tau)}{D(0)} \; , \qquad D(\tau) = \frac{\partial(x, y, z)}{\partial(\xi, \eta, \tau)} \; , \tag{2.1.13}$$

$\mathrm{d}a$ represents the cross section of the ray tube, and the quantities n^0, $\mathrm{d}a^0$ and $D(0)$ relate to the initial surface Q. The quantity $D(\tau)$ is the Jacobian of the transition from the ray coordinates ξ, η, τ to the Cartesian coordinates x, y, z.

In addition to the usual change of phase along the ray path, the method of geometrical optics takes into account the caustic phase shift S_c, which we reiterate in Sect. 2.3. In a formal way a phase shift S_c may be associated with the reversal of sign by the divergence $\mathscr{J} = D(\tau)/D(0)$ after the Jacobian $D(\tau)$ has vanished on the caustic. Therefore, it would be convenient to factor out the phase multiplier $\exp(\mathrm{i}S_c)$ from the amplitude, so that (2.1.12a) rewrites as

$$U_0 = U_0^0(\xi, \eta)\mathrm{e}^{\mathrm{i}S_c}/|\mathscr{J}|^{1/2} \; . \tag{2.1.12b}$$

Thus, in the zeroth approximation of geometrical optics, the wave field is given by the expression

$$\begin{aligned} u(\mathbf{r}) &= U_0(\mathbf{r})\mathrm{e}^{\mathrm{i}k\psi(\mathbf{r})} \\ &= u_0^0(\xi, \eta)|\mathscr{J}|^{-1/2}\exp\left(\mathrm{i}k\int_0^\tau n^2\,\mathrm{d}\tau + \mathrm{i}S_c\right) \; . \end{aligned} \tag{2.1.14}$$

If several rays come in one point **r**, then the fields should be summed as

$$u(\mathbf{r}) = \sum_{j=1}^{N} U_{0j}\exp(ik\psi_j)$$
$$= \sum_{j=1}^{N} u^0(\xi_j, \eta_j)|\mathscr{J}_j|^{-1/2}\exp\left(ik\int_0^{\tau_j} n^2 \mathrm{d}\tau + \mathrm{i}S_{c,j}\right) \quad (2.1.15)$$

and the ray parameters ξ_j, η_j, and τ_j should be evaluated from equations (2.1.9) for the given coordinates of the observation point

$$\xi_j = \xi_j(\mathbf{r}), \qquad \eta_j = \eta_j(\mathbf{r}), \qquad \tau_j = \tau_j(\mathbf{r}) \ . \quad (2.1.16)$$

2.1.2 Electromagnetic Waves in an Isotropic Medium

All results of scalar theory related to the eikonal, rays, and the magnitude of the electric vector $\mathbf{E}_0$ are valid, also for EM waves in isotropic media. New facets associated with polarization of the field were elucidated by *Rytov* [2.3] in 1938. The modern state of the art is covered in our previous volume [2.2].

In the zeroth approximation of geometrical optics, the vector amplitudes of **E** and **H** are transverse with the ray and can be resolved into components along the normal **n** and binormal **b** to the ray, namely,

$$\mathbf{E} = \Phi_n\mathbf{n} + \Phi_b\mathbf{b}, \qquad \varepsilon = n^2 \ ,$$
$$\mathbf{H} = \mathbf{p}\times\mathbf{E} = \sqrt{\varepsilon}(\Phi_n\mathbf{b} - \Phi_b\mathbf{n}) \ . \quad (2.1.17)$$

Hence, equations (2.1.17) fail to completely define polarization owing to polarization degeneracy of the field in the isotropic medium.

It has been learned that the quantity $\theta = \arctan(\Phi_n/\Phi_b)$ obeys the equation

$$\mathrm{d}\theta/\mathrm{d}\sigma = \kappa \ , \quad (2.1.18)$$

known as Rytov's law of field rotation about the Frenêt reference frame (**l**, **n**, **b**) associated with the ray; here κ stands for the ray torsion.

In the general case θ is complex valued. Its real part $\theta' = \mathrm{Re}\,\theta$ characterizes the angle between the large semiaxis of the polarization ellipse and the normal **n** to the ray, and the imaginary part $\theta'' = \mathrm{Im}\,\theta$ defines the ratio of the small semiaxis l_1 to the large semiaxis l_2 as $l_1/l_2 = |\tanh\theta''|$. The sign of θ'' controls the direction of rotation of vector **E** over the polarization ellipse. Since κ is real-valued, the quantity variable along the ray is θ' which governs the orientation of the axes rather than the shape of the ellipse.

Thus, having specified ϕ_n^0 and ϕ_b^0 one may find $\theta = \theta' + \mathrm{i}\theta''$ by (2.1.18) the value of

$$\theta = \theta^0 + \int_0^{\delta} \kappa\,\mathrm{d}\sigma = \theta^0 + \int_0^{\tau} n\kappa\,\mathrm{d}\tau \ , \quad (2.1.19)$$

which completely defines the polarization of the field. Like the modulus of the electric vector $|\mathbf{E}| = (|\phi_n|^2 + |\phi_b|^2)^{1/2}$ the amplitude of the field $\phi = (\phi_n^2 + \phi_b^2)^{1/2}$ obeys the first transport equation (2.1.4) (for details, see [2.2]).

2.1.3 Electromagnetic Waves in an Anisotropic Medium

For anisotropic media, the ray field is written in a somewhat different form. The eikonal equation breaks down into two independent equations to which there correspond two individual Hamiltonians

$$\mathscr{H}_j(p, r) = \tfrac{1}{2}[p^2 - n_j^2(\mathbf{r}, \mathbf{l})] = 0, \quad j = 1, 2 \ , \tag{2.1.20}$$

consistent with two normal waves. Here, n_j are the refractive indices depending on the direction of the wave normal $\mathbf{l} = \mathbf{p}/p$. Accordingly, the zero-approximation field is represented as the sum of two waves

$$\mathbf{E} = \phi_1 \mathbf{f}_1 \exp(ik\psi_1) + \phi_2 \mathbf{f}_2 \exp(ik\psi_2) \ , \tag{2.1.21}$$

where ϕ_j are the amplitudes of the normal waves, and f_j are the polarization vectors normalized to unity, $(\mathbf{f}_j \cdot \mathbf{f}_j^*) = 1$. These vectors depend on the properties of the medium and are defined at every point along the ray path as there is no polarization degeneracy.

For normal waves, the phases ψ_j obey the eikonal equation (2.1.20), where $\mathbf{p}$ should be taken as $\nabla\psi$, and can be represented as integrals along the rays. Unlike the situation in isotropic media, this time the direction of rays coincides with that of the group velocity and the Poynting vector $\mathbf{S}$, rather than with the direction of the wave normal $\mathbf{l} = \nabla\psi/|\nabla\psi| = \mathbf{p}/p$. As a result, the amplitudes ϕ_j can be found out of energy conservation as

$$\operatorname{div} \mathbf{S}_j = 0, \quad j = 1, 2 \ . \tag{2.1.22}$$

The concept of two independent normal waves (2.1.21) is valid so long as the difference of refractive indices $\Delta n = n_1 - n_2$ is sufficiently large,

$$\Delta n \gg \frac{1}{kL} \sim \mu \ , \tag{2.1.23}$$

where L is the characteristic length of medium inhomogeneity. This inequality breaks down in weakly anisotropic media where $\Delta n \lesssim \mu$. An effective formulism, to handle the fields in such media has been devised by *Kravtsov* [2.4] under the name of a quasi-isotropic approximation of geometrical optics. It treats the permittivity tensor components responsible for the weak anisotropy as small perturbations with the geometric-optical field in the isotropic medium as the zeroth approximation. This setting warrants that in the limit as $\Delta n \to 0$ the quasi-isotropic approximation yields the equations of geometrical optics for the isotropic medium. On the other hand, for $\Delta n \gg \mu$ the equations of quasi-isotropic approximation allow solutions in the form of independent normal waves. In other words, this approximation allows a continuous transition from the anisotropic medium (no degeneration) to the isotropic medium, i.e., to the case of polarization degeneracy. For an updated review of the state of the art, the reader is referred to [2.2, 5].

2.2 The Role of Rays in the Method of Geometrical Optics

2.2.1 The Locality Principle

In the wave theory, rays play a dual role. Primarily they are mathematical objects, specifically, the curves which are solutions to certain differential equations and possess some extremal properties.

The fundamental role of rays is that they serve to map the initial field $u^0(\xi, \eta)$ onto the entire space, to be more precise, onto the lit portion of the space where the rays penetrate. This mapping is described by the formulas (2.1.14, 5) and the very nature of the mapping $u^0(\xi, \eta) \to u(\mathbf{r})$ has come to be known as the principle of locality. It embodies the essence of the ray method implying that for a point $\mathbf{r}$ of observation lying on the ray $\mathbf{r} = \mathbf{R}(\xi, \eta, \tau)$ the field $u(\mathbf{r})$ depends only on the value $u^0(\xi, \eta)$ at the point where this ray begins.

The principle of locality is also characteristic of multiple modifications of the ray method describing the diffraction effects. In the ray method proper and in its quasi-ray modifications, the rays provide a "skeleton supporting the wave-field flesh".

2.2.2 Rays as Energy and Phase Trajectories

Rays are, no doubt, not only mathematical abstractions, but physical objects as well. Intuitively rays are paths along which energy flows, i.e., they play the part of energy trajectories. We shall treat the vector-valued quantity

$$\mathbf{I} = \frac{1}{2ik}(u^* \nabla u - u \nabla u^*) \tag{2.2.1}$$

as the density vecor of energy flux of the scalar field u. By virtue of the Helmholtz equations (2..1.1), this quantity is conserved, namely,

$$\operatorname{div} \mathbf{I} = \mathbf{0} \ . \tag{2.2.2}$$

In the geometrical optics approximation $\mathbf{I} \approx \mathbf{p} U_0^2 = \mathbf{I}_0$, and the conservation law becomes

$$\operatorname{div}(\mathbf{p} U_0^2) = 0 \ , \tag{2.2.3}$$

which is equivalent to the first line of the transfer equations (2.1.4) since $\mathbf{p} = \nabla \psi$ and $\operatorname{div} \mathbf{p} = \Delta \psi$. Hence, the approximation of geometrical optics has the energy flowing along the rays, i.e., $\mathbf{I}_0$ is parallel to $\mathbf{p}$.

Of course rays play the part of energy paths to a certain approximation only: higher terms deliver to **I** components transverse with the ray. At larger distances these components responsible for the diffraction processes may drastically change the spatial distribution of the energy flows. Consequently, the meaning of energy paths may be assigned to rays at limited distances where the applicability conditions of geometrical optics are not violated.

These words relate in full to the ability of a ray to serve as a phase trajectory, i.e., a line perpendicular to the phase fronts (in anisotropic media, rays lose this property, but retain the functions of energy paths). In areas where diffraction effects, which should be understood as deviations from geometrical optics, are substantial, the phase fronts behave differently from the manner predicted by ray theory.

2.2.3 Fresnel Volume of a Ray: The Physical Content of the Ray Concept

It is a common physical knowledge that wave fields (acoustic, electromagnetic, etc.) rather than rays are physical reality. None the less, the traditions to endow rays with certain physical properties, traced back to Descartes times, have been deeply enrooted in natural science. Rays are discussed as if they were real objects. This handling of rays is justified when rays can be localized in space, say with the aid of pinholes or Gaussian windows.

A deep insight into this problem indicates that a natural domain of localization of a ray treated as a physical object is its Fresnel volume. This volume is introduced as the envelope of all first Fresnel zones "threaded" on the ray, as illustrated in Fig. 2.2, see also [2.2, 6, 7].

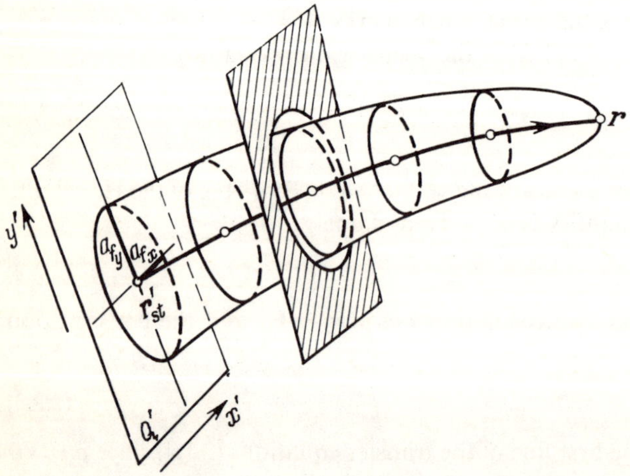

Fig. 2.2. Fresnel volume of a curved ray contains the first Fresnel zones threaded on the ray

If we contract a hole in a screen to dimensions smaller than the first Fresnel zone, we shall see at the point **r** of Fig. 2.2 a substantial distortion of the field associated with this ray. Conversely, with a rather large hole we avoid marked distortions of the field, but the information about where the ray is within the wide aperture will be lost. A pinhole about Fresnel radius size offers a reasonable "diffraction" tradeoff between the localization of the region that forms the field and the perturbation of the field.

Now, the mathematical ray is an "indefinitely thin" line satisfying the ray equations, whereas the physical ray has a finite thickness defined by the Fresnel volume and indicates the direction in which energy flows. Hence, we shall identify the physical content of the ray concept with the Fresnel volume of the ray.

In an inhomogeneous medium, the Fresnel volume can be readily derived on the basis of the Fresnel–Huygens representations on the formation of the field at a point **r** of observation. Let $\psi_Q(\mathbf{r}')$ be the eikonal of the initial wave at some surface Q and $\psi_g(\mathbf{r}, \mathbf{r}')$ be the eikonal of the Green's function associated with the virtual ray $\mathbf{r}' \rightarrow \mathbf{r}$ leaving Q to arrive at **r**. The resultant eikonal on the virtual ray will be $\psi_{\text{virt}}(\mathbf{r}, \mathbf{r}') = \psi_Q(\mathbf{r}') + \psi_g(\mathbf{r}, \mathbf{r}')$. With reference to Fig. 2.2, the point $\mathbf{r}'_{\text{st}}$ where ψ_{virt} assumes a stationary value corresponds to the reference ray $\mathbf{r}'_{\text{st}} \rightarrow \mathbf{r}$.

Denote

$$\begin{aligned}\tilde{\psi}(\mathbf{r}') &= \psi_{\text{virt}}(\mathbf{r}, \mathbf{r}') - \psi_{\text{ref}}(\mathbf{r}) \\ &= [\psi_Q(\mathbf{r}') + \psi_g(\mathbf{r}, \mathbf{r}')] - [\psi_Q(\mathbf{r}'_{\text{st}}) + \psi_g(\mathbf{r}, \mathbf{r}'_{\text{st}})] \ .\end{aligned} \tag{2.2.4}$$

If we expand $\tilde{\psi}(\mathbf{r}')$ into a Tailor series in powers of the deviation from the stationary point $\boldsymbol{\rho}' = \mathbf{r}' - \mathbf{r}'_{\text{st}}$, we find out that its leading term is quadratic. The value of $\tilde{\psi}$ at $\mathbf{r}'_{\text{st}}$ is zero which immediately follows from the definition (2.2.4), and linear terms vanish because of stationarity.

For convenience we take Q to be a plane perpendicular to the reference ray at $\mathbf{r}'_{\text{st}}$ and the coordinates $\boldsymbol{\rho}' = (x', y')$ be such that the mixed partial derivative, $\tilde{\psi}_{xy}$, should vanish. Then the eikonal assumes the Morse form

$$\tilde{\psi}(\mathbf{r}') = \tfrac{1}{2}\tilde{\psi}_{xx}x'^2 + \tfrac{1}{2}\tilde{\psi}_{yy}y'^2 + \cdots \ . \tag{2.2.5}$$

If the derivatives $\tilde{\psi}_{xx}$ and $\tilde{\psi}_{yy}$ are of the same sign ($\psi_{xx}\psi_{yy} > 0$) then the lines $\tilde{\psi} = \text{const.}$ form in Q figures close to elliptical shapes. To the first Fresnel zone there corresponds an ellipse for which the magnitude of $\tilde{\psi}$ equals $\lambda/2$, viz.,

$$\tfrac{1}{2}|\tilde{\psi}_{xx}|x'^2 + \tfrac{1}{2}|\tilde{\psi}_{yy}|y'^2 - \tfrac{1}{2}\lambda = 0 \ . \tag{2.2.6}$$

The semiaxes a_{fx} and a_{fy} of the Fresnel ellipse (see Fig. 2.2) are as follows:

$$\begin{aligned} a_{fx} &= \lambda^{1/2}|\tilde{\psi}_{xx}|^{-1/2} = (2\pi/k)^{1/2}|\tilde{\psi}_{xx}|^{-1/2} \ , \\ a_{fy} &= \lambda^{1/2}|\tilde{\psi}_{yy}|^{-1/2} = (2\pi/k)^{1/2}|\tilde{\psi}_{yy}|^{-1/2} \ . \end{aligned} \tag{2.2.7}$$

Equation (2.2.6) in which the derivatives $\tilde{\psi}_{xx}$ and $\tilde{\psi}_{yy}$ depend on σ', i.e., on the position of Q on the reference ray may be viewed as the equation of a surface

$F(x', y', \sigma') = 0$ combining the contours of all Fresnel zones for the given ray. This surface embraces the Fresnel volume of a ray that plays the decisive role in the formation of the field at the point of observation $\mathbf{r}$.

The term Fresnel volume suggested in the early works of the authors [2.2, 6] (also [2.7, 8]) to define the domain of actual localization of a ray seems to be more descriptive than the terms "region essential for diffraction" [2.9], three-dimensional Fresnel zone [2.10], and spatial Fresnel zone [2.11].

If the derivatives $\tilde{\psi}_{xx}$ and $\tilde{\psi}_{yy}$ are of different sign, then the lines $\tilde{\psi}$ = constant in plane Q are shaped as hyperbolas with infinite trails. Analysis indicates, however, that the trails play only an insignificant role in the formation of the field and that the Fresnel volume is quite satisfactorily described by equation (2.2.6) where $\lambda/4$ should be substituted for $\lambda/2$ [2.12, 13]. A detailed discussion of Fresnel volumes may be found in [2.8].

Looking a little bit ahead we remark that if the point of observation approaches the caustic, then one or both derivatives $\tilde{\psi}_{xx}$ and $\tilde{\psi}_{yy}$ tend to zero and the Fresnel scales a_{fx} and a_{fy} go to infinity. In these circumstances one should keep higher terms in x and/or y in the Taylor series of eikonal $\tilde{\psi}$. The estimates of Fresnel scales characterizing the domain essential for the field at the point of observation will change accordingly.

2.2.4 Heuristic Criteria of Applicability for Ray Theory

In accordance with the Huygens–Fresnel principle, the field at the observation point is formed as a result of the interference of secondary wavelets induced at every point of the prime wave front. The first Fresnel zone is the key contributor because the secondary wavelets from this zone differ in phase at most by π (by $\lambda/2$ in the eikonal) and cannot cancel each other, whereas the combined contribution of the higher Fresnel zones will be rather small because of many oscillatory pairs in antiphase.

From the mathematical standpoint, the geometrical-optics approximation occurs as the result of applying the method of stationary phase to the Huygens–Fresnel–Kirchhoff integral with the first Fresnel zone corresponding to the π neighborhood of a stationary point. In view of the analogy with the applicability conditions for the stationary-phase method, a heuristic criterion of applicability of geometrical optics has been suggested [2.2, 6] as follows:

> The parameters of the medium and the wave (amplitude and phase gradient) should not vary significantly over the cross section of the Fresnel volume, i.e., in all Fresnel zones belonging to the ray.

This condition implies that

$$a_f |\nabla_Q U_0| \lesssim |U_0|, \qquad a_f |\nabla_Q n| \lesssim n, \qquad a_f |\nabla_Q p_N| \lesssim p_N \, , \tag{2.2.8}$$

where ∇_Q is the differentiation operator associated with plane Q and transverse with the ray, $p_N = \partial\psi/\partial N$ is the component of the momentum in the direction of the normal **N** to plane Q, and a_{f} is the transverse size of the Fresnel volume. Choice of the direction of differentiation in (2.2.8) should be consistent with the orientation of the semiaxes $a_{\mathrm{f}x}$ and $a_{\mathrm{f}y}$. Similar requirements are imposed on the polarization of the wave when an electromagnetic field is concerned.

2.2.5 Distinguishability of Rays

The concept of Fresnel volume may be helpful in formulating a condition for actual distinguishability of rays, namely, the Fresnel volumes of two adjacent rays should not markedly overlap, i.e., should not intrude into one another at least in one transverse section plane. When this condition is met, the rays can be observed as separate entities, by having them pass through an aperture or a window in an opaque screen. Recognizing that the phases of virtual rays at the outside of Fresnel volume differ from the phase of the reference ray by more than π, the distinguishability condition may be rewritten as

$$|S_j - S_k| \gtrsim \pi \quad \text{or} \quad |\psi_j - \psi_k| \gtrsim \lambda/2 \ , \tag{2.2.9}$$

where S_j and S_k are the phases, and ψ_j and ψ_k the eikonals of the rays to be distinguished.

2.3 Physical Characteristics of Caustics

2.3.2 Caustics as Envelopes of Ray Families

Mathematically, caustic surfaces are envelopes of the family of rays. Physically, these surfaces or lines are distinctive in that the field intensity increases on them sharply as compared with the adjacent space. The rise of field is best of all seen at the focal point where all the rays corresponding to the converging spherical wave-front intersect.

In the geometrical optics approximation, at a caustic the amplitude of the field (2.1.12) becomes infinite since by the definition of the envelope.

$$D(\tau)|_{\text{caustic}} = 0 \ . \tag{2.3.1}$$

When the Jacobian $D(\tau)$ is zero, the equation of the ray family $\mathbf{r} = \mathbf{R}(\xi, \eta, \tau)$ cannot be uniquely resolved for the ray coordinates ξ, η, and τ. This implies approach of the rays however close to one another, vanishing of the ray tube section da, and consequently, singularity of the ray fields at caustics.

Moving across a caustic gives birth or annihilation of a pair of rays at a time because real-valued roots of (2.1.9) in ξ, η, and τ occur in pairs. The jumpwise variation of the number of rays across a caustic is qualified as a catastrophe. This new and fruitful approach to caustics, developed in recent years, allows a universal classification of the typical caustics. We will revisit this topic in the next section.

An important role of caustics is associated with the fact that they characterize the family of rays as a whole because every ray of the family touches the caustic. The position of caustics is described by the equation of ray family (2.1.9) and (2.3.1), that is, by the set

$$\mathbf{r} = \mathbf{R}(\xi, \eta, \tau), \qquad D(\xi, \eta, \tau) = 0 \ , \tag{2.3.2}$$

where the Jacobian D is a function of all three ray coordinates ξ, η, and τ. Eliminating τ between the lines of (2.3.2) we write the equation of caustic in parametric form

$$\mathbf{r} = \mathbf{R}[\xi, \eta, \tau(\xi, \eta)] \equiv \mathbf{r}_c(\xi, \eta) \ . \tag{2.3.3}$$

According to this equation the position of a point on a caustic may be characterized by the same coordinates ξ, η as the point where the ray emanates from the initial surface. In some situations as, say, in reconstructing the family of rays by the form of a caustic it would be reasonable to endow the caustic with its own coordinates, that along with the parameter l_n specifying the separation from the caustic would give the position in space (on the caustic, $l_n = 0$). We refer to such coordinates as caustic.

Reconstruction of a family of rays by the form of a caustic presents a formidable problem. In addition to the form of the caustic, it is required that the field of directions of the rays tangent to the caustic should be given. If, for example, a plane wave incident on a half space filled with a plane-stratified medium is known to produce a plane caustic surface, then having specified such a plane we may claim that the rays forming it correspond to a plane wave, but we cannot indicate unambiguously the direction of this wave.

A popular meaning that the form of a caustic is enough for a unique reconstruction of the ray family is based on familiar examples, all of which without exclusion relate to two-dimensional problems (some 3-D examples, e.g., axially symmetric caustics, also reduce to two-dimensions). Indeed, in two dimensions caustics unambiguously dictate the course of rays. This fact has engendered the known property of a homogeneous two-dimensional space where the phase fronts are evolvments of caustics and the caustics are evolutes of phase fronts. We cannot exclude solving the problem of reconstruction of rays by the form of caustics with the formalism of catastrophe theory. Moreover, one may think of a law used to transform a given caustic to the normal form (to be discussed in Chap. 3) as the transformation of the family of rays.

2.3.2 Caustic Phase Shift

From specific examples allowing exact solutions, it has been known that the phase of the field changes by $-\pi/2$ upon touching a nonsingular caustic and by $-\pi$ after passing a three-dimensional focus; however, a universal rule on the additional phase shift at a caustic has been formulated only in the comparatively recent works of *Maslov* [2.14] and *Lewis* [2.15]. It is based on calculating the integral by the method of stationary phase and tells that after touching the caustic the ray field acquires the phase factor

$$\exp(\mathrm{i}S_{\mathrm{c}}) = \exp(-\mathrm{i}\pi\beta_{\mathrm{c}}/4) \; . \tag{2.3.4}$$

Here, S_{c} is the caustic phase shift, and β_{c} is the caustic variation of signature of the matrix $\hat{D} = [\partial x_i/\partial \xi_j]$ that characterizes the transition from the ray coordinates ($\xi_1 = \xi$, $\xi_2 = \eta$, $\xi_3 = \tau$) to Cartesian ($x_1 = x$, $x_2 = y$, $x_3 = z$). The Jacobian $D(\tau)$ is the determinant of this matrix.

The phase factor (2.3.4) has been already included in the amplitude of the zeroth approximation (2.1.12b) and in the sum of ray fields (2.1.15). This factor should be taken into account in analysis of interference patterns of the field under the conditions of multipath propagation.

If a ray touches a few caustics, additional phase shifts $S_{\mathrm{c}}^{(m)}$ acquired at each caustic sum up and the resultant field has the phase factor

$$\exp\left(\mathrm{i}\sum_m S_{\mathrm{c}}^{(m)}\right) = \exp\left(-\mathrm{i}\pi/4\sum_m \beta_{\mathrm{c}}^{(m)}\right) \; . \tag{2.3.5}$$

The quantity $q = -1/2\sum_m \beta_{\mathrm{c}}^{(m)}$ is called the Maslov trajectory index. It is integer-valued since the signature of matrix $\hat{D}$ varies in even-numbered increments. The Maslov index occurs in many problems, specifically in the evaluation of eigenmodes in open resonators.

A widespread conviction of the caustic phase shift assuming only negative values, such as $-\pi/2$, $-\pi$, etc., has recently received a proof to the opposite. *Orlov* [2.16] found a positive phase shift of $S_{\mathrm{c}} = \pi/2$ corresponding to a negative Maslov index. This topic will be revisited in Chap. 10.

2.3.3 Caustic Zone and Caustic Volume

Because the amplitude U_0 becomes infinite at a caustic, the geometrical optics solution at a caustic and in the close neighborhood is inapplicable as actual wave fields are always finite.

Available exact and approximate solutions for some wave problems involving caustics indicate that a substantial concentration of the field takes place near a caustic. This phenomenon is more profound within some finite region which has not yet received a uniformly adopted name. We shall refer to this region as a *caustic zone* or a *caustic volume*.

In the general case, the width of the zone where the ray approximation is inapplicable may be estimated by violation of inequalities (2.2.8). Introduce the parameter

$$\delta \sim a_{\mathrm{f}} \max\left\{\frac{|\nabla_Q U_0|}{|U_0|}, \frac{|\nabla_Q n|}{n}, \frac{|\nabla_Q p_N|}{|p_N|}\right\} \tag{2.3.6}$$

and rewrite (2.2.8) as $\delta \lesssim 1$. Clearly, the opposite inequality $\delta \gtrsim 1$ will identify the domain where the geometrical optics is inapplicable. It follows that the approximate equality

$$\delta \approx 1 \qquad (2.3.7)$$

can be served as a condition defining the boundary of the applicability domain, Γ. In a particular case of a caustic domain of inapplicability, the growth of δ on approaching the caustic occurs predominantly on account of the unbounded growth of the Fresnel scale a_f (in Sect. 2.2 we noted that a_f increases because the derivatives $\tilde{\psi}_{xx}$ or $\tilde{\psi}_{yy}$ tend to zero).

The boundary Γ of a caustic inapplicability zone may be derived on other grounds. We recognize that in the vicinity of a caustic every point of observation $\mathbf{r}$ is hit by a few rays belonging to the same initial wave front. Consider the simplest situation of two rays with reference to Fig. 2.3. When the observation point is at a sufficient distance from the caustic, Fig. 2.3a, the Fresnel zones of these rays do not overlap in any intermediate surface Q. When $\mathbf{r}$ is moved closer to the caustic, at some time instant there will occur an overlap of the Fresnel zones, Fig. 2.3b. Such an overlap is inadmissible in the ray method as it implies that one and the same segment contributes twice in the resultant field.

In order to preclude overlapping of the Fresnel zones and, hence, Fresnel volumes, of two rays coming in the neighborhood of the caustic, we should require that $|S_j - S_k| \gtrsim \pi$. This inequality has already occurred in connection with the condition of distinguishability of rays (2.2.9). If we assume that the boundary of the caustic zone Γ will be identified by replacing the symbol $\gtrsim$ with the approximate equality $\approx$, then the boundary may be defined from the equation

$$|S_1 - S_3|_\Gamma \approx \pi \quad \text{or} \quad |\psi_1 - \psi_2|_\Gamma \approx \lambda/2 \ , \qquad (2.3.8)$$

which for multipath propagation gives way to

$$\min |S_j - S_k|_\Gamma \approx \pi \quad \text{or} \quad \min |\psi_j - \psi_k|_\Gamma \approx \lambda/2 \ , \qquad (2.3.9)$$

where j and k label the rays associated with the caustic under consideration.

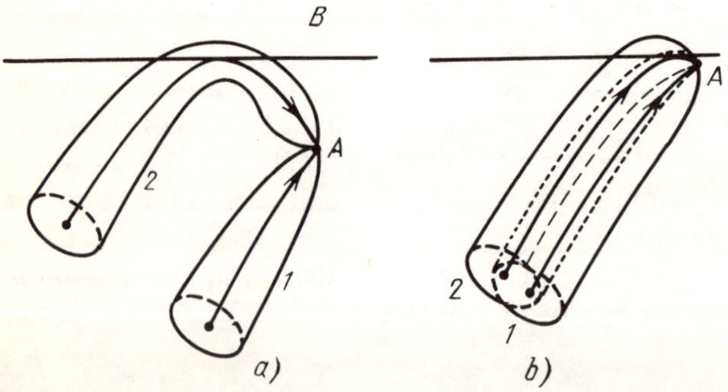

Fig. 2.3. (**a**) Distinguishable and (**b**) indistinguishable rays. The Fresnel volumes are separated in (**a**) and substantially overlap in (**b**)

Many particular cases suggest that the estimators (2.3.7) and (2.3.8) of boundary Γ agree well with one another and that the calculations with (2.3.8) are neater. Therefore, in what follows we shall stick to the latter.

By way of illustration we estimate the width of a caustic zone near a nonsingular segment of a caustic. Near a caustic the difference of the eikonals of two rays has found to be [2.17–19] (also see Sect. 5.1)

$$|\psi_1 - \psi_2| \approx \tfrac{4}{3}\gamma^{1/2}|v|^{3/2}, \quad \gamma = 2n_c^2|K_{rel}| , \tag{2.3.10}$$

where v is the distance normal to the caustic, n_c the refractive index at the caustic, K_{rel} the relative curvature of ray and caustic that in a two-dimensional problem is combined as $K_R \pm K_c$ of the ray curvature K_R and the caustic curvature K_c with the plus sign for the ray and caustic lying on either side of a common tangent. In three dimensions, $K_{rel} = |K_c \cos\vartheta - K_R|$ with K_c being the curvature of the caustic in the plane of the ray, and the angle between the normals to the ray, **n**, and to the caustic, N_c.

Substituting (2.3.10) into (2.3.5) yields the width of caustic zone as

$$\begin{aligned}\Delta v_c &\approx \left(\frac{3\lambda}{8}\right)^{2/3}\gamma^{-1/3} = \left(\frac{3\pi}{4k}\right)^{2/3}\gamma^{-1/3}\\ &= 1.77\,k^{-2/3}\gamma^{-1/3} \equiv 1.77\,\Lambda .\end{aligned} \tag{2.3.11}$$

This estimate agrees well with the results of exact and approximate calculations of field intensity near nonsingular segments of caustics. In Sect. 5.1 we shall see that the quantity $\Lambda = k^{-2/3}\gamma^{-1/3}$ describes the scale of field variations near a simple caustic where the field is defined by an Airy function,

$$u \propto \mathrm{Ai}(-v/\Lambda) \equiv \mathrm{Ai}(\zeta) . \tag{2.3.12}$$

The plot of this function is shown in Fig. 2.4. In the light region, $\zeta < 0$, the first maximum is at $\zeta = -1.02$, i.e., at $v = 1.02\Lambda$, and the first zero at $\zeta = -2.34$, i.e., at $v = 2.34\Lambda$. Thus, the estimate (2.3.11) corresponds to $\zeta = -1.77$ which lies midway between the first zero and first maximum of the Airy function.

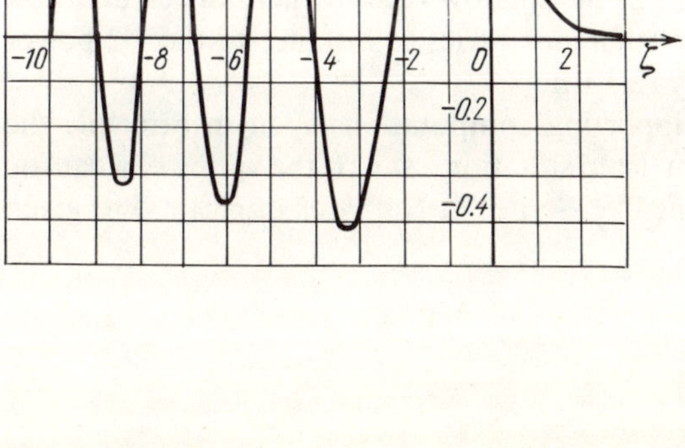

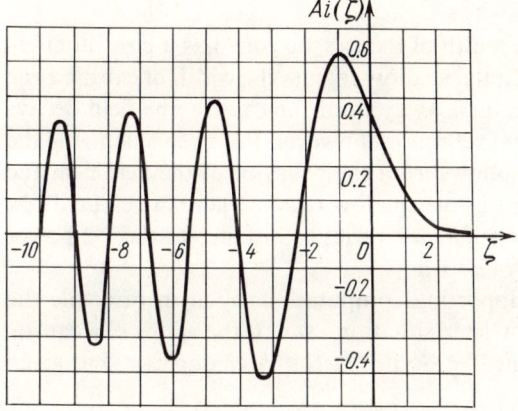

Fig. 2.4. Plot of the Airy function

An insignificant modification of the above consideration allows the distance to the first interference zero and to the first interference maximum to be given with amazing accuracy. We take into account the caustic phase shift and substitute into (2.3.8) the total difference

$$S_{tot} = S_{2,tot} - S_{1,tot} = (S_2 + S_{c2}) - (S_1 + S_{c1}) \ , \tag{2.3.13}$$

for the purely geometric phase difference $S_2 - S_1$. For a simple caustic where the second ray has $S_{c2} = -\pi/2$ and the first ray (does not touch the caustic) has $S_{c1} = 0$, we get in place of (2.3.8)

$$\Delta S_{tot} = \left| S_2 - \frac{\pi}{2} - S_1 \right|_r = \pi \ , \tag{2.3.14a}$$

whence

$$|S_2 - S_1| = k|\psi_2 - \psi_1| = 3\pi/2 \ . \tag{2.3.14b}$$

In view of (2.3.10) this condition is satisfied by the distance $\nu = 2.32\Lambda$ which is almost the distance to the first zero of the Airy function $\nu = 2.34\Lambda$.

Now, the requirement that the two ray fields be in phase becomes with allowance for the caustic phase shift

$$S_{tot} = 0 \ , \tag{2.3.15a}$$

whence

$$|S_2 - S_1| = \pi/2 \quad \text{or} \quad k|\psi_2 - \psi_1| = \pi/2 \ . \tag{2.3.15b}$$

This condition is met by the distance $\nu = 1.11\Lambda$ which differs only insignificantly from the distance to the first maximum of the Airy function $\nu = 1.02\Lambda$.

There is some uncertainty concerning the width of the caustic zone $\Delta\nu_c$ (from 1.02Λ to 2.34Λ). It may be diminished if we refine the purpose of the estimate. Our convention henceforth is as follows: In the problem addressed, the resolution of the branches of caustics we take $\Delta\nu_c = 2.34\Lambda$, i.e., the distance to the first zero of the Airy function because the zero level of the field provides a clear-cut separation of the branches, whereas for ray estimation of the fields at caustics we take $\Delta\nu_c \cong \Lambda$ because this meets the condition of constructive interference $\Delta S_{tot} = 0$ of the two ray fields.

It should be obvious that the width of the caustic zone has a clear interference sense in the lit region only. In the shadow region, the width of caustic zone is controlled already by other factors. Away from the caustic, the field decays exponentially, like in total internal reflection; however, this time again Λ is the rate of decay. This immediately follows from the behavior of the field near the caustic and also from the fact that in the shadow region where the eikonals ψ_1 and ψ_2 are complex-valued, their difference varies as distance to the 3/2 power like in the lit region described by (2.3.10).

Similar consideration and supporting computations elicit in principle the width of a caustic zone in more intricate situations, say, at the vertex of a caustic cusp, near a caustic pocket, and the like. Some illustrative examples will be given below.

2.3.4 Ray Estimates of Fields at Caustics and in Focal Spots

Having established the limit of applicability for ray theory we are in a position to derive, although rough but correct in the order of magnitude, an estimate of the field in the region of inapplicability, using for this merely the ray considerations. We indicate the available possibilities with respect to the focal and caustic domains of inapplicability, although this reasoning (first outlined in [2.2, 6]) is applicable in equal measure to other situations, e.g., to penumbral fields. Estimation of fields in far shadow regions would require, however, a different approach.

First of all, the focal field u_{foc} may be estimated by the ray field (2.1.14) calculated at the boundary Γ of the focal domain of inapplicability as

$$|u_{\text{foc}}| \cong |U_0|_\Gamma \cong (|U_0^0/\sqrt{\mathscr{J}}|)_\Gamma \ . \tag{2.3.16}$$

Secondly, the reasoning may be based on the conservation of the flow of energy in the ray tube assuming in addition that the initial flow $\Delta\Pi^0 = n^0|U^0|^2\Delta a^0$ is distributed more or less uniformly over the focal (caustic) zone,

$$\Delta\Pi_{\text{c}} = n_{\text{c}}|u_{\text{foc}}|^2\Delta a_{\text{foc}} \approx \Delta\Pi^0 = n^0|U^0|^2\Delta a^0 \ , \tag{2.3.17}$$

where Δa^0 is the initial cross section of the ray tube corresponding to a section of finite width Δa_{foc} in the focal zone, as illustrated in Fig. 2.5. It follows the estimate

$$|u_{\text{foc}}| \approx |U_0^0|\left|\frac{n^0\Delta a^0}{n\Delta a_{\text{foc}}}\right|^{1/2} \ , \tag{2.3.18}$$

which is close to (2.3.16) as on the boundary of caustic zone

$$\mathscr{J}|_\Gamma \approx n_{\text{c}}\Delta a_{\text{foc}}/n^0\Delta a^0 \ .$$

As an example, consider the field in the vicinity of a simple caustic. Represent the cross section of the ray tube $\Delta a_{\text{foc}} = \Delta v_{\text{c}}\Delta l_\perp$ as the product of the width of caustic zone Δv_{c} by the width of the ray tube $\Delta l_\perp$ measured in the direction parallel to the caustic. Putting $\Delta v_{\text{c}} = \Lambda$, which has been agreed upon in Sect. 2.3.3, yields

$$|u_{\text{foc}}| \sim |U_0^0|\left(\frac{n^0\Delta a^0}{n_{\text{c}}\Delta l_\perp\Lambda}\right)^{1/2} \ . \tag{2.3.19}$$

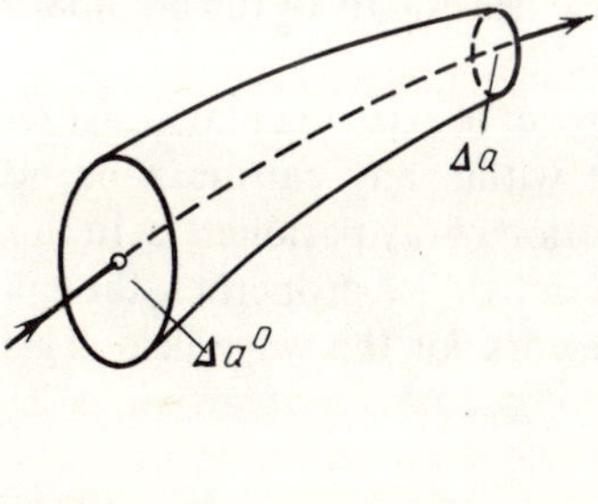

Fig. 2.5. Conservation of an energy flow in a ray tube

General consideration judgement might lead one to think that this estimation technique for focal and caustic fields can be accurate at most in the order of magnitude. However, as we shall see below, comparison of the heuristic estimates derived with (2.3.16, 18) against the computations with the diffraction formulas indicates that they seldom differ by more than 1.5 times. In particular, the local asymptotic expansions yield a caustic field value lying within a factor of 0.89 of (2.3.19), and 1.37 times the field at the first interference maximum. More examples to the point will be given in the following chapters.

2.3.5 Indistinguishability of Rays in a Caustic Zone

Indistinguishability of rays in a caustic zone follows already from (2.3.8), for this definition of caustic-zone boundary obviously contradicts the distinguishability condition (2.2.9).

In Sect. 2.2 we discussed distinguishability of rays from the standpoint of their separability with pinholes and windows. There exist, however, other techniques to distinguish rays, specifically, by the angle of incidence, which may mislead one into believing that incapability of one method does not imply incapability of another. One may readily verify with a simple caustic that inside a caustic zone, i.e., at $\lesssim \Lambda$, the rays cannot be distinguished by their angle of incidence either. Indeed, on approach to the caustic, the angle $\Delta\theta$ between the rays decreases as $v^{1/2}$ and from (2.3.10) it follows that $\Delta\theta \approx (\gamma v)^{1/2}/n_c$. To resolve such rays with an antenna (or cylindrical lens) measuring l_a across the aperture the width of its radiation pattern $\theta_a \sim \lambda/l_a n_c$ must be narrower than $\Delta\theta$, viz.,

$$\lambda/l_a n_c \lesssim (\gamma v)^{1/2}/n_c \ . \tag{2.3.20}$$

At first glance it might seem that increasing the length l_a of the antenna one could satisfy this inequality at any distance v, including locations within the caustic zone where $v \leq \Lambda$. However, in actual systems the length of the antenna cannot be shorter than the width of the caustic zone $\Delta v_c \approx \Lambda$, or else the field within the antenna will be appreciably inhomogeneous (segment $|\zeta| < 1$ in Fig. 2.4) and can no longer be viewed as a superposition of two locally plane waves – a requirement to distinguish the rays.

To estimate the minimal distance from the caustic at which the rays can still be resolved we let $l_a = 2v$, i.e., take an antenna twice as large as the distance from the caustic to the observation point. Equating both sides of (2.3.20) gives $v_{min} \approx \pi^{2/3}\Lambda \approx 2.2\Lambda$, that is the minimum distance lies exactly at the boundary of the caustic zone.

Looking at this particular example from the aspect of experimental measurements we may draw a general conclusion that within any caustic zone no physical devices are capable of separate determination of ray parameters. In this sense, in a caustic zone rays loose their physical individual properties, though continue to play the role of the geometric framework for the wave field. This conclusion has been borne out by the results of a numerical modeling performed by *Asatryan* et al. [2.20] on the basis of measurements with a Gaussian window. The field passing through such a window senses its presence once the effective width of the window has reduced down to a value of the order of Λ.

2.3.6 Reality of Caustics

In his book *Stavrodis* [2.21] remarked that in contrast to rays and wave fronts, the caustic is one of a few objects in geometrical optics that can be observed in reality. This remark emphasizing the role of caustics, of course, has its own range of validity. What we actually "see" is the image of objects on the retina of our eye. The image is formed by the laws of wave optics, but if we are disinterested in minute details, then it would be quite accurate to say that the image is formed by rays. Rays can form in the retina the image of a caustic, but the same caustic may be recorded by physical instruments. Outside the optical range we have to relegate measurements to instrumentation which can record and sometimes visualize invisible images, including those of caustics.

On these lines, the above remark on the physical distinguishability and "reality" of caustics is true only to the point that in the close vicinity of caustics one can observe or measure a concentration of the field. Is a caustic real in the above sense in all situations, i.e., is the effect of field buildup on a caustic appreciable enough for instruments to reveal, separate and identify the caustics? This question may be answered with heuristic criteria to be discussed below for specific conditions.

A caustic may be deemed real, i.e., observed or recorded, if two conditions are met:

(i) the amplitude on the caustic is at least a few times the field value elsewhere;

(ii) the nearcaustic zones of adjacent caustics do not overlap completely (partial overlapping is admissible).

Some other relevant conditions of practical character should be satisfied of course, namely, noise should not be high, resolution and sensitivity of the instruments should be sufficient, and the like.

With these conditions in sight, let us consider the caustic zones of swallow-tail sections obtained at different wavelengths (Fig. 2.6). The diagram in Fig. 2.6a shows a caustic that occurs in a wave field with very short wavelength. As the wavelength increases, the caustic zones of the branches exhibit a marked overlapping. The caustic zones of all three branches merge and cannot be distinguished any longer beginning from the time instant when the width Δv_c of the caustic zone exceeds the distance between two adjacent branches, l_b. The situation of $\Delta v_c > l_b$ is presented in Fig. 2.6b. As a result, the caustic pocket becomes "unreal", i.e., unobservable. It is virtually impossible to reveal the presence of such a pocket with the aid of physical devices and a looped caustic

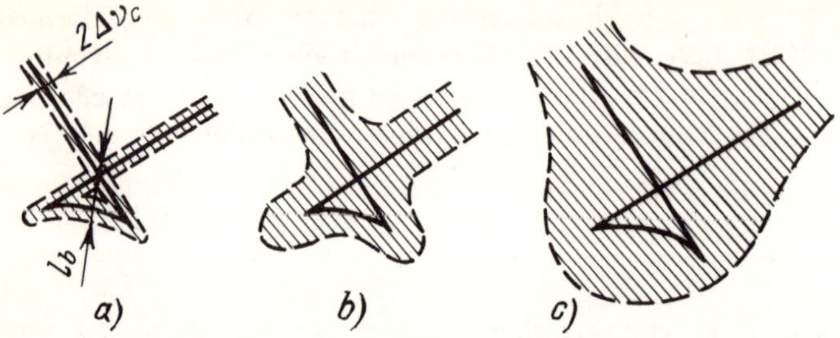

Fig. 2.6a–c. Distinguishable, coalescent, and indistinguishable branches of a caustic loop. Caustic zones are emphasized by hatching

becomes indistinguishable from a smooth portion of a caustic, as indicated in Fig. 2.6c.

Thus, speaking about the caustic as physical object we deal not with the caustic proper but with the caustic volume surrounding it. This circumstance aggravates the problem of caustic identification, for different types of caustic can have caustic volumes of similar shape.

2.3.7 A Remark on Multipath Propagation

The term multipath propagation has gained a strong foothold in the field (see, e.g., [2.22]) despite its conditional nature. It is used to denote the rough surface of the phase front for each of the "macrorays" oncoming in the observation point in a random inhomogeneous medium. Unlike macrorays, microrays disappear when random inhomogeneities are "turned off".

One may speak of reality, i.e., physical distinguishability, of microrays provided that their Fresnel volumes do not overlap. In a random inhomogeneous medium with large-scale fluctuations, this condition is met only in the region of unsaturated fluctuations, i.e., where relative fluctuations of intensity are small: $\Delta I/\bar{I}$, and where no caustics are formed [2.23]. Well pronounced caustics occur in the focusing range (Fig. 2.7) where strong fluctuations of intensity with $\Delta I/\bar{I} \gtrsim 1$ take place. The principal claim in the work of *Kravtsov* [2.23] is that a marked (say, 30–50%) probability of occurrence of caustics is observed just in the approach to the strong fluctuation range.

In the range of saturated fluctuations, caustics fill the space with so high a density that there is no way to speak of distinguishability of individual rays and caustics any longer (Fig. 2.7b). Here, the microray representation cannot be associated with real rays, and can be treated only as a conditional reflection of the complex nature of the fluctuation field.

It is worth noting that random caustics occur not only in a random inhomogeneous medium, but also in the reflection of rays from rough surfaces. The density of caustics, critical indices and other important aspects of this problem have been analyzed by *Berry* and *Upstill* [2.24].

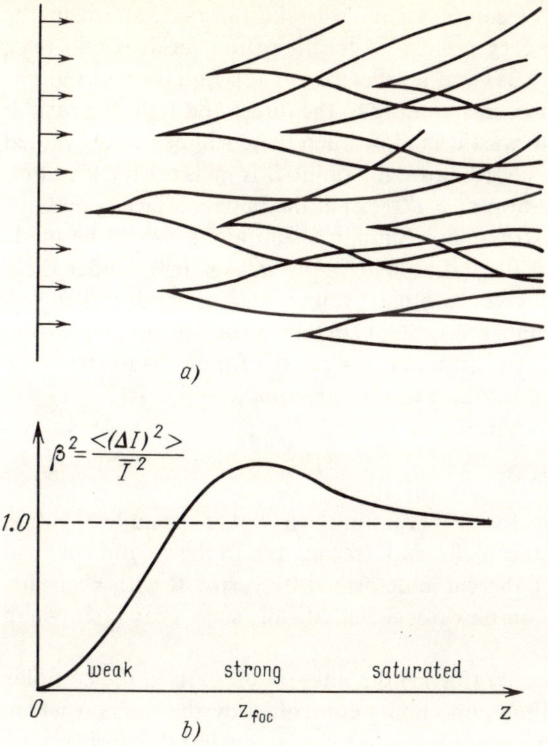

Fig. 2.7. (**a**) Random caustics in a turbulent medium. (**b**) Behavior of the scintillation index $\langle \Delta I^2 \rangle / \langle I \rangle^2$ with distance in the ranges of weak, strong and saturated fluctuations

2.4 Complex Rays

2.4.1 Main Properties of Complex Rays

Complex geometrical optics deals with complex eikonals and complex rays. Complex solutions occur at least in three situations: in absorbing media, in caustic shadow, and in description of inhomogeneous waves. Complex rays are introduced as the complex solutions of the ray equations (2.1.5), therefore, their analysis needs no additional mathematical formalism except that outlined in Sect. 2.1. On the other hand, the properties of complex rays are very specific and deserve separate and attentive consideration.

The concept of complex rays seems to have been first introduced by *Keller* [2.25] and later refined and generalized by *Seckler* and *Keller* [2.26], *Keller* and *Karal* [2.27], *Maslov* [2.28], *Babich* [2.29], and *Kravtsov* [2.30]. Among the recent contributions to the field we draw attention to the works of Felsen and co-workers who addressed waves with sources of large aperture, specifically [2.31]. Below we shall stick to the treatment of complex rays that has been outlined in the paper of *Kravtsov* [2.30].

To illustrate the concept of complex rays we take the ray pattern in the vicinity of a simple caustic. Every point $\mathbf{r}$ of the lit region receives two rays, a direct ray (labeled 1 in Fig. 2.3a) and a reflected ray (2). Equation (2.1.9) has two real-valued roots in ξ, η, τ corresponding to the direct and reflected rays. If the observation point is at B in caustic shadow, then there will be no real valued solutions of (2.1.9) for the ray coordinates, for point B is inaccessible for light.

Suppose now that the equation $\mathbf{r}^0 = \mathbf{r}^0(\xi, \eta)$ of the initial surface Q remains valid under complex-valued parameters. Similar assumptions should be made about the permittivity $\varepsilon(\mathbf{r})$ and the initial field values $u^0 = u^0(\mathbf{r}^0)$. Under these assumptions, equation (2.1.9) gives complex values of ξ, η, and τ that will correspond to the complex points $\mathbf{r}^0 = \mathbf{r}^0(\xi, \eta, \tau)$ where the ray emanates from the initial surface Q. Consequently, as parameter τ varies from zero to $\tau(\mathbf{r})$ of the observation point $\mathbf{r}$, the formal solution to the ray equations $\mathbf{r} = \mathbf{R}(\xi, \eta, \tau)$ will vary from the *initial complex* vector

$$\mathbf{r}^0[\xi(\mathbf{r}), \eta(\mathbf{r})] = \mathbf{R}[\xi(\mathbf{r}), \eta(\mathbf{r}), 0]$$

to the real-valued final vector $\mathbf{r}$ assuming en route complex values $\mathbf{R}(\tau) = \mathbf{R}'(\tau) + i\mathbf{R}''(\tau)$, where $'$ means Re and $''$ means Im. In the six-dimensional space $(X', X''; Y', Y''; Z', Z'')$, the curve described by vector $\mathbf{R}$ with τ varying from zero to $\tau(\mathbf{r})$ represents a complex ray. Schematically such a ray is shown in Fig. 2.8.

In calculating the eikonal ψ by (2.1.11) the integration path in the complex plane τ may be deformed within some limits controlled by the domain where functions $\varepsilon(\mathbf{r})$, $\mathbf{r}^0(\xi, \eta)$ and some other functions characterizing the problem are analytic. This deformation will certainly alter the configuration of the rays which have only their initial and final points fixed. It follows that the extremum of the optical path ψ will be attained at a continuum of complex paths. This might seem to contradict the Fermat principle which says that the time of light propagation between two points is extremal at a finite number of paths. However, the classic formulation of this principle does not apply to complex rays because here the very notion of propagation time becomes meaningless. None the less another more abstract variational principle remains valid, namely, the complex eikonal ψ attains extrema at complex trajectories defined by equations (2.1.5).

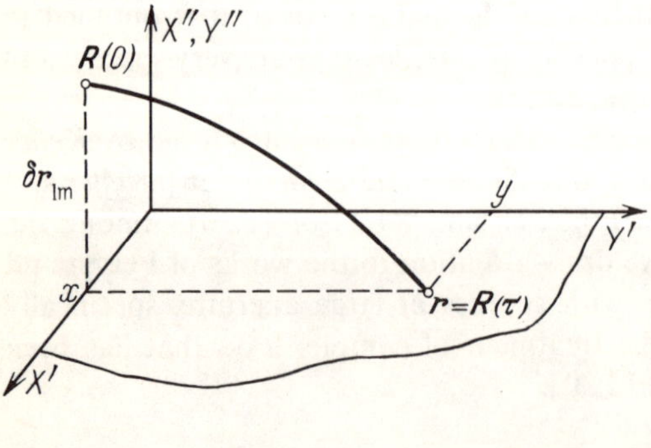

Fig. 2.8. Trajectory of a complex ray begins at a complex point $\mathbf{R}(0)$ and terminates at a real observation point $\mathbf{r}$

In the considered example of a simple caustic where any point in the lit region is hit by two rays, one may expect that through an arbitrary point in the shadow region there will pass at least two complex rays. From physical considerations it is clear that the field must fall off inward toward the shadow region. Therefore, of the two rays under consideration we should exclude one for which the imaginary part of the eikonal, ψ'', is negative and increases in magnitude inward toward the caustic shadow.

This example indicates that for complex rays, certain selection rules should be set up that would ensure that the physical constraints imposed on the wave field are met, including the requirement that the field should fall off toward the deep caustic shadow. Such a selection rule may be devised by analysis of available exact and asymptotic solutions which allow interpretation in terms of complex rays. The work in this direction is at its infancy.

2.4.2 Reflection of a Plane Wave from a Linear Slab

Consider a layer of linear permittivity $\varepsilon(z) = \varepsilon_0 - \varepsilon_1 z$. Let an initial field $u^0 = U^0 \exp(ikn_0 x \sin\theta^0)$ be given in the initial plane Q: $z = z^0$ (θ^0 is the angle of incidence at the level $z = z^0$, and $n^0 = [\varepsilon(z^0)]^{1/2}$ the refractive index at this level). The oncoming wave is assumed to propagate toward positive z, i.e., toward decreasing permittivity, as illustrated in Fig. 2.9. For simplicity we confine ourselves to the two-dimensional case and take as a ray labeling coordinate $x^0 \equiv \xi$ where the ray leaves the initial plane $z = z^0$ so that $\psi^0 = n^0 \xi \sin\theta^0$.

In these circumstances the equations of characteristics (2.1.5) become

$$\frac{\mathrm{d}x}{\mathrm{d}\tau} = p_x, \qquad \frac{\mathrm{d}z}{\mathrm{d}\tau} = p_z, \qquad \frac{\mathrm{d}p_x}{\mathrm{d}\tau} = 0, \qquad \frac{\mathrm{d}p_z}{\mathrm{d}\tau} = -\frac{\varepsilon_1}{2}.$$

The solution of this system of equations subject to the initial conditions $x(0) = \xi$, $z(0) = z^0$, $p_x(0) = n^0 \sin\theta^0$, and $p_z(0) = n^0 \cos\theta^0$ is as follows:

$$\begin{aligned} x(\tau) &= \xi + \tau n^0 \sin\theta^0, \qquad z = z^0 + \tau n^0 \cos\theta^0 - \varepsilon_1 \tau^2/4\,, \\ p_x(\tau) &= n^0 \sin\theta^0, \qquad p_z(\tau) = n^0 \cos\theta^0 - \varepsilon_1 \tau/2\,. \end{aligned} \tag{2.4.1}$$

This solution yields two "distances" τ_1 and τ_2 and two coordinates ξ_1 and ξ_2 for emanation points of rays on the initial surface Q, viz.,

$$\begin{aligned} \tau_{1,2} &= \frac{2}{\varepsilon_1}\{n^0 \cos\theta^0 \mp [(n^0 \cos\theta^0)^2 - \varepsilon_1 (z - z^0)]^{1/2}\} \\ &= (2/\varepsilon_1)(\alpha \mp \beta^{1/2}), \end{aligned} \tag{2.4.2}$$

$$\begin{aligned} \alpha &= n^0 \cos\theta^0, \qquad \beta = \alpha^2 - \varepsilon_1 (z - z^0) \equiv \varepsilon_1 (z_c - z)\,, \\ \xi_{1,2} &= x - \tau_{1,2} n^0 \sin\theta^0\,. \end{aligned} \tag{2.4.3}$$

Using these parameters yields the eikonal and amplitude by (2.1.11, 13) in the form

$$\psi_{1,2} = n^0 x \sin\theta^0 + \frac{2}{3\varepsilon_1}\{(n^0\cos\theta^0)^3 \mp [(n^0\cos\theta^0)^2 - \varepsilon_1(z - z^0)]^{3/2}\}$$

$$= n^0 x \sin\theta^0 + \frac{2}{3\varepsilon_1}(\alpha^3 \mp \beta^{3/2}) , \tag{2.4.4}$$

$$U_1 = (n^0\cos\theta^0)^{1/2}[(n^0\cos\theta^0)^2 - \varepsilon_1(z - z^0)]^{-1/4} = \alpha^{1/2}\beta^{-1/4} ,$$
$$U_2 = -\mathrm{i}U_1 . \tag{2.4.5}$$

The factor $-\mathrm{i} = \mathrm{e}^{-\mathrm{i}\pi/2}$ relating U_2 with U_1 is caused by the caustic phase shift on a ray that has touched the caustic.

When the observation points are below the caustic surface, $z < z_c \equiv z^0 + (n^0\cos\theta^0)^2/\varepsilon_1$, the quantities τ_j and ξ_j, $j = 1, 2$, are real-valued, they correspond to two real rays arriving at the point of observation. The field u is a sum of two components representing the direct and reflected waves, namely,

$$u = U_1\exp(\mathrm{i}k\psi_1) + U_2\exp(\mathrm{i}k\psi_2)$$
$$= \alpha^{1/2}\beta^{-1/4}\exp(\mathrm{i}kn^0\xi\sin\theta^0 + 2\mathrm{i}k\alpha^3/3\varepsilon_1)$$
$$\times[\exp(2\mathrm{i}k\beta^{3/2}/3\varepsilon_1) - \exp(-2\mathrm{i}k\beta^{3/2}/3\varepsilon_1)] . \tag{2.4.6}$$

Above the caustic, i.e., at $z > z_c$, the quantities τ_j and ξ_j along with the other parameters of the ray field become complex valued, because the parameter $\beta = \alpha^2 - \varepsilon_1(z - z_0)$ takes on a negative value, viz.,

$$\tau_{1,2} = \frac{2}{\varepsilon_1}(\alpha \mp \mathrm{i}|\beta|^{1/2}), \qquad \psi_{1,2} = n^0\xi\sin\theta^0 + \frac{2}{3\varepsilon_1}(\alpha^3 \pm \mathrm{i}|\beta|^{3/2}) .$$

Of the two values of eikonal only one (ψ_1) has a positive real part, which corresponds to an exponential decay of the field beyond the caustic. Therefore, the selection rules require that above the caustic the field should be described by a monomial formula,

$$u = U_1\exp(\mathrm{i}k\psi_1)$$
$$= \alpha^{1/2}|\beta|^{-1/4}\exp\left(\mathrm{i}kn^0\xi\sin\theta^0 + \frac{2\mathrm{i}k\alpha^3 - 2k|\beta|^{3/2}}{3\varepsilon_1}\right) . \tag{2.4.7}$$

This procedure of discarding of extraneous complex rays may be justified by comparison with the available exact solution.

2.4.3 Nonlocal Nature of Complex Rays

From the very method of construction of complex solutions it follows that complex rays treated as lines in complex space cannot be localized in principle

– we simply cannot measure at complex values of coordinates. However, localization of complex rays may be endowed with a limited physical meaning implying the evaluation of a real-valued domain of the field associated with complex rays.

The local nature of the wave field (2.1.14) coupled with real rays manifests itself in that the field at an observation point $\mathbf{r}$ depends only upon the values of the field at the point $\mathbf{r}_0$ where the ray leaves the initial surface Q. In contrast, the fields associated with complex rays are expressed not only through the initial values of the field in real domain, but also through the derivatives of the initial values. As an illustration, represent the complex initial point in the form

$$\mathbf{r}^0 = \mathrm{Re}(\mathbf{r}^0) + \mathrm{i}\,\mathrm{Im}(\mathbf{r}^0) \tag{2.4.8}$$

and expand the initial field $u^0(\mathbf{r}^0)$ entering the ray solution (2.1.14) in a series in $\mathrm{Im}(\mathbf{r}^0)$:

$$\begin{aligned} u^0(\mathbf{r}^0) &= u^0[\mathrm{Re}(\mathbf{r}^0) + \mathrm{i}\,\mathrm{Im}(\mathbf{r}^0)] \\ &= u^0[\mathrm{Re}(\mathbf{r}^0)] + \mathrm{i}[\mathrm{Im}(\mathbf{r}^0), \nabla]u^0[\mathrm{Re}(\mathbf{r}^0)] \\ &\quad + \frac{\mathrm{i}^2}{2}[\mathrm{Im}(\mathbf{r}^0), \nabla]^2 u^0[\mathrm{Re}(\mathbf{r}^0)] + \cdots . \end{aligned} \tag{2.4.9}$$

This expansion indicates that at $\mathbf{r}$ the field (2.1.14) depends not only upon the values of the field u^0 at real-valued points $\mathrm{Re}(\mathbf{r}^0)$, but also upon the derivatives of this field. This fact reflects the diffraction nature of the formation of the field at the point of observation.

A lucid proof of the diffraction effects associated with complex rays is delivered by the properties of a Gaussian beam. *Kravtsov* [2.30] has demonstrated that the field of a Gaussian beam is similar to a spherical wave focused into a complex point. Expressed differently, we may treat Gaussian beams as diffused traces of singularities situated in the complex space (also see [2.32, 33]).

Another evidence of non-locality is provided by the direction of flows of energy. Whereas for real rays the density vector of an energy flow is directed along the ray, $\mathbf{I} \parallel \mathbf{p} = \nabla\psi$ (Sect. 2.1), for the field $u = U\exp(ik\psi)$ with complex-valued U and ψ, the expression for $\mathbf{I}$ reduces to

$$\mathbf{I} = \left[|U|^2\nabla\psi' + \frac{1}{k}(U'\nabla U'' - U''\nabla U')\right]\exp(-2k\psi'') .$$

The second term in this expression may be neglected on account of a slow variation of the amplitude U, then, accurate to the terms of the order of $\mathrm{O}(1/k)$, we have

$$\mathbf{I} \approx |U|^2\exp(-2k\psi'')\mathbf{p}', \qquad \mathbf{p}' \equiv \nabla\psi' . \tag{2.4.10}$$

This formula indicates that the flow of energy is associated only with the real-valued part $\mathbf{p}'$ of vector $\mathbf{p}$ tangent to the complex ray and that the magnitude of the density of energy flow I decays exponentially with the growth of ψ''. The direction of the fastest decay of I is controlled by the imaginary part

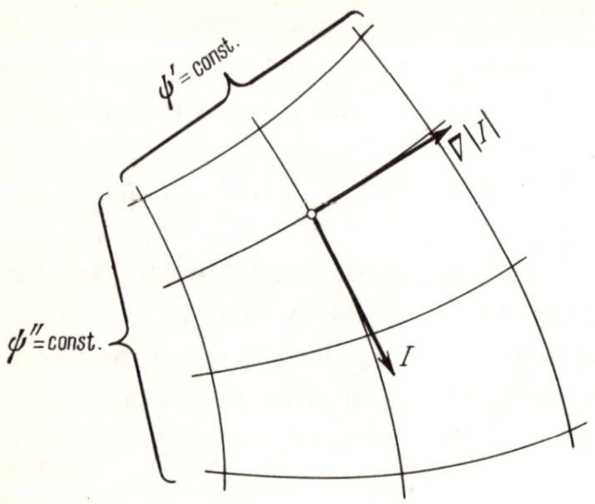

Fig. 2.9. In a nonabsorbing medium, the lines $\psi' =$ constant and $\psi'' =$ constant are orthogonal to each other

$\mathbf{p}'' = \nabla\psi''$. Indeed, recognizing that both U and $\mathbf{p}'$ vary slowly over the wavelength we get

$$\nabla I = [\nabla(|U|^2 p') - 2k\mathbf{p}''|U|^2 p']\exp(-2k\psi'')$$
$$\approx -[2k|U|^2 p' \exp(-2k\psi'')]\mathbf{p}'' \ .$$

For real-valued ε (nonabsorbing medium), vectors $\mathbf{p}'$ and $\mathbf{p}''$ (as well as the surfaces $\psi' =$ constant and $\psi'' =$ constant) are orthogonal to one another, $\mathbf{p}' \cdot \mathbf{p}'' = 0$. The vectors $\mathbf{I}$ and $\nabla|\mathbf{I}|$ are also almost orthogonal, as shown in Fig. 2.9. This pattern is similar to the one occurring for plane inhomogeneous waves where the directions of the flow of energy and of the fastest decay of intensity are orthogonal. This analogy is by no means accidental. Just as the geometric-optical fields associated with real rays are similar to locally plane waves, the fields associated with complex rays play the role of locally plane inhomogeneous waves. This feature of complex geometrical optics has been reflected by *Choudhary* and *Felsen* [2.34].

Thus, for inhomogeneous waves we are not in a position to indicate where the energy is coming from, as the transfer of disturbances is essentially distributed and delocalized.

In absorbing media, the behavior of complex rays is somewhat different than in transparent media. First of all, in absorbing media all rays become complex-valued. The surfaces $\psi' =$ constant and $\psi'' =$ constant are no longer orthogonal, because from the eikonal equation $p^2 = \varepsilon = \varepsilon' + i\varepsilon''$ it follows that $\mathbf{p}' \cdot \mathbf{p}'' = \varepsilon''/2 \neq 0$. If absorption is not very strong, $\varepsilon'' \ll \varepsilon'$, then the complex rays differ from the real rays only insignificantly and the domain of influence is governed primarily by the phase, "Fresnel" considerations. Under strong absorption, $\varepsilon'' \approx \varepsilon'$, the phase relations give way to the amplitude formulas.

2.4.4 Domain of Localization of Complex Rays

The aforementioned features of complex solutions complicate the evaluation of the domain of influence over ordinary geometrical optics. Now the interference phenomena caused by the phase relationships of Fresnel type give way to more complicated phase-amplitude phenomena where attenuation of waves and non-local transfer of wave disturbances are essential. In general, the domain where the field is formed becomes more extended than in the case of real-valued rays.

Analysis of this problem has been carried out in the papers of *Asatryan* and *Kravtsov* [2.35] and *Kravtsov* [2.7]. They assumed the domain of localization to be the width of Gaussian window. Beginning from this width, the field at the observation point "sensed" the presence of the window. It is worth noting that transmission of waves through a Gaussian window to localize the domain of influence has been suggested by *Felsen* and co-workers [2.10].

Analysis of a few examples (total reflection of a wave from the interface between two homogeneous media, reflection from a linear slab) has suggested that the width w of the domain of influence is controlled by the largest of two quantities, the shift δr_{Im} of the emanation point of a ray from the plane of observation to the complex space (i.e., $\delta r_{\mathrm{Im}} = |\mathrm{Im}(\mathbf{r}^0)|$) and the saddle scale a_s characterizing a saddle path in the vicinity of a complex emanation point. This scale is introduced as the coefficient of the square term in the development of the eikonal in powers of the deviation from the complex stationary point and goes to substitute for the Fresnel scale a_f. Thus,

$$w \sim \max(\delta r_{\mathrm{Im}}, a_s) \ . \tag{2.4.11}$$

For the simple caustic, considered in Sect. 2.4.2, in a linear layer, the shift δr_{Im} exceeds the saddle scale, therefore an estimate of w may be obtained as follows:

$$w \sim \delta r_{\mathrm{Im}} = \mathrm{Im}(\xi) = n_0 \sin\theta^0 (z - z_c)^{1/2} \varepsilon_1^{-1/2} \ .$$

The quantity $1/\varepsilon_1$ is the scale of thc lincar slab, which appreciably exceeds the wavelength. Accordingly, the width of the domain of influence $w \sim \sqrt{(z - z_c)/\varepsilon_1}$ is markedly larger than the Fresnel scale $a_f \sim \sqrt{(z - z_c)\lambda}$ which would characterize the size of the influence region for a path $|z - z_c|$. It is quite obvious that this example is rather specific, but it illustrates the general trend: the domain of influence for complex rays is substantially larger than that for real rays.

This circumstance imposes severe constraints on the distinguishability of complex rays, namely, if a few complex paths lead to a given point, their zone of influence will overlap and the rays can no longer be separated. A condition for distinguishability of such rays may be written in the form

$$|\Delta\,\mathrm{Re}(\mathbf{r}^0)| > w \sim \max(\delta r_{\mathrm{Im}}, a_s) \ , \tag{2.4.12}$$

where $|\Delta\,\mathrm{Re}(\mathbf{r}^0)|$ is the distance between the real-valued projections of the complex emanation points.

The applicability conditions of complex geometrical optics may be derived from a requirement that the wave parameters should vary insignificantly within a saddle scale a_s from the complex stationary point, i.e.,

$$a_s |\nabla U| \ll |U| \ . \tag{2.4.13}$$

This condition now replaces the inequalities of the type (2.2.8) related to real rays.

3 Caustics as Catastrophes

A principally new treatment of caustics as catastrophes has been a relatively recent proposition, but the significance of this innovation could hardly be overestimated. The new approach has allowed one to classify caustics, to select among them structurally stable species, and to establish a subordinance for caustics of different complexity. It has been the historically first breakthrough to understanding the nature of caustics.

3.1 Mappings Induced by Rays

3.1.1 The Ray Surface and Lagrange's Manifold

In this section we wish to elicit how the theory of caustics relates with the theory of mappings. To this end we write the equation of ray family leaving an initial surface (see Fig. 2.1) in parametric form:

$$x = x(\xi, \eta, \tau), \qquad y = y(\xi, \eta, \tau), \qquad z = z(\xi, \eta, \tau) \ .$$

Joining the ray coordinates into a multicomponent parameter $\boldsymbol{\xi} = (\xi, \eta, \tau)$ rewrites this relation as

$$\mathbf{r} = \mathbf{R}(\boldsymbol{\xi}) \ . \tag{3.1.1}$$

At $\tau = 0$ these equations describe the initial surface Q. The functions characterizing this problem include the equation of initial surface Q, initial conditions for the eikonal on this surface, and a law representing the variation of medium parameters in space. If all these functions are continuous along with all their derivatives, then (3.1.1) involves infinitely differentiable functions that describe a smooth three-dimensional hypersurface $S_\mathbf{R}$ in the extended six-dimensional space $\{\mathbf{r}, \boldsymbol{\xi}\} = \{x, y, z, \xi, \eta, \tau\}$. This surface $S_\mathbf{R}$ formed by rays will be referred to as a *ray surface*.

Projecting (mapping) a ray surface $S_\mathbf{R}$ from the extended space $\{\mathbf{r}, \boldsymbol{\xi}\}$ onto the three dimensional (configurational) space $\mathbf{r} = \{x, y, z\}$ brings about singularities which would be natural to identify with caustics, for these singularities are where the Jacobian D of the transition from Cartesian to ray coordinates vanishes. A transition over a caustic, i.e., across a locus where the map of $S_\mathbf{R}$ onto $\{x, y, z\}$ has singularities corresponding to birth or annihilation of some even number of rays reaching the observation point $\mathbf{r} = \{x, y, z\}$. From the standpoint of mapping theory a jumpwise appearance or disappearance of pairs of rays is interpreted as a catastrophe. In this case it is interpreted as a qualitative change in the ray pattern in moving from one point to another. For example, moving across a simple caustic from the shadow to lit region gives birth to two rays (Fig. 1.4).

Because we cannot represent the six-dimensional space $\{\mathbf{r}, \boldsymbol{\xi}\}$ on a plane, we illustrate the consideration by taking a simplified model of a ray surface in the three-dimensional space of parameters. Let ξ be the coordinate of the emanation point on the x axis, and $\theta(\xi)$ the angle, measured from the y axis, at which the ray is launched in the x, y plane, as shown in Fig. 3.1. With this notation the equation of ray family in the x, y plane takes on the form

$$x = \xi + y \tan\theta(\xi) \ . \tag{3.1.2}$$

In the extended 3-D space $\{x, y, \xi\}$ this equation describes a two-dimensional ray surface S_R: $\xi = \xi(x, y)$. Figure 3.2 shows schematically such a surface for ray slope varying by the law $\tan\theta = \beta\xi/(\xi^2 + a^2)$. Projecting the ray surface S_R onto the physical plane $\{x, y\}$ gives rise to caustics corresponding to the

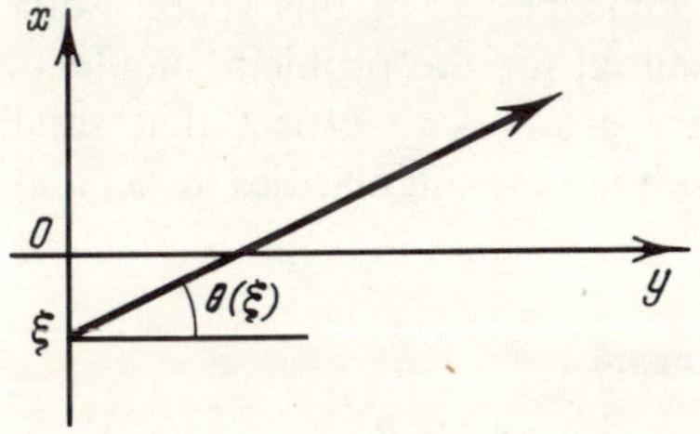

Fig. 3.1. Ray path in the (x, y) plane of a homogeneous medium

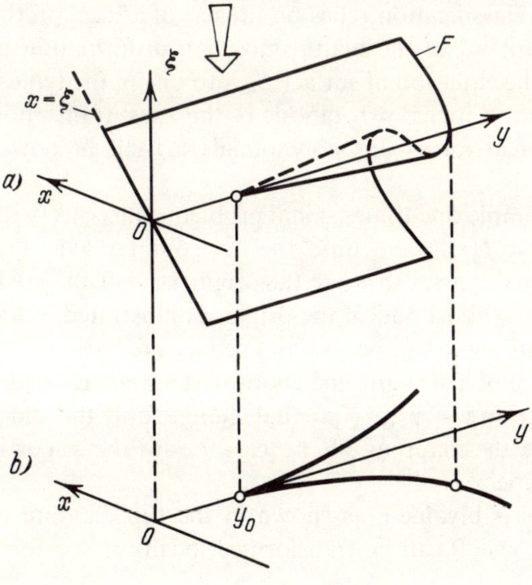

Fig. 3.2. (**a**) Ray surface F in the extended space (x, y, ξ). (**b**) Bifurcation set (cusp) in mapping surface F onto the physical (x, y) plane

singularities of the mapping. In Fig. 3.2 the caustic has the form of a cusp and the corresponding singularity of $S_{\mathbf{R}}$ is known as a fold. Crossing over the caustic inside the cusp increases the number of rays from one to three.

A similar approach to caustics will result from considering the rays in the six-dimensional space $\{\mathbf{r}; \mathbf{p}\} = \{x, y, z; p_x, p_y, p_z\}$, where $\mathbf{p} = \nabla\psi$ is the gradient of the eikonal playing the role of momentum. What was a ray surface becomes in this case a Lagrange's manifold [3.1]. The parametric equation of a Lagrangian manifold $\mathbf{r} = \mathbf{R}\{\xi, \eta, \tau\}$, $\mathbf{p} = \mathbf{P}\{\xi, \eta, \tau\}$ may be obtained from the ray equations (2.1.5) written in Hamiltonian form.

A more general treatment of caustics is achieved by specifying the ray surface in the extended space of parameters $\{\mathbf{r}, \boldsymbol{\alpha}\}$. The dimension of this space is increased by including all the parameters essential for the problem. By way of example, these parameters α_k, $k = 1, \ldots, K$, may characterize the position of the source; the point where the ray leaves the initial surface Q; the derivatives of the refractive index in certain directions; the form of the initial surface Q; etc.

The total number of variables H forms along with three Cartesian coordinates an extended space $\mathbf{w} = \{\mathbf{r}, \boldsymbol{\alpha}\}$ of dimension $K + 3$, where caustics are singularities of a mapping of a ray surface $S_{\mathbf{R}}$ onto a subspace of lower dimensionality. More often than not researchers are interested in mappings of $S_{\mathbf{R}}$ onto the physical space $\{x, y, z\}$ or on some planes in this space. Some situations require that some parameters essential for the problem should be mapped onto a plane, especially if there are grounds to expect that small variations of one of the parameters might result in new singularities in $\{x, y, z\}$.

3.1.2 Classification of Structurally Stable Caustics

Catastrophe theory has been a useful formalism enabling a classification of structurally stable caustics. The classification relies on the use of a local (in the neighborhood of a singular point $\mathbf{w}_0$ of the map), smooth transformation of variables, $\mathbf{w} - \mathbf{w}_0 \to \boldsymbol{\zeta}$, to carry the equation of surface $S_{\mathbf{R}}$ into one of the typical (normal) forms. The transformation includes rotations of the axes, translation and change of scales. The typical forms are polynomials arising in power expansions.

As an illustration we take a simple one-dimensional problem concerned with the projection of a smooth curve $f(x, y) = 0$ onto the horizontal x axis. For simplicity we assume that this curve passes through the origin, i.e., $f(0, 0) = 0$. If $y = y(x)$ is monotonous in the neighborhood of the origin, as illustrated in the top diagram of Fig. 3.3a, then this curve can be straightened, i.e., reduced to the equation $f_1(x_1, y_1) = y_1 - x_1 = 0$ of a straight line [bottom diagram at (a)] by a smooth one-to-one transformation $y \to y_1$, $x \to x_1$ that changes only the scales in the axes x and y. In this case the mapping of $f(x, y) = 0$ onto the x axis is one-to-one and has no singularities.

If the dependence of y on x is bivalued, as shown in the top diagram of Fig. 3.3b, then the function $f(x, y) = 0$ can be transformed locally to the form

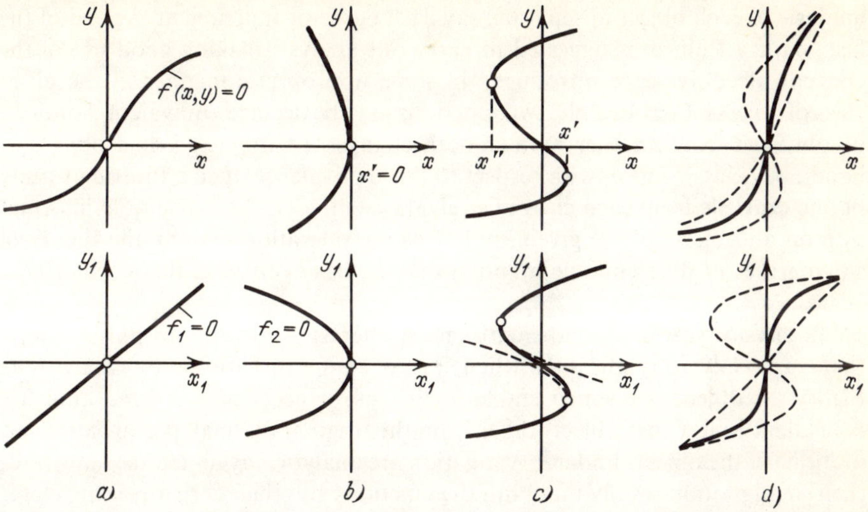

Fig. 3.3. Simple typical situations in mapping a smooth curve $f(x, y) = 0$ onto the horizontal axis. (**a**) One-to-one map. (**b**) Map with one critical point. (**c**) Map with two critical points. (**d**) Structurally unstable (nontypical) situation (solid lines) which is reduced by small perturbations (dashed lines) to either (a) or (c)

$f_2(x_1, y_1) = x_1 + y_1^2 = 0$ (bottom diagram). In this mapping of $f_2(x_1, y_1) = 0$ on the x_1 axis there is only one singular point, $x_1 = 0$, where $|dy_1/dx_1| = \infty$, i.e., the curve $f_2(x_1, y_1) = 0$ has a vertical tangent, much like the original curve $f(x, y) = 0$.

If the dependence of y on x is three-valued, as in the top diagram of Fig. 3.3c, the mapping has two singular points, x_0' and x_0'', in which $|dy/dx| = \infty$. Moving through these points gives rise to catastrophes which are jumpwise variations of the number of branches uniquely projected onto the x axis. In the neighborhood of each of the two singular points, a smooth transformation reduces the function $f(x, y) = 0$ locally to a second-order polynomial (bottom diagram).

The linear ($f_1 = y_1 - x_1$) and square ($f_2 = x_1 + y_1^2$) polynomials render examples of the normal forms to which virtually all curves in the x, y plane may be reduced. Exclusions are some degenerate cases corresponding to structurally unstable maps.

An example of a structurally unstable mapping is displayed in the top diagram of Fig. 3.3d. It shows a curve in which the point with the vertical tangent is simultaneously a point of inflection. Locally this curve reduces to the form $f_3(x_1, y_1) = x_1 - y_1^3 = 0$, in the bottom of Fig. 3.3d. If small variations of the problem parameters carry the curve $f(x, y) = 0$ to one of two dashed plots in the top diagram of Fig. 3.3d, then this curve may be reduced either to a linear function like in Fig. 3.3a, or to two square polynomials locally fit to the singular points, as in the bottom diagram of Fig. 3.3c.

Mathematics deals with objects of different nature: curves, functions, families of functions, maps, caustics, and the like. Instead of having to attempt an analysis of each object in separate, say each curve or function, at the end of the last century Poincaré suggested to carry out analysis making good use of the concept of equivalence introduced in some appropriate manner in the given class of objects. For example, two functions may be deemed equivalent if one can be obtained from another by a smooth change of variables. When objects at hand obey classification with respect to this equivalence, then a thorough study of one example from each class of equivalence may yield considerable information on all objects of the given kind. This consideration underly the theory of singularities of differentiable mappings which has evolved in the recent 10–15 years.

In physics as well as mathematics most interest is attracted to stable objects that vary only insignificantly when forced by perturbations. In catastrophe theory an object is deemed stable if all close objects of the same kind are equivalent to it. Close objects differ from the original by small perturbations or motions. If the objects under investigation are analytic, say functions, maps, etc., then small motions imply that both the functions and their derivatives are close.

The concept of *structural stability*, fundamental to catastrophe theory, is allied to the concept of roughness introduced into dynamic system theory by *Andronov* [3.2]. Transitions from one normal form to another via a structurally unstable state correspond to bifurcations in dynamic system theory. For example, a crossover from one-ray mode to three-ray mode observed in increasing y in Fig. 3.2b occurs at the point of bifurcation y_0.

While all singularities that occur in mapping $f(x, y) = 0$ onto the x axis can be classed by common analysis, in a multidimensional case, classification is markedly complicated by the fact that the typical polynomials are also multidimensional. Therefore, sequential classification of structurally stable singularities called for invoking the ideas from topology and differential geometry. *Arnold* [3.3] took advantage of the relation of the classification with the Lie group theory. For the history of the topic and early works in the field, the reader is referred to [3.3–9].

By the time of the writing of this book all typical forms of caustics in multidimensional spaces up to dimension $m = 10$ have been revealed and analytical expressions have been obtained for the structurally stable singularities of mappings onto spaces of lower dimension. Below we outline the procedure without having to dwell on mathematical niceties. The universal classification developed by *Arnold* [3.9] will be adhered to.

A notion worth making here concerns the contents of the term "small perturbations as used in mathematics and in physics. The difference consists that in mathematics this term implies that both the perturbation $\tilde{f}$ and its derivatives are small. To demonstrate this peculiarity of small stirring we point out that the sine perturbation

$$\tilde{f} = a \sin(\kappa y_1) \tag{3.1.3}$$

may be a small movement provided that simultaneously both a and κ are small so that the derivative, $|\tilde{f}'| \sim a\kappa$, be small. If the parameter $a\kappa$ is too large,

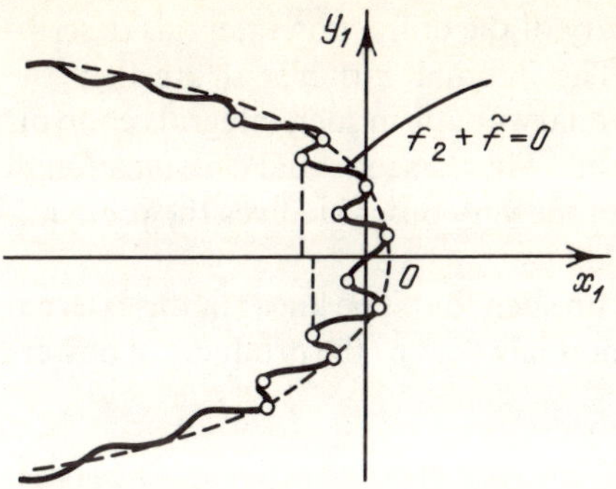

Fig. 3.4. Mapping a smooth curve $f_2 = x_1 + y_1^2 = 0$ perturbed by a small sinusoidal component $\tilde{f} = a \sin \kappa y_1$ onto the axis gives rise to a multitude of degenerate points

$a\kappa \gg 1$, then the perturbation (3.1.3) can no longer be treated as small stirring. If we add the perturbation (3.1.3) to the parabola $f_2 = x_1 + y_1^2$, then the perturbed curve

$$f_2(x_1, y_1) + \tilde{f}(x_1, y_1) = x_1 + y_1^2 + a\sin(\kappa y_1) = 0$$

may acquire many additional critical points, as illustrated in Fig. 3.4. The number of these points is estimated to be κa. This example demonstrates that small perturbations and associated representations concerning structural stability are treated in mathematics somewhat differently than in physics.

3.2 Classification of Typical Caustics

3.2.1 Generating Function: Codimension and Corank

A most economical method of classification of caustics relies upon the polynomial generating functions

$$\Phi(\boldsymbol{\zeta}, \mathbf{t}) = \Phi_0(\mathbf{t}) + \sum_{p=1}^{m} \Phi_p(\mathbf{t})\zeta_p \tag{3.2.1}$$

where $\boldsymbol{\zeta} = \{\zeta_1, \zeta_2, \ldots, \zeta_m\}$ are the *external* parameters obtained by applying a local transformation (rotations, translations and rescaling) to the initial physical parameters $\mathbf{w} = \{\mathbf{r}, \boldsymbol{\alpha}\}$, and $\mathbf{t} = \{t_1, t_2, \ldots, t_l\}$, i.e., the *internal* or *state* variables, which are introduced to obtain a manageable description of typical singularities of caustics. The number of external variables, m, essential for the problem at hand is called the *codimension* of caustics, and the number of internal variables, l, is called the *corank* of a caustic or its *internal dimensionality*.

Generating functions (3.2.1) are selected such that they may serve as the phase functions in diffraction integrals of the form

$$\int \exp[\mathrm{i}\Phi(\boldsymbol{\zeta}, \mathbf{t})]\, \mathrm{d}^l t \ .$$

We shall revisit such integrals in Chap. 4.

Corank l corresponds to the multiplicity of the diffraction integrals describing the field in the vicinity of caustics. The internal variables t_k are then the integration variables. The number of stationary points in such integrals controls the number of rays for a given type of caustic. Thus, a sequential classification of caustics as purely geometrical objects from the very onset involves the prerequisites for the construction of the wave field.

The generating function (3.2.1) is built up such that it be linear in the external variables ζ_p. The coefficients of ζ_p are monomials Φ_p, i.e., the products of powers of t_k, namely,

$$\Phi_p(\mathbf{t}) = C_p \Pi t_k^{s_{pk}}, \quad s_{pk} \geq 0, \; p \geq 1 \; . \tag{3.2.2}$$

Each type of caustic is characterized by its own values of l and m, and their universal functions $\Phi_p(t)$ of which the most important role is played by the function $\Phi_0(t)$ called the *unfolding* of the catastrophe.

Typical (normal) forms of ray surface S_R result by differentiating the generating function Φ with respect to t:

$$\frac{\partial \Phi}{\partial t_k} = 0, \quad k = 1, 2, \ldots, l \; . \tag{3.2.3}$$

Setting these derivatives to zero corresponds to the evaluation of stationary points of the phase function $\Phi(\zeta, \mathbf{t})$ over the space $\mathbf{t} = \{t_1, \ldots, t_l\}$. The position of stationary points $\mathbf{t}_\mathrm{st}$ and the values of the phase function $\Phi(\boldsymbol{\zeta}, \mathbf{t}_\mathrm{st})$ at these points depend on the external variables $\boldsymbol{\zeta} = \{\zeta_1, \ldots, \zeta_m\}$.

Equations (3.2.3) set up l ties between parameters $\mathbf{t}$ and $\boldsymbol{\zeta}$, so that in the external space of external and internal variables $(\boldsymbol{\zeta}, \mathbf{t})$ of dimension $m + l$ the ray surface S_R has $(m + l) - l = m$ dimensions. Singularities of this surface occur when it is projected (mapped) from the extended space $\{\zeta_1, \ldots, \zeta_m; t_1, \ldots, t_l\}$ of all problem variables, where S_R is given by (3.2.3), onto the space $\{\zeta_1, \ldots, \zeta_m\}$ of external variables.

An equation for the locus of the map singularities may be obtained by setting to zero the determinant composed of the second order partial derivatives of Φ, known as the Hessian,

$$h \equiv \det \left[\frac{\partial^2 \Phi}{\partial t_j \partial t_k} \right] . \tag{3.2.4}$$

Segments where the Hessian is nonzero allow local unique projection onto the space of external parameters, whereas segments where it vanishes give rise to singularities or catastrophes identified as caustics.

3.2.2 Caustic Surfaces of Low Codimension

Table 3.1 summarizes the typical generating functions Φ and equations of ray surface S_R for the simplest types of singularities, known as the Thom's seven elementary catastrophes. A lucid illustration of the occurrence of singularities in

Table 3.1. Simplest caustics (seven elementary catastrophes)

Name and symbol	Generating function $\Phi(\boldsymbol{\zeta}, \mathbf{t})$	Equation of ray surface	Indexes of caustic zone $\alpha_1, \ldots, \alpha_m$	Focusing index, σ_{foc}
Fold (A_2)	$\frac{1}{3}t_1^3 + \zeta_1 t_1$	$t_1^2 + \zeta_1 = 0$	2/3	1/6
Cusp (A_3)	$\pm\frac{1}{4}t_1^4 + \zeta_1 t_1 + \frac{1}{2}t_1^2\zeta_2$	$\pm t_1^3 + \zeta_1 + \zeta_2 t_1 = 0$	3/4, 1/2	1/4
Swallowtail (A_4)	$\frac{t_1^5}{5} + \zeta_1 t_1 + \frac{\zeta_2 t_2^2}{2} + \frac{\zeta_3 t_1^3}{3}$	$t_1^4 + \zeta_1 + \zeta_2 t_1 + \zeta_3 t_1^2 = 0$	4/5, 3/5, 2/5	3/10
Hyperbolic umbilic (D_4^-) Elliptic umbilic (D_4^+)	$\pm t_1 t_2^3 + t_1^2 t_2 + \zeta_1 t_1 + \zeta_2 t_2 + \zeta_3 t_2^2$	$\pm 3t_2^2 - \zeta_2 + 2\zeta_3 t_2 = 0$ $2t_1 t_2 + \zeta_1 = 0$	2/3, 2/3, 1/3	1/3
Butterfly (A_5)	$\pm\frac{t_1^6}{6} + \zeta_1 t_1 + \frac{\zeta_2 t_1^2}{2} + \frac{\zeta_3 t_1^3}{3} + \frac{\zeta_4 t_1^4}{4}$	$\pm t_1^5 + \zeta_1 + \zeta_2 t_1 + \zeta_3 t_1^2 + \zeta_4 t_1^3 = 0$	5/6, 2/3, 1/2, 1/3	1/3
Parabolic umbilic (D_5)	$\pm t_2^4 + t_1^2 t_2 + \zeta_1 t_1 + \zeta_2 t_2 + \zeta_3 t_2^2 + \zeta_4 t_2^3$	$2t_1 t_2 + \zeta_1 = 0$ $\pm 4t_2^3 + t_1^2 + \zeta_2 + 2\zeta_3 t_2 + 3\zeta_4 t_2^2 = 0$	5/8, 3/4, 1/2, 1/4	3/8

mappings can be given only for two examples, namely, a fold (type A_2, $m = 1$, $l = 1$) and a cusp catastrophe (type A_3, $m = 2$, $l = 1$). Singularities of the fold type correspond to a non-degenerate caustic (Fig. 3.5), while the surface in Fig. 3.6 corresponds to a caustic cusp with a single point of reflection. We have already met the last type of caustic in Fig. 3.1.

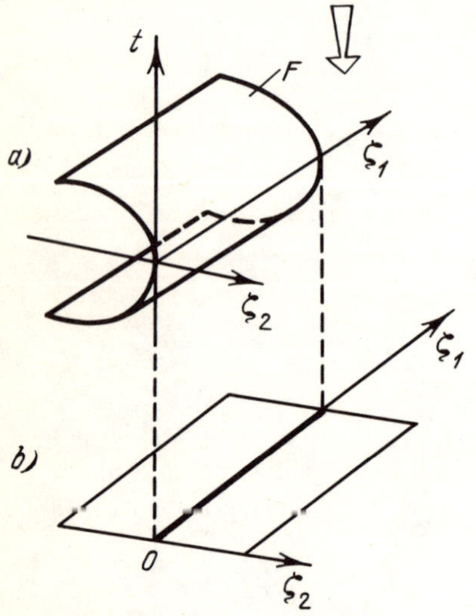

Fig. 3.5. Simplest singularity (fold type A_2, $m = 1$, $l = 1$) corresponding to a nondegenerate caustic

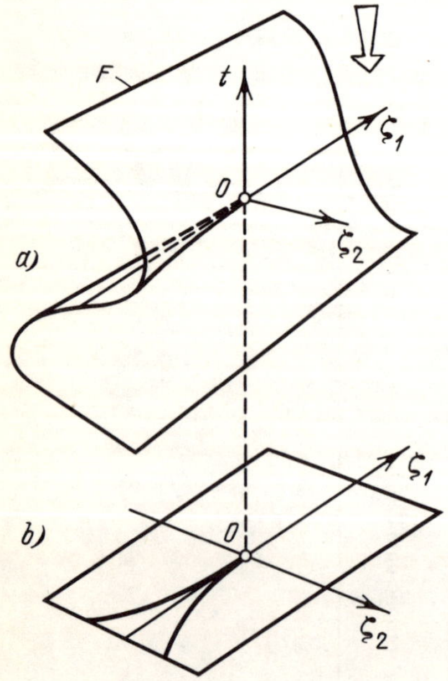

Fig. 3.6. Ray surface and the bifurcation set of the cusp catastrophe (type A_3, $m = 2$, $l = 1$)

The next complex singularity, swallowtail (type A_4, $m = 3$, $l = 1$) is already a three-dimensional surface (Fig. 3.7). The umbilic types of caustic are also represented by three-dimensional surfaces. These are the hyperbolic umbilic (D_4^+) shown in Fig. 3.8, and the elliptic umbilic (D_4^-), shown in Fig. 3.9. These caustics have codimension $m = 3$ and corank $l = 2$.

Caustics of higher codimension with $m > 3$ may be illustrated only by sections with planes. By way of example, Fig. 3.10a shows a three-dimensional section of an A_5 type caustic ($m = 4$, $l = 1$) called the butterfly, and the diagram at (b) shows two-dimensional cross sections of this bifurcation set. For all caustics the maximum number of ray is dictated by codimension and equals $m + 1$.

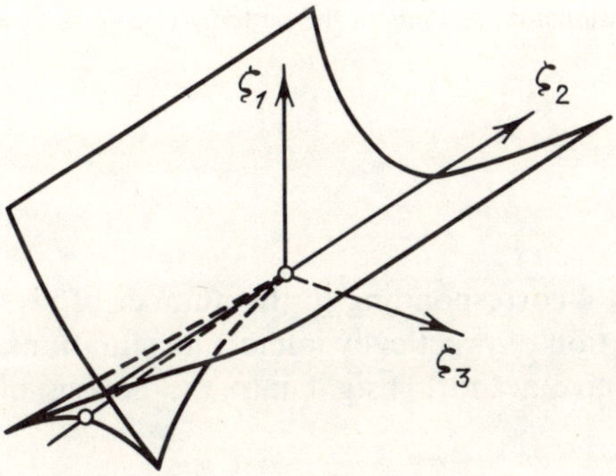

Fig. 3.7. Bifurcation set of the swallowtail (type A_4, $m = 3$, $l = 1$)

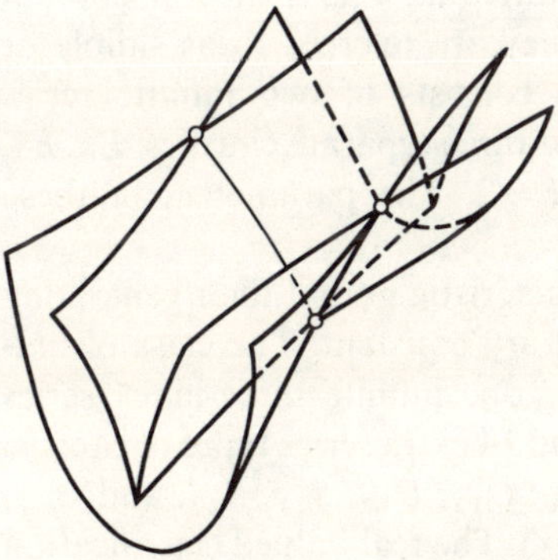

Fig. 3.8. Bifurcation set of the hyperbolic umbilic (type D_4^+, $m = 3$, $l = 2$)

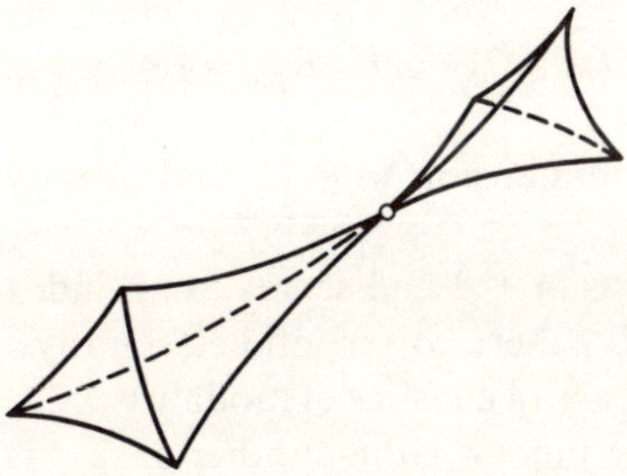

Fig. 3.9. Bifurcation set of the elliptic umbilic (D_4^-, $m = 3$, $l = 2$)

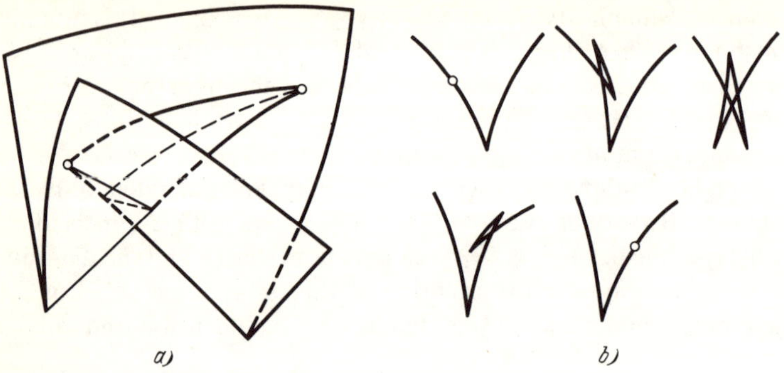

Fig. 3.10. Three-dimensional (**a**) and two-dimensional (**b**) sections of the butterfly caustic surface (A_5, $m = 4$, $l = 1$)

3.2.3 Caustics of High Codimension

The listed caustics of codimension $m \leq 4$ corresponding to the number of rays $m + 1 \leq 5$ cover a considerable proportion of practically interesting situations. None the less it would be useful to have a short insight into the genesis of caustics having higher codimension.

a) Simple Caustics. The caustics described in Sect. 3.2.2 have one feature in common, namely, the associated generating functions $\Phi(\zeta, \mathbf{t})$ do not contain arbitrary constants or "modules". Therefore they are referred to as simple or null-modal. A complete list of such caustics consists of two infinite series A_{m+1} ($m \geq 1, l = 1$) and D_{m+1} ($m \geq 3, l = 2$) and three separate caustics: E_6, E_7, and E_m having respectively $m = 5, 6, 7$ and $l = 2$. The parameters of these caustics are summarized in Table 3.2.

b) Unimodal Caustics. These caustics are characteristic in that their generating functions contain only one uneliminated arbitrary constant. The class of unimodal caustics [3.9] is completely covered by one infinite three-index series $T_{p,q,r}$ ($4 \leq p \leq q \leq r, m = p + q + r - 3 \geq 9$) and 14 extra series types of caustic: K_{12}, K_{13}, K_{14}, Z_{11}, Z_{12}, Z_{13}, W_{12}, W_{13}, Q_{10}, Q_{11}, Q_{12}, S_{11}, S_{12}, and U_{12}. These symbols relate to the complex classification. The real-valued classification brings forth the parabolic singularities $X_9 = T_{2,4,4}$, $J_{10} = T_{2,3,6}$, $P_8 = T_{3,3,3}$ and a series of hyperbolic singularities J_{n+2} ($m \geq 9$), X_{m+2} ($m \geq 8$), $Y_{p,q}$ ($5 \leq p < q$, $m = m + q - 1 \geq 9$), P_{m+2} ($m \geq 7$), and $R_{p,q}$ ($4 \leq p \leq q$, $m = p + q \geq 8$).

Examples of unimodal caustics having codimension $m \leq 11$ and corank $l = 2$ are compiled in Table 3.3.

For unimodal caustics, the number of rays is $m + 2$ and serves as an index for most caustics. For example, to the caustic J_{11} there correspond eleven rays.

c) Caustics of Higher Modality. In the general case of caustics of modality μ, i.e., with μ uneliminated constants in the generating function, the number of rays is

Table 3.2. Classification of simple caustics

Type	m	$(\boldsymbol{\zeta}, \mathbf{t})$	$\alpha_1, \ldots, \alpha_m$	σ_{foc}
A_{m+1}	≥ 1	$\pm \frac{1}{m+2} t^{m+2} + \sum_{p=1}^{m} \zeta_p \frac{t^p}{p}$	$\frac{m+1}{m+2}, \frac{m}{m+2}, \ldots, \frac{2}{m+2}$	$\frac{1}{2} - \frac{1}{m+2}$
D_{m+1}	≥ 3	$\pm t_2^m + t_1^2 t_2 + \zeta_1 t_1 + \sum_{p=2}^{m} \zeta_p t_2^{p-1}$	$\frac{1}{2} + \frac{1}{2m}, \frac{m-1}{m}, \ldots, \frac{1}{m}$	$\frac{1}{2} - \frac{1}{2m}$
E_6	5	$t_1^3 \pm t_2^4 + \zeta_1 t_1 + \zeta_2 t_2 + \zeta_3 t_2^2 + \zeta_4 t_1 t_2 + \zeta_5 t_1 t_2^2$	$\frac{2}{3}, \frac{3}{4}, \frac{1}{2}, \frac{5}{12}, \frac{1}{6}$	$\frac{5}{12}$
E_7	6	$t_1^3 + t_1 t_2^3 + \zeta_1 t_1 + \sum_{p=2}^{5} \zeta_p t_2^{p-1} + \zeta_6 t_1 t_2$	$\frac{2}{3}, \frac{7}{9}, \frac{5}{9}, \frac{1}{3}, \frac{1}{9}, \frac{4}{9}$	$\frac{4}{9}$
E_8	7	$t_1^3 + t_2^5 + \zeta_1 t_1 + \sum_{p=2}^{4} \zeta_p t_2^{p-1} + \zeta_5 t_1 t_2$ $+ \zeta_6 t_1 t_2^2 + \zeta_7 t_1 t_2^3$	$\frac{2}{3}, \frac{4}{5}, \frac{3}{5}, \frac{2}{5}, \frac{7}{15}, \frac{4}{15}, \frac{1}{15}$	$\frac{7}{15}$

Table 3.3. Unimodal caustics of $m \leq 11$ and corank 2

Type	m	$\Phi_0(\mathbf{t})$	$\Phi_{\mathrm{p}}(\mathbf{t})$	σ_{foc}
X_{m+2}	≥ 8	$t_1^4 + t_1^2 t_2^2 + a t_2^{m-3}$	$t_1, t_1^2, t_1^3, t_1 t_2, t_2, \ldots, t_2^{m-4}$	1/2
J_{m+2}	≥ 9	$t_1^3 + t_1^2 t_2^2 + a t_2^{m-2}$	$t_1, t_1, t_2, t_1 t_2^2, t_2, \ldots, t_2^{m-3}$	1/2
$Y_{p,q}$	$p + q - 1 \geq 9$	$t_1^p + t_1^2 t_2^2 + a t_2^q$	$t_1, \ldots, t_1^{p-1}, t_1 t_2, t_2, \ldots, t_2^{q-1}$	1/2
Z_{11}	9	$t_1^3 t_2 + t_2^5 + a t_1 t_2^4$	$t_1, t_1^2, t_1 t_2, \ldots, t_1 t_2^3, t_2, \ldots, t_2^4$	8/15
Z_{12}	10	$t_1^3 t_2 + t_1 t_2^4 + a t_1^2 t_2^3$	$t_1, t_1^2, t_1 t_2, \ldots, t_1 t_2^3, t_1^2 t_2^2,$ $t_2, \ldots, t_2^8$	6/11
Z_{13}	11	$t_1^3 t_2 + t_2^6 + a t_1 t_2^5$	$t_1, t_1^2, t_1 t_2, \ldots, t_1 t_2^4, t_2, \ldots, t_2^5$	5/9
W_{12}	10	$t_1^4 + t_2^5 + a t_1^2 t_2^3$	$t_1, t_1^2, t_1 t_2, \ldots, t_1 t_2^3, t_1^2 t_2,$ $t_2, \ldots, t_2^3$	11/20
W_{13}	11	$t_1^4 + t_1 t_2^4 + a t_2^6$	$t_1, t_1^2, t_1 t_2, t_1 t_2^2, t_1^2 t_2, t_1^2 t_2^2,$ $t_2, \ldots, t_2^5$	9/16
K_{12}	10	$t_1^3 + t_2^7 + a t_1 t_2^5$	$t_1, t_1 t_2, \ldots, t_1 t_2^4, t_2, \ldots, t_2^5$	11/21
K_{13}	10	$t_1^3 + t_1 t_2^5 + a t_2^8$	$t_1, t_1 t_2, \ldots, t_1 t_2^3, t_2, \ldots, t_2^7$	8/15

$m + \mu + 1$. Beginning with codimension $m = 10$ the list of singularities acquires the five-modal caustic O_{16}, and at $m = 11$ the bimodal caustics Q_{14}, S_{14}, U_{14} and the three-modal caustic V_{15}.

Table 3.4 summarizes caustics of codimension $m \leq 11$ along with corank and modality. Caustics of corank $l = 3$ are seen to occur at $m = 6$ and upwards. Unimodal caustics also appear beginning with $m = 6$, and the caustics of higher modality ($l > 1$) occur at $m = 10$, the number of rays ($m + \mu + 1$) being at 8 and 16, respectively. Ray patterns with this number of rays are rare in practical research and very hard to analyze. Quite reasonably, no application has yet been reported with identified caustics of corank $m = 3$, nor unimodal caustics to say

Table 3.4. Classification of caustics having codimension $m \leq 11$ and corank l

m																					
1	A_2																				
2	A_3																				
3	A_4	$D_4^{\pm}$																			
4	A_6	D_6																			
5	A_6	D_6	E_6																		
6	A_7	D_7	E_7								P_8										
7	A_8	D_8	E_8			X_9					P_9										
8	A_9	D_9		J_{10}		X_{10}					P_{10}	$R_{4,4}$				Q_{10}					
9	A_{10}	D_{10}		J_{11}		X_{11}	$Y_{5,5}$		Z_{11}		P_{11}	$R_{4,5}$		$T_{4,4,4}$		Q_{11}	S_{11}				
10	A_{11}	D_{11}		J_{12}	K_{12}	X_{12}	$Y_{5,6}$		Z_{12}	W_{12}	P_{12}	$R_{4,6}$	$R_{5,5}$	$T_{4,4,5}$		Q_{12}	S_{12}	U_{12}			O_{16}
11	A_{12}	D_{12}		J_{13}	K_{13}	X_{13}	$Y_{5,7}$	$Y_{6,6}$	Z_{13}	W_{13}	P_{13}	$R_{4,7}$	$R_{6,6}$	$T_{4,4,6}$	$T_{4,5,5}$				Q_{14} S_{14} U_{14}	U_{15}	
	Simple			Unimodal															Bimodal	3 mod	5 mod
	$l = 1$					$l = 2$										$l = 3$					$l = 4$

Table 3.5. Number of types of caustics of given codimension

Codimension	1	2	3	4	5	6	7	8	9	10	11
Total number	1	1	2	2	3	4	5	7	11	16	18
No with corank $l \leq 2$	1	1	2	2	3	3	4	4	6	8	9

nothing of higher modality caustics. Table 3.4 gives us a hope for a more focused search for caustic structures predicted by catastrophe theory. For a given codimension m, the number of caustic types increases, not very fast, at least up to $m = 7$ or 8. This fact is evident from Table 3.5 compiled on the basis of Table 3.4. Indeed, at $m = 7$ (8 rays) there exist only 5 types of essentially different caustic structures, and at $m = 8$ (9 rays) there are seven types.

Analysis of the data summarized in Tables 3.3–5 indicates the free parameters a_λ, $\lambda = 1, \ldots, \mu$, enter the unfoldings linearly, like the external variables ζ_p. Reasons why a_λ should not be identified with external variables ζ_p evolve from rather delicate reasoning and are conditional to a considerable extent.

In applications the difference between ζ_p and a_λ is inessential, and the sum $m + \mu = M$ may be viewed as the generalized codimension of a caustic. Then, the number of rays is $N = M + 1$.

3.2.4 Subordinance Relations

Caustics of different complexity are tied by certain relations which can be represented in the form of dominance sets or adjacency graphs. These graphs are useful vehicles in predicting how a given type of caustic may degenerate when some or other parameters are changed. Lists of important adjacencies are available in many books, see, e.g., [3.5–10].

4 Typical Integrals of Catastrophe Theory

This chapter is devoted to standard integrals associated with typical caustics. We present the simplest integrals corresponding to the seven Thom's catastrophes and also some more complex integrals associated with caustics that occur in series. A novel point in this exposition is the one of Fresnel's criteria for passing over from some normal forms to others when external parameters vary.

4.1 Standard Caustic Integrals

4.1.1 Use of Generating Functions as Phase Functions

Catastrophe theory has not only elicited the typical forms of caustics but also radically cleared the problem of the typical or standard integrals describing the field near caustics. The type of such integrals, also known as wave catastrophe functions, is defined by generating functions $\Phi(\boldsymbol{\zeta}, \mathbf{t})$ which depend on external ($\boldsymbol{\zeta}$) and internal ($\mathbf{t}$) variables. Integration of the oscillating function $\exp[\mathrm{i}\Phi(\boldsymbol{\zeta}, \mathbf{t})]$ with respect to the "superfluous" parameters $t_1, \ldots, t_l$ leads to what is known as the typical caustic integral [4.1–4]

$$I(\zeta) = (2\pi)^{-1/2} \int \mathrm{d}^l t \exp[\mathrm{i}\Phi(\boldsymbol{\zeta}, \mathbf{t})] \; . \tag{4.4.1}$$

As a result, the multiplicity of typical caustic integrals is governed by the number of the internal parameters, i.e., by the corank l of the caustic, and the number of independent external variables $\zeta_1, \ldots, \zeta_m$ is controlled by the codimension m of the caustic.

In the case of modal caustics, the exponent also includes μ arbitrary coefficients (modules) $a_1, \ldots, a_\mu$ which, like the external variables ζ_m, enter the generating function Φ as linear terms (Sect. 3.2.3). Making use of this circumstance and thinking of modules a_λ conditionally as independent variables ζ_p with $p = m+1, m+2, \ldots, m+\mu$, then the total number of variable parameters is $M = m + \mu$, that is, the generalized codimension. In agreement with the general statements of catastrophe theory, for a given caustic the maximum number of rays N is the generalized codimension plus one: $N = M + 1 = m + \mu + 1$.

In many circumstances it is convenient to deal with another typical integral

$$\tilde{I}(\boldsymbol{\zeta}) = (k/2\pi)^{1/2} \int \exp[\mathrm{i}k\Phi(\tilde{\zeta}, \tilde{\mathbf{t}})]\mathrm{d}^l \tilde{t} \; , \tag{4.1.2}$$

whose phase function Φ is equipped with the dimensional factor $k = \omega/c$ which is the wavenumber of the problem at hand. Incorporating this factor we have to substitute new rescaled dimensional variables $\tilde{\boldsymbol{\zeta}}$, $\tilde{\mathbf{t}}$ for dimensionless variables,

$$\begin{aligned} \tilde{\zeta}_p &= k^{-\alpha_p}\zeta_p, \quad p = 1, \ldots, m, m+1, \ldots, M ; \\ \tilde{t}_q &= k^{-\gamma_q} t_q, \quad q = 1, \ldots, l . \end{aligned} \tag{4.1.3}$$

The possibility of this rescaling transformation stems from the power-function behavior of all functions entering the generating function $\Phi(\boldsymbol{\zeta}, \mathbf{t})$. The exponents α_p and γ_q are determined from the invariance of the phase functions in coming from $\boldsymbol{\zeta}$, $\mathbf{t}$ to $\tilde{\boldsymbol{\zeta}}$, $\tilde{\mathbf{t}}$, namely,

$$\Phi(\boldsymbol{\zeta}, \mathbf{t}) = k\Phi(\tilde{\boldsymbol{\zeta}}, \tilde{\mathbf{t}}) . \tag{4.1.4}$$

Specifically, γ_q are obtained from the invariance of the unfolding Φ_0. For example, for simple caustics of type A_{m+1}, the unfolding $\Phi_0(t_1) = t_1^{m+2}/(m+2)$, so that from

$$\Phi_0(t_1) = \frac{t_1^{m+2}}{m+2} = k\Phi_0(\tilde{t}_1) = k\frac{(k^{-\gamma_1}t_1)^{m+2}}{m+2}$$

it follows:

$$\gamma_1 = 1/(m+2) . \tag{4.1.5}$$

Indexes γ_q may be connected with the dimensions of the domain essential for the integration with respect to $\mathbf{t}$ or $\tilde{\mathbf{t}}$, to be more specific, they define the frequency dependence of the Fresnel zones responsible for the formation of the "fundamental" part of the field at the point of interest. Accordingly, the indexes γ_q might be called Fresnel. How they relate to the actual dimensions of the wave-field formation zone will be illustrated below with specific examples.

As to α_p, they are readily found from the invariance of the terms $\Phi_p(t_1)\zeta_p$ linear with respect to ζ_p. For simple A_{m+1} caustics, when $\Phi_p(t_1) = t_1^p/p!$, from

$$p!\Phi(t_1)\zeta_p = t_1^p\zeta_p = k\Phi_p(\tilde{t}_1)\tilde{\zeta}_p = k(k^{-\gamma_1}t_1)^p(k^{-\alpha_p}\zeta_p)$$

it follows:

$$\alpha_p = 1 - p\gamma_1 = 1 - \frac{p}{m+2} . \tag{4.1.6}$$

These exponents α_p are called *caustic indices.* They define the frequency dependence of spacings between interference fringes formed near caustics [4.1–4] and the frequency dependence of the width of caustic zone [4.5–7]. For lower types of caustics, the values of α_p are summarized in Tables 3.1, 2. Some specific estimates based on these indexes have been reported [4.5–7] and will be displayed in the following sections.

Clearly, crossover to dimensional variables $\tilde{\boldsymbol{\zeta}}$, $\tilde{\mathbf{t}}$ renders a typical integral I also dimensional.

The integrals (4.1.1) and (4.1.2) are related by

$$\tilde{I}(\ldots, \tilde{\zeta}_p, \ldots) = k^{1/2-\gamma_\Sigma} I(\ldots, \tilde{\zeta}_p k^{\alpha}p, \ldots) , \tag{4.1.7a}$$

where $\gamma_\Sigma = \sum_1^l \gamma_q$ along with $l/2$ defines the index of focusing or singularity:

$$\tfrac{1}{2} - \gamma_\Sigma = \tfrac{1}{2} - \sum_{q=1}^{l} \gamma_q = \sigma_{\mathrm{foc}} \ . \tag{4.1.7}$$

This index introduced by *Arnold* [4.8] indicates how sharp the field focusing is at the critical point [4.1–4]. Use of this index has been reported [4.5–7].

Each of the two forms (4.1.1, 2) has its own advantages. The dimensionless integrals $I(\boldsymbol{\zeta})$ are convenient to tabulate. Their characteristic values and typical scale of variation over various variables are comparable with unity (in the neighborhood of the critical point $\zeta = 0$). On the other hand, many computations associated with the crossover to geometrical optics and with the construction of uniform asymptotics are easier to handle with the dimensional integrals $\tilde{I}$. Where necessary we shall add to I and $\tilde{I}$ subscripts according to the type of caustic, for example, $\tilde{I}_{A,m+1}(\tilde{\boldsymbol{\zeta}})$ for simple A_{m+1} caustics.

Representing a caustic field as an integral of an oscillating function $\exp(ik\Phi)$ is a natural way where comparison with geometrical optics is necessary. The stationary points of integral (4.1.2) corresponding to the geometrical optics approximation lie in the space $(\tilde{\boldsymbol{\zeta}}, \tilde{\mathbf{t}})$ on the ray surface S_R given by equations

$$\frac{\partial \Phi(\tilde{\boldsymbol{\zeta}}, \tilde{\mathbf{t}})}{\partial \tilde{t}_q} = 0, \quad q = 1, 2, \ldots, l \ , \tag{4.1.8}$$

that are completely analogous to (3.2.3). If t_j, $j = 1, \ldots, M+1$, are $M+1$ solutffions of (4.1.8), then these stationary values of internal parameters depend on the external variables of the problem, i.e., $\tilde{t}_j = \tilde{t}_j(\tilde{\boldsymbol{\zeta}})$.

Two or more stationary points merge on caustics where the Hessian determinant

$$\tilde{H} = \det[\tilde{H}_{\alpha\beta}] = \det[\partial^2 \Phi(\tilde{\boldsymbol{\zeta}}, \tilde{\mathbf{t}})/\partial \tilde{t}_\alpha \partial \tilde{t}_\beta] \tag{4.1.9}$$

vanishes.

A formal calculation of the integrals (4.1.1) and (4.1.2) with the aid of the multidimensional method of stationary phase [4.9, 10] yields the sum of ray components

$$I(\boldsymbol{\zeta}) \cong \sum_{j=1}^{N} |H_j|^{-1/2} \mathrm{e}^{\mathrm{i}\frac{\pi}{4}\beta_j} \exp[\mathrm{i}\Phi(\boldsymbol{\zeta}, \mathbf{t}_j)] \ , \tag{4.1.10}$$

$$\tilde{I}(\tilde{\boldsymbol{\zeta}}) \cong \sum_{j=1}^{N} |\tilde{H}_j|^{-1/2} \mathrm{e}^{\mathrm{i}\frac{\pi}{4}\beta_j} \exp[\mathrm{i}k\Phi(\tilde{\boldsymbol{\zeta}}, \tilde{\mathbf{t}}_j)] \ , \tag{4.1.11}$$

where summation is over all stationary points that total at most at $N = M + 1$. The quantity $\tilde{H}_j$ corresponds to the value of (4.1.9) at a stationary point $\tilde{t}_j$, and β_j implies the signatures of matrices $[H_{\alpha\beta}]$ and $[\tilde{H}_{\alpha\beta}]$ at this point. Complex roots of equations (4.1.8), corresponding to the region of caustic shadow, will not be touched on at the moment.

We introduced the factor $(k/2\pi)^{1/2}$ in (4.1.2) for convenience: at non-caustic points it "neutralizes" the multipliers $(2\pi/k)^{1/2}$ that occur in integration with respect to each of the parameters $\tilde{t}_j$, thus rendering the amplitudes of ray field components in (4.1.11) independent of wavenumber k. In the vicinity of caustics where $\tilde{H} \to 0$ the order of the integral with respect to k will be a different, moreover positive, number which reflects the concentration of the field at caustics.

4.1.2 Reducing Integrals to Normal Form

The standard integrals (4.1.1, 2) are thought of as normal forms to which one can reduce many caustic integrals with infinite integration limits, i.e., the integrals of the form

$$u(\mathbf{r}, \boldsymbol{\alpha}) = \int \ldots \int d^l s \, F(\mathbf{r}, \boldsymbol{\alpha}, \mathbf{s}) \exp[ik\Psi(\mathbf{r}, \boldsymbol{\alpha}, \mathbf{s})] \ , \tag{4.1.12}$$

where $\boldsymbol{\alpha}$ are parameters essential for the problem, $\mathbf{s} = (s_1, \ldots, s_l)$ are the integration variables, and l is the multiplicity of the integration. To this end one has to perform smooth transformations (translations, turns, and rescaling) such that the real-valued phase function $\Psi(\mathbf{r}, \boldsymbol{\alpha}, \mathbf{s})$ would be reduced locally, in the neighborhood of a given critical point, to one of the normal forms $\Phi(\zeta, \mathbf{t})$, as delineated in Sect. 3.2.1.

As a matter of fact, the search for the transformation

$$\Psi(\mathbf{r}, \boldsymbol{\alpha}, \mathbf{s}) \ \to \Phi(\zeta, \mathbf{t}) + \text{Morse form} \tag{4.1.13}$$

is the hard core of the problem. Rigorous derivations of the necessary transformations obtained in an explicit analytical form have been successful for only simple problems. Most situations, however, have to be tackled with approximate computations based on the expansion of phase function Ψ into power series near a critical point. The key point is to identify the type of the caustic and to reveal the respective unfolding $\Phi_0(\zeta, \mathbf{t})$. The coefficients of the remaining terms of the power expansions may be determined by a more or less regular procedure of elimination of the terms that do not occur in the normal form. Elimination consists in the selection of coefficients in the polynomials that carry out the transformation $\{\mathbf{r}, \boldsymbol{\alpha}, \mathbf{s}\} \to \{\zeta, \mathbf{t}\}$. This procedure treats the nonlinear terms of the transformation as small corrections to the fundamental linear transformation.

Turning to the amplitude factor $F(\mathbf{r}, \boldsymbol{\alpha}, \mathbf{s})$ we say that the transformation (4.1.13) of the phase function to a normal form is most important but not the unique operation needed to reduce (4.1.12) to the typical form (4.1.2) – one has to move $F(\mathbf{r}, \boldsymbol{\alpha}, \mathbf{s})$ outside the integral symbol. To be more precise, we speak about the amplitude factor F_1 that occurs in (4.1.12) after the transformation of variables, $\mathbf{s} \to \mathbf{t}$, and differs from F by the factor $D(s)/D(t)$ equal to the Jacobian of the transformation $(s_1, \ldots, s_l) \to (t_1, \ldots, t_l)$, i.e., $F_1 = FD(\mathbf{s})/D(\mathbf{t})$. However, owing to an almost linear relation between $\mathbf{s}$ and $\mathbf{t}$ the Jacobian is almost a constant. Therefore, in many situations we may safely take the initial factor F rather than F_1. In the general case, of course, the behavior of F_1 should be also analyzed.

For the diffraction integral (4.1.12) to be reducible to a typical form (4.1.2), the pre-exponent F in its integrand should remain practically invariant over the essential domain of integration. The size of this domain may be estimated qualitatively on the grounds of "Fresnel type" considerations [4.5–7].

Let $\mathbf{s}_{st}$ be a stationary point. In agreement with the Fresnel criterion we assume that at the boundary of the respective Fresnel region $\mathbf{s}_{st} + \delta\mathbf{s}_f$ the phase shift relative to the stationary value $k\Psi(\mathbf{r}, \boldsymbol{\alpha}, \mathbf{s}_{st})$ is about π, viz.,

$$k\delta\Psi = k\Psi(\mathbf{r}, \boldsymbol{\alpha}, \mathbf{s}_{st} + \delta\mathbf{s}_f) - k\Psi(\mathbf{r}, \boldsymbol{\alpha}, \mathbf{s}_{st}) \sim \pi \ . \tag{4.1.14}$$

Solving this equation for $\delta\mathbf{s}_f$ we may write the condition of practical constancy of $F(\mathbf{r}, \boldsymbol{\alpha}, \mathbf{s})$ in the Fresnel region $\delta s \leq \delta s_f$ in the form of an inequality

$$\delta s_f \left|\frac{\partial F}{\partial s}\right| \ll |F| \ , \tag{4.1.15}$$

where $F(\mathbf{r}, \boldsymbol{\alpha}, \mathbf{s})$ is taken at the stationary point $\mathbf{s}_{st}$.

No detailed treatment of respective inequalities has been reported, but it should be clear without analysis that this requirement is similar to the applicability condition for the stationary-phase method and stronger than the latter in general, because for caustic points the Fresnel region δs_f is, as a rule, greater than a common Fresnel zone in the proportion that is larger the more complicated the caustic is.

In the s parameter space, the condition $\delta\Psi \sim \pi$ confines a Fresnel zone of a rather complicated shape that can have infinite trails like the phase front of negative curvature.

It has been demonstrated [4.7, 11] that the actual region where the field is formed can never be infinite. Moreover, for functions F of "general form", i.e., having no special form emphasizing the in-phase regions of the oscillating factor $\exp(ik\Psi)$, as estimates of δs_f it is convenient to take values obtained from (4.1.14) by expanding $\delta\Psi$ into a Taylor series in $\delta\mathbf{s}$. The leading term in this series

$$\delta\Psi = \frac{1}{2}\left(\delta\mathbf{s}, \frac{\partial}{\partial\mathbf{s}}\right)^2 \Psi + \frac{1}{6}\left(\delta\mathbf{s}, \frac{\partial}{\partial\mathbf{s}}\right)^3 \Psi + \ldots$$

is quadratic because Ψ is stationary at $\mathbf{s} = \mathbf{s}_{st}$. Thus, if $\partial^2\Psi/\partial s^2 \equiv \Psi_{ss} \neq 0$, then for δs_f from (4.1.14) we obtain the estimate (for the one-dimensional case)

$$\delta s_f \sim (\lambda/\Psi_{ss})^{1/2} \ . \tag{4.1.16}$$

If $\Psi_{ss} = 0$ but $\Psi_{sss} \neq 0$, then

$$\delta s_f \sim (3\lambda/\Psi_{sss})^{1/3} \ . \tag{4.1.17}$$

Finally, if the first $n - 1$ derivatives vanish, but $\partial^n\Psi/\partial s^n \neq 0$, then,

$$\delta s_f \sim \left(\frac{n!\lambda/2}{\partial^n\Psi/\partial s^n}\right)^{1/n} \ . \tag{4.1.18}$$

Weak inconsistency of the amplitude factors within the Fresnel region essential for integration may be taken into account by expanding the pre-exponent F (or F_1 in the general case) into a Taylor series in $\delta\mathbf{s} = \mathbf{s} - \mathbf{s}_{st}$. Integration of the oscillating component $\exp[ik\Psi(\mathbf{r}, \boldsymbol{\alpha}, s)]$ with multipliers $(\delta\mathbf{s})^k$ will lead, as commonly happens in such situations, to the appearance of derivatives of the standard integral $\tilde{I}(\tilde{\boldsymbol{\zeta}})$ with respect to some or other external parameters $\boldsymbol{\zeta}$. So long as the inequalities (4.1.15) are valid, these correction terms will be sufficiently small and may be neglected similar to how higher terms of the expansion are neglected in the stationary phase method. When these inequalities are invalidated, the higher terms will increase sharply, thus rendering the asymptotic expansion of the field inapplicable.

4.1.3 Multiplicity of Standard Integrals

All available integral representations have a rather low multiplicity: more often $l = 1$ or $l = 2$ and very seldom $l = 3$. Double integrals occur, for example, in using the Huygens–Kirchhoff principle when the integration is taken over a two-dimensional surface. By way of illustration, if the scalar monochromatic field $u^0(x', y')$ is specified in the plane $z = 0$, then the field at a point (x, y, z) of the homogeneous half-space $z > 0$ is defined by

$$u(x, y, z) = \frac{1}{2\pi} \iint_{-\infty}^{\infty} u^0(x', y') \frac{\partial}{\partial z}\left(\frac{e^{ikR}}{R}\right) dx'\, dy' , \tag{4.1.19}$$

where $R = \sqrt{z^2 + (x - x')^2 + (y - y')^2}$.

Decomposing in plane waves gives rise to integrals of Rayleigh type. For example, for the above problem, the following spectral decomposition will hold

$$u(x, y, z) = \iint_{-\infty}^{\infty} d\kappa_1\, d\kappa_2\, u^0(\kappa_1, \kappa_2) e^{i\kappa_1 x + i\kappa_2 y + i\sqrt{k^2 - \kappa_1^2 - \kappa_2^2}\, z} , \tag{4.1.20}$$

where $u^0(\kappa_1, \kappa_2)$ is the spatial spectral density of the initial field $u^0(x', y')$ in the plane $z = 0$, given by

$$u^0(\kappa_1, \kappa_2) = \frac{1}{4\pi^2} \iint dx'\, dy'\, u^0(x', y') e^{-i\kappa_1 x' - i\kappa_2 y'} . \tag{4.1.21}$$

It should be obvious that (4.1.19, 20) are equivalent.

In three space, the double integral family also includes Maslov's representation because there the integration is over two components of momentum tied with some or other coordinates. In all listed situations, two-dimensional integrals can be reduced to single integrals.

Triple integrals occur in three-dimensional problems when time-dependent processes are considered. If we recover the argument ω in (4.1.19, 20) and consider the field $u(\mathbf{r})$ as the spectral amplitude $u(\mathbf{r}, \omega)$, then for the field $u(\mathbf{r}, t)$ in a stationary medium we get the integral expression

$$u(\mathbf{r}, t) = \int_{-\infty}^{\infty} u(\mathbf{r}, \omega) e^{-\omega t}\, d\omega , \tag{4.1.22}$$

which becomes a triple integral upon substituting (4.1.13) or (4.1.20).

In view of the fact that most useful integral representations have the form of single, double, or more seldom, triple integrals, it would be reasonable to ask what might cause more than three integrations corresponding to caustics with corank $l \geq 3$? Expressed differently, what physical processes may lead to the appearance of caustics having corank $l \geq 3$?

As you will recall complex caustics can appear in inhomogeneous media with complex structure. In this case one should no longer use simple formulas like (4.1.19, 20) relying on the Green's functions of free space $\sim e^{ikR}/R$ and plane waves $\exp(i\mathbf{kr})$. In inhomogeneous media, Green's functions and initially plane waves can themselves acquire a caustic structure, i.e., may be represented as multiple integrals.

If we wish to confine ourselves to simple Green functions of the form e^{ikR}/R valid for homogeneous space, we have to divide the entire space by cutting planes into limited almost homogeneous regions and apply diffraction integrals, say, of Kirchhoff's type to recalculate the field from one surface to another. We finally obtain multiple integrals of dimension exceeding that of the configuration space.

If the Huygens concept of field formation is applied consecutively to ever decreasing volumes ΔV, then in the limit as $\Delta V \to 0$ we arrive at the Feynman path integrals with infinite multiplicity of integration, $l = \infty$. Certainly, for caustics of moderate complexity the path integrals may be reduced to finite-dimensional integrals of the type (4.1.1) or (4.1.2).

Similar reasoning suggests that one may obtain caustics of corank $l = 3$ and higher by multiple reflection of light from curved mirrors. Indeed, by Huygens–Kirchhoff's principle each reflection is equivalent to an integration over a surface. Therefore, sequential reflection of a light beam from two mirrors has a potential to induce caustics of corank $l = 4$ and upwards.

In this example, additional parameters of integration are coordinates on the mirror surfaces, but in the general case the parameter of integration may have another physical sense. In the literature the multiplicity of caustic integrals under typical conditions is inadequately reported.

4.2 The Airy Integral

4.2.1 Basic Properties

The concept of simplest standard integrals has evolved earlier than catastrophe theory, suffice it to recall the Airy function (integral), Pearsey integral, generalized Airy functions and such. The merit of catastrophe theory consists not only in compiling the list of typical catastrophes, but also in proving the completeness of this list for given m and l.

To the simplest caustic, fold (A_2), there correspond the phase function $\Phi = \zeta t + t^3/3$ and the standard integral

$$I_{A,2}(\zeta) = \frac{1}{\sqrt{2\pi}} \int\limits_{-\infty}^{\infty} \exp[\mathrm{i}(\zeta t + t^3/3)]\mathrm{d}t \ . \tag{4.2.1}$$

This integral is expressible through the standard Airy function [4.10, 12, 13]

$$\begin{aligned} \mathrm{Ai}(\zeta) &= \frac{1}{2\pi} \int\limits_{-\infty}^{\infty} \exp[\mathrm{i}(\zeta t + t^3/3)]\mathrm{d}t \\ &= \frac{1}{\pi} \int\limits_{0}^{\infty} \cos(\zeta t + t^3/3)\mathrm{d}t \ , \end{aligned} \tag{4.2.2}$$

which is also called the Airy integral, and through the Airy function in the Fock normalization [4.14]

$$\begin{aligned} v(\zeta) &= \frac{1}{2\sqrt{\pi}} \int\limits_{-\infty}^{\infty} \exp[\mathrm{i}(\zeta t + t^3/3)]\mathrm{d}t \\ &= \frac{1}{\sqrt{\pi}} \int\limits_{0}^{\infty} \cos(\zeta t + t^3/3)\mathrm{d}t \ . \end{aligned} \tag{4.2.3}$$

Thus,

$$I_{A,2}(\zeta) = \sqrt{2\pi}\,\mathrm{Ai}(\zeta) = \sqrt{2}\,v(\zeta) \ . \tag{4.2.4}$$

In what follows we shall rely predominantly on the Airy function $\mathrm{Ai}(\zeta)$. It is plotted in Fig. 2.4. The caustic $\zeta = 0$ divides the axis ζ in the shadow region $\zeta > 0$, and the light region $\zeta < 0$. In the lit region, the Airy integral describes two-ray interference, and equation (4.1.3) for stationary points assumes the form

$$\frac{\partial \Phi}{\partial t} = \zeta + t^2 = 0 \ . \tag{4.2.5}$$

For $\zeta < 0$, both stationary values

$$t_{1,2} = \pm(-\zeta)^{1/2}, \quad \zeta < 0 \tag{4.2.6}$$

are real-valued.

Calculation of the Airy integral in the neighborhood of these points by the stationary phase method yields an asymptotic expression as follows:

$$\begin{aligned} \mathrm{Ai}(\zeta) &= \frac{1}{2\sqrt{\pi}}(-\zeta)^{-1/4}\left[\exp\left\{-\mathrm{i}\left[\frac{2}{3}(-\zeta)^{3/2} - \frac{\pi}{4}\right]\right\} + \exp\left\{\mathrm{i}\left[\frac{2}{3}(-\zeta)^{3/2} - \frac{\pi}{4}\right]\right\}\right] \\ &= \frac{1}{\sqrt{\pi}}(-\zeta)^{-1/4}\cos\left[\frac{2}{3}(-\zeta)^{3/2} - \frac{\pi}{4}\right], \quad \zeta < 0, \quad |\zeta| \gg 1 \ . \end{aligned} \tag{4.2.7a}$$

In the shadow region, $\zeta > 0$, there are no real-valued stationary points and the integral exhibits an exponential asymptotic behavior

$$\mathrm{Ai}(\zeta) = \frac{1}{2\sqrt{\pi}}\zeta^{-1/4}\exp\left(-\frac{2}{3}\zeta^{3/2}\right), \quad \zeta > 0, \quad |\zeta| \gg 1 . \tag{4.2.7b}$$

The asymptotic dependances (4.2.7a, b) are valid at large values of ζ, i.e., at $|\zeta| \gg 1$, but practically they ensure a suitable accuracy (better than 5%) already for $\zeta > 1.3$ in the shadow region, and for $|\zeta| > 1.4$ in the lit region.

The derivative of the Airy integral

$$\mathrm{Ai}'(\zeta) = \mathrm{i} \int_{-\infty}^{\infty} t \exp[\mathrm{i}(\zeta t + t^3/3)]\mathrm{d}t$$

has asymptotic expressions for $|\zeta| \gg 1$ as follows:

$$\mathrm{Ai}'(\zeta) \sim \frac{1}{\sqrt{\pi}}(-\zeta)^{1/4}\sin\left[\frac{2}{3}(-\zeta)^{3/2} - \frac{\pi}{4}\right], \quad \zeta < 0 , \tag{4.2.8a}$$

$$\mathrm{Ai}'(\zeta) \sim -\frac{1}{2\sqrt{\pi}}\zeta^{1/4}\exp\left(-\frac{2}{3}\zeta^{3/2}\right), \quad \zeta > 0 . \tag{4.2.8b}$$

The dimensional Airy integral is introduced as

$$\tilde{I}_{A,2}(\tilde{\zeta}) = \sqrt{\frac{k}{2\pi}} \int_{-\infty}^{\infty} \exp[ik(\tilde{\zeta}\tilde{t} + \tilde{t}^3/3)]\mathrm{d}\tilde{t} . \tag{4.2.9}$$

In agreement with (4.1.5–7) at $m = 1, l = 1$ the caustic indexes α_1 and γ_1 and the focusing index are

$$\gamma_1 = 1/3, \quad \alpha_1 = 1 - 1/3 = 2/3, \quad \sigma_{\mathrm{foc}} = \tfrac{1}{2} - \tfrac{1}{3} = \tfrac{1}{6} \tag{4.2.10}$$

so that $\tilde{I}_{A,2}(\tilde{\zeta})$ can be rewritten in terms of other integral forms as

$$\begin{aligned}\tilde{I}_{A,2}(\tilde{\zeta}) &= k^{1/6} I_{A,2}(k^{2/3}\tilde{\zeta}) = k^{1/6}\sqrt{2\pi}\,\mathrm{Ai}(k^{2/3}\tilde{\zeta}) \\ &= k^{1/6}\sqrt{2}\,v(k^{2/3}\tilde{\zeta}) .\end{aligned} \tag{4.2.11}$$

The asymptotic values of this integral are as follows:

$$\tilde{I}_{A,2}(\tilde{\zeta}) \cong \sqrt{2}(-\tilde{\zeta})^{-1/4}\cos\left[\frac{2k}{3}(-\tilde{\zeta})^{3/2} - \frac{\pi}{4}\right], \quad \text{for } k^{2/3}|\tilde{\zeta}| \gg 1, \tilde{\zeta} < 0 , \tag{4.2.12a}$$

$$\tilde{I}_{A,2}(\tilde{\zeta}) \cong 2^{-1/2}(\tilde{\zeta})^{-1/4}\exp(-\tfrac{2}{3}k\tilde{\zeta}^{3/2}), \quad \text{for } k^{2/3}\tilde{\zeta} \gg 1, \tilde{\zeta} > 0 . \tag{4.2.12b}$$

The derivative $\tilde{I}'_{A,2}(\tilde{\zeta}) = \partial\tilde{I}_{A,2}/\partial\tilde{\zeta}$ is expressed via $I'_{A,2}(\zeta)$, $\mathrm{Ai}'(\zeta)$ and $v'(\zeta)$ as follows:

$$\begin{aligned}\tilde{I}'_{A,2}(\tilde{\zeta}) &= k^{5/6}\sqrt{2\pi}\,\mathrm{Ai}'(k^{2/3}\tilde{\zeta}) = k^{5/6} I'_{A,2}(k^{2/3}\tilde{\zeta}) \\ &= k^{5/6}\sqrt{2}\,v'(k^{2/3}\tilde{\zeta}) .\end{aligned} \tag{4.2.13a}$$

The asymptotic representation of this derivative has the form

$$\tilde{I}_{A,2}(\tilde{\zeta}) \cong \sqrt{2}(-\tilde{\zeta})^{1/4} k \sin\left[\frac{2k}{3}(-\tilde{\zeta})^{3/2} - \frac{\pi}{4}\right], \quad \tilde{\zeta} < 0 ,$$
$$\tilde{I}_{A,2}(\tilde{\zeta}) \cong \frac{k}{2}\tilde{\zeta}^{1/4} \exp(-\tfrac{2}{3}k\tilde{\zeta}^{3/2}), \quad \tilde{\zeta} > 0 . \tag{4.2.13b}$$

4.2.2 The Airy Differential Equation

Like the functions $\mathrm{Ai}(\zeta)$ and $v(\zeta)$, the integral $I_{A,2}(\zeta)$ satisfies the Airy equation

$$\frac{\partial^2 y(\zeta)}{\partial \zeta^2} - \zeta y(\zeta) = 0 , \tag{4.2.14}$$

which differs from the harmonic equation $y'' + y = 0$ in having a linear coefficient $(-\zeta)$ of y rather than a constant. Therefore, in going from negative to positive values the behavior of the solution changes drastically: the oscillatory profile of $\mathrm{Ai}(\zeta)$ for $\zeta > 0$ gives way to a monotonously falling profile for $\zeta > 0$ (Fig. 2.4). Equation (4.2.14) facilitates greatly the calculation of the Airy integral [4.12, 13].

The second linearly independent solution to (4.2.14) is usually denoted by $\mathrm{Bi}(\zeta)$. *Fock* [4.14] dealt with the function $u(\zeta) = \sqrt{\pi}\,\mathrm{Bi}(\zeta)$. For $\zeta < 0$, this second solution oscillates as $\sin[\frac{2}{3}(-\zeta)^{3/2} - \pi/4]$, and for $\zeta > 0$ it rises exponentially as $\zeta^{-1/4}\exp(\frac{2}{3}\zeta^{3/2})$.

The integral $\tilde{I}(\tilde{\zeta})$ satisfies the equation

$$\frac{\mathrm{d}^2 Y(\tilde{\zeta})}{\mathrm{d}\tilde{\zeta}^2} - k^2\tilde{\zeta} Y(\tilde{\zeta}) = 0 , \tag{4.2.15}$$

which, unlike (4.2.14), has the coefficient k^2 in the second term. This equation may be obtained from (4.2.14) by the change of variable $k^{2/3}\tilde{\zeta} = \zeta$.

4.2.3 An Exact Airy-Integral Solution to the Wave Problem

It is an easy matter to write an exact solution to the problem of plane wave incidence with the aid of the Airy integral. The wave

$$u^0(x, z^0) = U^0 \exp(ikxn^0 \sin\theta^0) ,$$

given in the plane $z^0 =$ constant is incident on a linear slab of permittivity $\varepsilon(z) = \varepsilon_0 - \varepsilon_1 z$. The geometrical solution of this problem has been outlined in Sect. 2.4.2. If we separate the variables by writing the solution as $u(x, z) = \exp(ikxn^0 \sin\theta^0)\, W(z)$, then for $W(z)$ we get

$$\frac{\mathrm{d}^2 W}{\mathrm{d}z^2} + k^2[\varepsilon_0 - \varepsilon_1 z - (n^0)^2 \sin^2\theta^0]\, W = 0 . \tag{4.2.16}$$

The substitution $\zeta = k^{2/3}\varepsilon_1^{1/3}(z - z_c)$ with $z_c = (\varepsilon_0 - \sin^2\theta^0)^2\varepsilon_1$ reduces this equation to (4.2.14).

The amplitude coefficient of the Airy function can be determined by equating the coefficient of the first term in (4.2.7) (precisely this term corresponds to the wave approaching the point of return $\zeta = 0$) to the amplitude of the incident wave, u^0. Now, the solution of the problem under consideration takes on the form

$$u(x, z) = CU^0\left(\frac{k}{\varepsilon_1}\right)^{1/6} \mathrm{Ai}[k^{2/3}\varepsilon_1^{1/3}(z - z_c)]\mathrm{e}^{ikx\sin\theta^0} , \tag{4.2.17}$$

where

$$C = 2\sqrt{\pi}\left(\frac{k}{\varepsilon_1}\right)^{1/6} \exp\left\{\frac{2ik}{3\varepsilon_1}[\varepsilon_1(z_c - z^0)]^{3/2} - \frac{i\pi}{4}\right\} .$$

In view of (4.2.7) this expression reduces to the geometrical solution (4.2.6) for $z < z_c$, whereas for $z > z_c$ it becomes (4.2.7) which has been obtained by the method of complex geometrical optics. To establish relation between (4.2.17) and the geometric optical asymptotics it would pay to keep in sight that $\varepsilon_1(z - z_c) = \varepsilon_1(z - z^0) - (n^0\cos\theta^0)^2$ and that $\varepsilon_1(z^0 - z_c) = -(n^0\cos\theta^0)^2$.

Seemingly simple on the surface the outlined solution involves some non-trivial points. First of all, this solution justifies the caustic phase shift $-\pi/2$ that appears in the reflected wave. Secondly, the solution (4.2.17) serves as a basis for the rules of selection of complex rays in the region of caustic shadow. Thirdly, it illustrates the natural way of appearance of the caustic index $\alpha_1 = 2/3$ (power of the wavenumber in the argument of the Airy function) and the focusing index $\sigma_{foc} = 1/6$ (the power of k in the coefficient of Ai) in wave problems associated with simple caustics of the fold type. An interpretation of the Fresnel index $\gamma = 1/3$ is deferred until Sect. 4.4.1.

4.2.4 The Airy Integral as a Standard Function for the One-Dimensional Wave Equation

Consider the one-dimensional wave equation

$$u''(z) + k^2\varepsilon(z)u(z) = 0, \quad k = \omega/c . \tag{4.2.18}$$

If the permittivity $\varepsilon(z)$ is a slowly varying, positive, and nowhere vanishing function, then the solution to (4.2.18) is well approximated by a sum of two opposite waves (WKB approximation), namely,

$$u(z) \approx U_1\varepsilon^{-1/4}\exp\left(ik\int_{z_0}^{z}\varepsilon^{1/2}\,dz\right) + U_2\varepsilon^{-1/4}\exp\left(-ik\int_{z_0}^{z}\varepsilon^{1/2}\,dz\right) . \tag{4.2.19}$$

This solution is invalidated in the vicinity of turning points z_c where $\varepsilon(z)$ vanishes. As a way out, in the neighborhood of a zero (deemed to be simple), the permittivity may be replaced with the linear term of the power expansion, $\varepsilon(z) \approx \varepsilon'(z_c)(z - z_c)$, then, locally, at $z \sim z_c$ (4.2.18) reduces to the Airy equation (4.2.14).

Under certain conditions, the domain of applicability of the local Airy asymptotic partially overlaps the applicability domain of the WKB approximation (4.2.19). This gives us hope for a uniform Airy asymptotic that would correctly describe the behavior of the wave function near the turning point z_c and become the WKB approximation at a distance from this point. The idea underlying this reasoning has been set forth by *Langer* [4.15] and developed in subsequent works including those given in [4.16, 17].

Here we give the result of Langer without proof (in Chap. 5 we shall derive it from a more general theory). Looking for an asymptotic solution to (4.2.18) as an Airy integral with undefined argument $\zeta(z) = k^{2/3}\tilde{\zeta}(z)$ and undefined amplitude $A(z)$

$$u(z) = A(z)\mathrm{Ai}[k^{2/3}\tilde{\zeta}(z)] \tag{4.2.20}$$

one may verify that $\tilde{\zeta}(z)$ is related to $\varepsilon(z)$ as

$$\tilde{\zeta}(z) = \begin{cases} -\left(\dfrac{3}{2}\displaystyle\int_z^{z_c} \sqrt{\varepsilon(z)}\,\mathrm{d}z\right)^{2/3}, & z < z_c \ (\varepsilon > 0) \\ \left(\dfrac{3}{2}\displaystyle\int_{z_c}^{z} \sqrt{-\varepsilon(z)}\,\mathrm{d}z\right)^{2/3}, & z > z_c \ (\varepsilon < 0), \end{cases} \tag{4.2.21}$$

whereas the amplitude factor is proportional to $[\zeta'(z)]^{-1/2}$:

$$A(z) = \frac{C}{\sqrt{\tilde{\zeta}'(z)}}, \qquad \tilde{\zeta}' = \frac{\mathrm{d}\tilde{\zeta}}{\mathrm{d}z} = \left(-\frac{\varepsilon}{\tilde{\zeta}}\right)^{1/2}. \tag{4.2.22}$$

Thus, $u(z) = \mathrm{const}\,(-\zeta/\varepsilon)^{1/4}\mathrm{Ai}(\zeta)$.

Near the turning point, $\tilde{\zeta}$ is a linear function of z, indeed, if $\varepsilon(z) \approx \varepsilon'(z_c)(z - z_c)$. As a result, the amplitude factor is always finite at turning points, $A(z_c) \approx C[\varepsilon'(z_c)]^{-1/6}$. Away from the turning point, a suitable selection of C carries the uniform Airy asymptotic (4.2.20) to the WKB asymtotic (4.2.9).

It is worth noting that for a linear function $\varepsilon(z)$, the solution (4.2.20) becomes exact and with the aid o the substitution $\varepsilon(z) = \varepsilon(z) - (n^0 \sin\theta^0)^2$ it may be adapted to the description of a plane wave $u^0 = U^0 \exp(ikxn^0 \sin\theta^0)$ incident on a plane stratified medium of permittivity $\varepsilon(z)$.

4.2.5 Applicability Conditions of the Uniform Airy Asymptotic in One-Dimensional Problems

For Langer's solution, the applicability domain may be derived from heuristic considerations [4.18]. Let Λ be the width of the caustic zone, i.e., the distance from the turning point z_c to the first maximum of the Airy function, defined from the condition $k^{2/3}\tilde{\zeta}(\Lambda) = -1.02$.

If we resort to the linear approximation $\tilde{\zeta} = |\varepsilon'|^{1/3}(z - z_c)$, then the estimate for Λ is as follows:

$$\Lambda \sim k^{-2/3}|\varepsilon'|^{-1/3} \ .$$

If we require that within the Λ neighborhood of the turning point the deviation of $\varepsilon(z)$ from a linear law $\varepsilon(z) \approx \varepsilon'(z_c)(z - z_c)$ be small, we have

$$\Lambda|\varepsilon''| \ll |\varepsilon'| \ . \tag{4.2.23}$$

Available examples, including those given in [4.17], prove that this inequality must hold.

Outside of the caustic zone, applicability of (4.2.20) is limited by a common condition of slow variation over the wavelength in the medium $\lambda\!\!\!^{-}/\sqrt{\varepsilon} = 1/(k\sqrt{\varepsilon})$, $\lambda\!\!\!^{-} = \lambda/2\pi = 1/k$:

$$\frac{\lambda\!\!\!^{-}}{\sqrt{\varepsilon}}|\varepsilon'| \ll \varepsilon \quad \text{or} \quad |\varepsilon'| \ll k\varepsilon^{3/2} \ . \tag{4.2.24}$$

Clearly this inequality breaks down on approach to point z. If we put $\varepsilon \approx \varepsilon'(z - z_c)$, then the condition (4.2.24) becomes

$$|z - z_c| \gg |\varepsilon'|^{-1/3}k^{-2/3} = \Lambda \ . \tag{4.2.25}$$

Hence, the ordinary condition of applicability of the WKB asymptotic (4.2.24) is valid outside of the caustic zone, whereas inside the Λ-neighborhood of a turning point the condition (4.2.23) must hold.

4.3 The Pearsey Integral

4.3.1 Properties

The standard integral

$$I_{A,3}(\zeta_1, \zeta_2) = \frac{1}{\sqrt{2\pi}} \int \exp\left[\mathrm{i}\left(\zeta_1 t + \frac{1}{2}\zeta_2 t^2 - \frac{1}{4}t^4\right)\right] \mathrm{d}t \tag{4.3.1}$$

described the wave field near an A_3 caustic, i.e., near a cusp caustic. One of the first analyses of such field was done by *Pearsey* [4.19] who tabulated the integral

$$I_p(X_1, X_2) = \int \exp[\mathrm{i}(X_1 t + X_2 t^2 + t^4)]\, \mathrm{d}t \tag{4.3.2}$$

now known as the Pearsey integral. The relation between (4.3.1) and (4.3.2) is given by

$$I_{A,3}(\zeta_1, \zeta_2) = \frac{1}{\sqrt{\pi}} I_p^*(-\sqrt{2}\zeta_1, -\zeta_2) \ . \tag{4.3.3}$$

The isolines of modulus and phase of I_p have been plotted by *Pearsey* [4.19] and by some Soviet authors [4.20, 21]. Reliefs of $|I_p(X_1, X_2)|$ have been computed by *Connor* and *Farrelly* [4.22]. *Connor* and *Curtis* [4.23] have tabulated I_p along with $\partial I_p/\partial X_1$ and $\partial I_p/\partial X_2$.

The caustic cusp corresponding to (4.3.1) and (4.3.2) is given by the equation

$$\zeta_1^2 = \tfrac{4}{27}\zeta_2^3 \quad \text{or} \quad X_1^2 = \tfrac{8}{27}(-X_2)^3 \; . \tag{4.3.4}$$

Inside the caustic cusp the field exhibits rather intense variations caused by the interference of the three ray fields, as shown in Fig. 1.5. A less intense interference is evident on the caustic branches proper where the field of one ray overlaps the caustic field associated with the two other rays. Outside the caustic cusp, the field is practically non-oscillating.

4.3.2 Focusing in the Presence of Cylindrical Aberration

Cylindrical aberration is a two-dimensional analog of spherical aberration. We take up the problem of cylindrical aberration to illustrate application of Pearsey's integral. Note that historically it was this problem that gave birth to the Pearsey integral (4.3.2).

Let the initial field

$$u^0(x') = U^0 \exp(-ikx'^2/2F) \tag{4.3.5}$$

be given on the line $z = 0$. The parabolic dependence of the phase $k\Psi^0(x') = -kx'^2/2F$ on the x' coordinate may be a good approximation for the central (paraxial) portion of the strictly cylindrical wave

$$u^0(x') = \frac{F}{(x'^2 + F^2)^{1/2}} \exp[ik(\sqrt{x'^2 + F^2} - F)] \; , \tag{4.3.6}$$

focused on a point $(0, F)$. For non-paraxial rays the parabolic dependence of the phase is stronger than required for an ideal focusing. As a result, the rays cross the axis $x' = 0$ ahead of the focal plane $z = F$, and form a caustic cusp, as shown in Fig. 4.1a.

We take as a point of departure the two-dimensional Kirchhoff integral

$$u(x_1, z) = \left(\frac{ik}{2\pi}\right)^{1/2} \int \frac{u^0(x')}{\sqrt{R}} \exp\sqrt{z^2 + (x - x')^2}\, dx' \tag{4.3.7}$$

which, unlike the 3-D integral (4.1.19), is valid only at a distance at least a few wavelengths from the line $z = 0$. (The point is that in the two-dimensional problem, the Green's function is expressed through a Bessel function $H_0(kR)$ that may be replaced with $e^{ikR}/\sqrt{R}$ only for $kR \gg 1$, where $R = \sqrt{z^2 + (x - x')^2}$.)

Assume that for the major portion of the field forming region, the values of x and x' are small compared with the focal length and represent

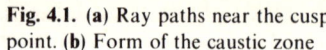

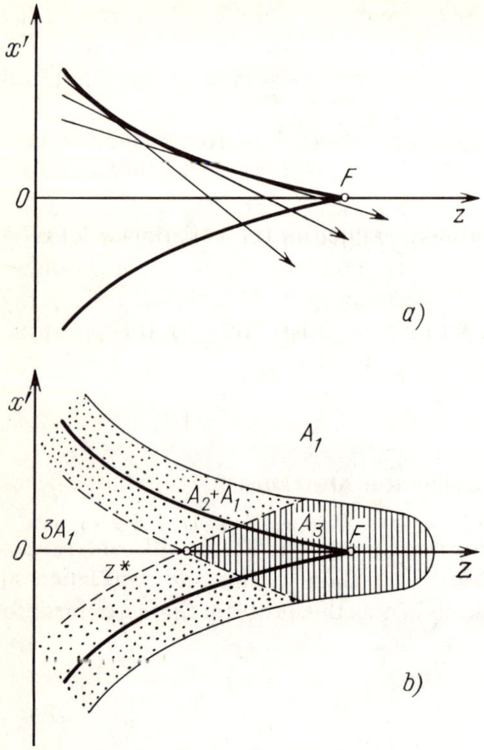

Fig. 4.1. (**a**) Ray paths near the cusp point. (**b**) Form of the caustic zone

$R = \sqrt{z^2 + (x - x')^2}$ in the form of an expansion

$$R = z + \frac{(x - x')^2}{2z} - \frac{(x - x')^4}{8z} + \cdots .$$

Write z as $F + (z - F)$ and assume the deviation $z_1 = z - F$ be small compared with F. Using the smallness of x/F, x'/F, and z_1/F compared to unity, we retain in the exponent only that terms which are essential for integration. Then

$$u(x, z) \approx \left(\frac{ik}{2\pi F}\right)^{1/2} \exp\left[ik\left(z + \frac{x^2}{2F}\right)\right] U^0 \times \int_{-\infty}^{\infty} \exp\left[ik\left(-\frac{xx'}{F} - \frac{x'^2 z_1}{2F^2} - \frac{x'^4}{8F^3}\right)\right] dx' . \tag{4.3.8}$$

The change of variables

$$\zeta_1 = -x\left(\frac{2k^3}{F}\right)^{1/4}, \quad \zeta_2 = (F - z)\left(\frac{2k}{F}\right)^{1/2}, \quad t = (k/2F^3)^{1/4} x' , \tag{4.3.9}$$

reduces this integral to the normal form (4.3.1):

$$u(x, z) = (2ikF)^{1/2} \exp\left[ik\left(z + \frac{x^2}{2F}\right)\right] I_{A,3}(\zeta_1, \zeta_2) \ . \qquad (4.3.10)$$

In dimensional variables x, z the caustic curve (4.3.4) becomes $x^2 = 8(F - z)^3/27F$.

4.3.3 Caustic Indices and Field Structure

By virtue of (4.3.10) the field at the focal point $x = 0$, $z = F$ is proportional to $(2kF)^{1/2} I_{A,3}(0, 0)$ which is about $(kF)^{1/2}$ times the field of a plane (unfocused) wave. This conclusion totally agrees with the singularity index $\sigma_{\text{foc}} = 1/2$ given in Table 3.2. The powers $\alpha_1 = 3/4$ and $\alpha_2 = 1/2$ at which the wavenumber enters the transformed variables ζ_1 and ζ_2 in (4.3.8) correspond exactly to the caustic indexes of Tables 3.1, 2.

We have already mentioned that the scales of variation of typical integrals over all variables are comparable with unity. This relates to the Pearsey integral as has been borne out by the numerical computations of this integral. Making good use of this fact, we get estimators for dimensional scales characterizing the caustic zone. Note that near the cusp point ($\zeta_1 = \zeta_2 = 0$) the phase function $\Phi(\zeta, t)$ is equal to $\Phi_0(t) = t^4/4$. Equating this function to π, we obtain a Fresnel estimate of the region of field formation with respect to t parameter: $t_f \sim (4\pi)^{1/4}$. Now, from the Fresnel conditions $\zeta_1 t_f \sim \pi$ and $\zeta_2 t_f^2/2 \sim \pi$ we obtain the estimates of the dimensionless parameters $\zeta_1 \sim (\pi^{3/4})^{1/4}$ and $\zeta_2 \sim \pi^{1/2}$ and, with the aid of (4.3.8), the estimates of dimensional scales

$$|x| \sim \zeta_1(F/2k^3) \cong (F\lambda^3)^{1/4}/2\sqrt{2} \cong l_\perp \ ,$$

$$z_1 = |F - z| \sim \zeta_2(F/2k)^{1/2} \cong (\lambda F)^{1/2}/2 = l_\parallel \ .$$

These expressions may be identified as the transverse $|x| \sim l_\perp$ and longitudinal $|F - z| \sim l_\parallel$ dimensions of the focal spot near the cusp [4.7, 24].

A similar reasoning applied to the integration variable x' yields the size of the domain essential for field formation on the axis $z = 0$:

$$|x'| \sim t_f(2F^3/k)^{1/4} \sim (4F^3\lambda)^{1/4} = a_f \ . \qquad (4.3.11)$$

The scale $(4F^3\lambda)^{1/4}$ is the Fresnel radius on the axis $z = 0$, which may be calculated in a more direct manner. At the cusp point, i.e., at $z = 0$ and $z_1 = 0$, the exponent in (4.3.8) includes the quantity $ikx'^4/(8F^3)$. If on the lines of the Fresnel criterion we equate the modulus of this quantity to π, $kx'^4/(8F^3) \sim \pi$, then we arrive at precisely the estimate (4.3.11) for x'. All these dependencies correspond to the Fresnel index $\gamma_1 = 1/4$.

While at the caustic cusp point $x = 0$, $z_1 = 0$ all three stationary points coincide, away from it along the caustic the phase function $\Phi(\zeta, t)$ retains only two coalesced points. This pair of stationary points induces an Airy function, whereas the third point associated with a ray tangent to another branch of the caustic induces a ray field. A transition from the full Pearsey integral to the asymptotic sum of an Airy function and a ray term occurs practically beyond the focal spot, i.e., at $|z_1| \gtrsim l$. This conclusion may be verified by comparing the Pearsey integral with its asymptotic approximations.

It can be demonstrated [4.7] that the caustic zones of separate branches of the caustic, approach one another near the cusp point so that their point of intersection z^* (Fig. 4.1b) marks exactly the longitudinal dimension of the focal spot: $|F - z^*| \sim l_\parallel$. This example indicates that the caustic index 2/3 corresponding to a simple caustic (caustic zone width is proportional to $k^{2/3}$) agrees completely with the longitudinal caustic index $\alpha = 1/2$ of the focal spot formed at the cusp point ($l_\parallel \sim k^{1/2}$). In other words, the caustic zones of individual caustic branches smoothly touch one another despite the different values of the indexes.

4.4 Other Typical Integrals

4.4.1 Generalized Airy Functions

The Airy and Pearsey integrals belong to the A_{m+1} series. Like these two, other typical integrals of this series belong to the family of generalized Airy functions considered by *Ludwig* [4.25] prior to the evolution of catastrophe theory:

$$I_{A,m+1}(\zeta) = \frac{1}{\sqrt{2\pi}} \int_{-\infty}^{\infty} \exp\left(\mathrm{i} \sum_{p=1}^{m} \zeta_p \frac{t^p}{p} \pm \frac{t^{m+2}}{m+2}\right) \mathrm{d}t \ . \tag{4.4.1}$$

Ludwig assumed the pre-integral coefficient equal to $1/2\pi$ in agreement with the standard definition of the Airy function (4.2.2).

Beginning with $m = 2$ for a cusp the A_{m+1} caustics are called cuspoid caustics or cuspoids for short, for all of them contain turning points or edges at which separate branches of caustics merge into a cusp. Cuspoid caustics may be realized by passing a plane wave through a two-dimensional lense suffering from aberrations or, what is the same, through a phase screen with a suitable phase variation law. The above Pearsey integral corresponds to a lens when it is near a focus eliminates the aberration terms up to the fourth order. The following integral in the order of increasing complexity is $I_{A,4}$, swallowtail, which eliminates aberrations up to the fifth order

$$I_{A,4}(\zeta) = \frac{1}{\sqrt{2\pi}} \int_{-\infty}^{\infty} \exp\left[\mathrm{i}\left(\zeta_1 t + \zeta_2 \frac{t^2}{2} + \zeta_3 \frac{t^3}{3} - \frac{t^5}{5}\right)\right] \mathrm{d}t \tag{4.4.2}$$

and the integral $I_{A,5}$, butterfly, eliminates the terms up to the sixth order,

$$I_{A,5}(\zeta) = \frac{1}{\sqrt{2\pi}} \int_{-\infty}^{\infty} \exp\left[\mathrm{i}\left(\zeta_1 t + \cdots + \zeta_4 \frac{t^4}{4} - \frac{t^6}{6}\right)\right] \mathrm{d}t \ . \tag{4.4.3}$$

Thus, given only values of caustic indexes we are in a position to assess most essential characteristics of the field at the focus of a lens suffering from aberrations. The easiest estimate is the factor of focusing K_{foc} which indicates how many times the field at the focus, $x = 0$, $z = F$, exceeds the field of the plane wave. So long as the field formation region does not reach the aperture, the focusing problem is essentially defined by one scale – the focal length F, therefore, the wavenumber should enter all estimates in combinations with F. Therefore,

$$K_{\text{foc}} \sim (kF)^{\sigma_{\text{foc}}} . \tag{4.4.4}$$

For an A_{m+1} caustic, the singularity index σ_{foc} is equal to $1/2 - \gamma_1 = 1/2 - 1/(m+2)$ so that

$$K_{\text{foc}} \sim (kF)^{\frac{1}{2}-\frac{1}{m+2}} = (kF)^{\frac{m}{2(m+2)}} . \tag{4.4.5}$$

When $m \to \infty$ and $\gamma_1 = 1/(m+2) \to 0$, the factor (4.4.5) corresponds to an ideal cylindrical focusing with $\sigma_{\text{foc}} \to 1/2$ and $K_{\text{foc}} \to (kF)^{1/2}$.

The Fresnel scale a_{f}, i.e., the radius of the zone (at the initial line $z = 0$) that forms the field in the focal plane is also described with the aid of the Fresnel index, viz.,

$$a_{\text{f}} \sim F(kF)^{-\gamma_1} \sim F^{1-\gamma_1}\lambda^{\gamma_1} \tag{4.4.6}$$

while the transverse, l_1, and longitudinal, l_2, scales of the focal spot are computed with the caustic indexes $\alpha_1 = 1 - \gamma_1$ and $\alpha_2 = 1 - 2\gamma_1$:

$$\begin{aligned} l_1 &\sim F(kF)^{-\alpha_1} \sim F^{\gamma_1}\lambda^{1-\gamma_1} , \\ l_2 &\sim F(kF)^{-\alpha_2} \sim F^{2\gamma_1}\lambda^{1-2\gamma_1} . \end{aligned} \tag{4.4.7}$$

As $m \to \infty$, the Fresnel scale a_{f} tends to the focal length F, while l_1 and l_2 tend to λ, but so that $l_1^2/l_2 = \lambda$. Note that the relationships $l_1 = (l_2\lambda)^{1/2}$ and $a_{\text{f}} l_1 \sim F_\lambda$ are characteristic of the focal spot at any values of m. For caustics A_3, A_4 and A_5 these parameters are summarized in Table 4.1.

Table 4.1. Parameters characterizing two-dimensional focusing in the presence of aberrations

m	Type		σ_{foc}	α_1	α_2	a_{f}	l_1	l_2
2	A_3	1/4	1/4	3/4	1/2	$(F^3\lambda)^{1/4}$	$(F\lambda^3)^{1/4}$	$(F\lambda)^{1/2}$
3	A_4	1/5	3/10	4/5	3/5	$(F^4\lambda)^{1/5}$	$(F\lambda^4)^{1/5}$	$(F^2\lambda^3)^{1/5}$
4	A_5	1/6	1/3	5/6	2/3	$(F^5\lambda)^{1/6}$	$(F\lambda^5)^{1/6}$	$(F\lambda^2)^{1/3}$

$\sigma_{\text{foc}} = 1/2 - \gamma$; $\alpha_1 = 1 - \gamma$; $\alpha_2 = 1 - 2\gamma$.

4.4.2 Fresnel Criteria for Transition to Subasymptotics

The above estimates provide an insight into the order of magnitude of the linear scales that occur in coming from the full typical integral $I_{A,m+1}$ to its subasymptotics at increasing distance from the critical point $\zeta = 0$. In dimensionless variables ζ_1, ζ_2, etc. the transition to simpler subordinate integrals occurs, roughly speaking, at $|\zeta_p| \sim 1$. A more accurate estimate may be derived with the Fresnel criterion $\zeta_p t_f^p/p \sim \pi$, where $t_f \sim [(m+2)\pi]^{1/(m+2)}$ so that

$$|\zeta_p| \sim \pi p/[(m+2)\pi]^{p/(m+2)}, \; p = 2, \ldots, m \; . \tag{4.4.8}$$

Errors caused by the substitution of a subordinate asymptotic for the integral (4.4.1) never exceeds 20–30% at the boundary of this region. Numerical calculations for the caustics A_4, A_5 and $D_4^{\pm}$, e.g., see [4.21], indicate that a satisfactory accuracy of 5–10% is achieved already at $|\zeta_{1,2}| \geq \pi$.

This reasoning suggests, for example, that in the case of a swallowtail (Fig. 4.2), one should resort to the full integral (4.4.2) in the region $|\zeta_{1,2,3}| \leq \pi$. With reference to the cross sections of the caustic by cutting planes $\zeta_2 =$ constant shown in Fig. 4.2, we see that moving beyond the neighborhood of the

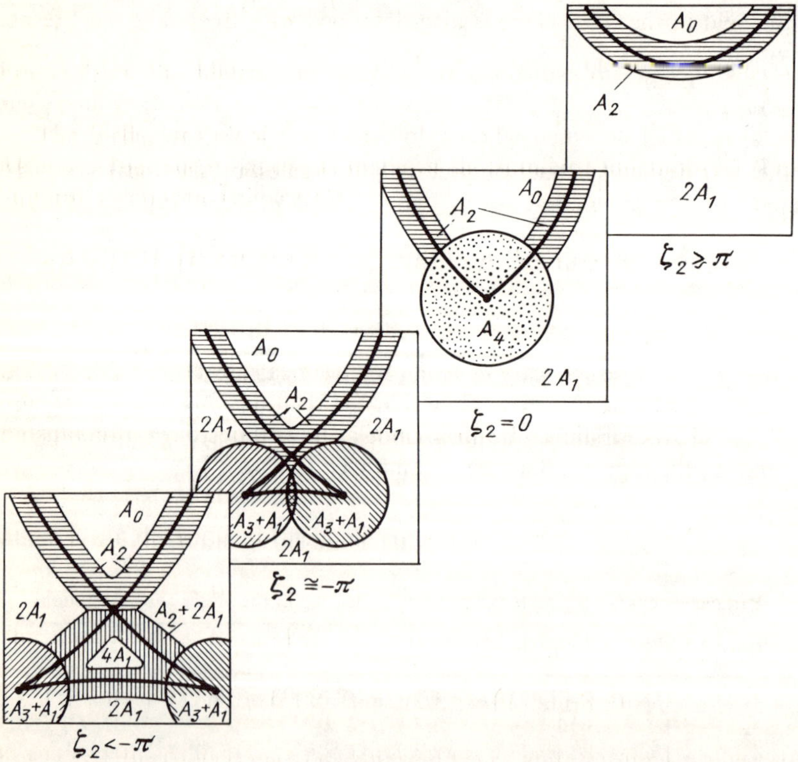

Fig. 4.2. Evolution of cross sections of the swallowtail (A_4) indicating subordinate and caustic branches

critical point $\zeta = 0$ in any direction, one may cross over to the subordinate asymptotics A_3, A_2, and A_1 (ray field). By way of example, in section $\zeta_2 = 0$, the integral $I_{A,4}$ is necessary at the vertex of the caustic, while on the side branches, one may use the Airy asymptotic $I_{A,2}$, and below the caustic the field is represented by a sum of two ray fields designated by $2A_1$. Above the caustic there is the zone of shadow.

In the section $\zeta_2 > \pi$, the caustic A_4 degenerates into the simple A_2 caustic over which there are no rays while beneath the field is represented by a sum of two ray fields $2A_1$.

In the section $\zeta_2 \leq -\pi$ the relationship with the subasymptotics is complicated. In the vicinity of each of two cusp points of the caustic loop, the field is the sum of the Pearsey integral and the ray field, $A_3 + A_1$. At a distance from the tail we deal with the sum of two ray fields $2A_1$, and on the outward branches of the caustic the field is described by the Airy integral $I_{A,2}$.

At still lower values of ζ_2, say $\zeta_2 \leq -4\pi$, when the cusp points are separated farther than by π, inside the caustic loop we have a sum of four ray fields (four stationary points of $I_{A,4}$ whose π-neighborhoods never overlap). On the internal branches of the tail, the field is described by the Airy functions with addition of two ray fields, at the vertices the field will be the superposition of fields A_3 and A_1, and below the tail the field will be the sum of two ray fields $2A_1$. The move to subordinate asymptotics in the case of A_5 caustics and cuspoids of higher orders occurs in a more complicated but similar manner.

In configuration space, to $|\zeta_p| \sim 1$ or π there correspond certain scales estimated by (4.4.7), namely,

$$l_p \sim F/(kF)^{\alpha_p} \sim F^{p\gamma_1} \lambda^{1-p\gamma_1} \,. \tag{4.4.9}$$

Specifically, the ζ_2 axis of a swallowtail is associated with the scale $l_2 \approx (F^2 \lambda^3)^{1/5}$ corresponding to the longitudinal size of the focal spot.

4.4.3 Field Structure in Different Areas of the External Variable Domain

The interference structure of a field depends essentially on the number of independent wavelets contributing to the interference pattern. The simplest pattern of absent interference is observed in the region of single-ray propagation, for example, outside of the caustic cusp for an A_3 caustic (of course a simpler pattern of an exponentially decaying field is where rays do not penetrate altogether). In the range of two-ray interference, regular interference fringes are evident near the A_2 caustic.

The interference pattern complicates inside the caustic cusp (Fig. 1.5) where three ray fields overlap simultaneously, and in a still greater extent this takes place inside the swallowtail where four ray fields sum up. These quantitative considerations have been borne out by model computations. For caustics of

$A_4, A_5, D_4^\pm$, and D_5 type, such computations have been performed by a number of workers [4.26–30].

4.4.4 Integrals of the D_{m+1} Series

The caustics of D_{m+1} series have corank $l = 2$ and the corresponding typical wave functions are described by double integrals of the form

$$I_{D,m+1}(\zeta) = \frac{1}{2\pi} \iint_{-\infty}^{\infty} d^2t \exp\left[i\left(\pm t_2^m + t_1^2 t_2 + \zeta_1 t_1 + \sum_{p=2}^{m} \zeta_p t_2^{p-1} \right) \right], \tag{4.4.10}$$

where $m \geq 3$, and $d^2t \equiv dt_1\, dt_2$. Specifically, at $m = 3$ we get

$$I_{D^+,4} = \frac{1}{\sqrt{2\pi}} \iint_{-\infty}^{\infty} d^2t \exp[i(t_2^3 + t_1^2 t_2 + \zeta_1 t_1 + \zeta_2 t_2 + \zeta_3 t_2^2)] \tag{4.4.11}$$

(elliptic umbilic) and

$$I_{D^-,4} = \frac{1}{\sqrt{2\pi}} \iint_{-\infty}^{\infty} d^2t \exp[i(-t_2^3 + t_1^2 t_2 + \zeta_1 t_1 + \zeta_2 t_2 + \zeta_3 t_2^2)] \tag{4.4.12}$$

(hyperbolic umbilic).

The Fresnel indexes of arbitrary codimension m are $\gamma_1 = (1 - 1/m)/2$ and $\gamma_2 = 1/m$, that is how we deal with two different Fresnel scales in two mutually perpendicular directions.

$$a_{f1} \sim F^{1-\gamma_1}\lambda^{\gamma_1}, \qquad a_{f2} \sim F^{1-\gamma_2}\lambda^{\gamma_2} .$$

Note that for D_4^+ and D_4^- caustics, these scales are identical, since at $m = 3$, $\gamma_1 = \gamma_2 = 1/3$, viz.,

$$a_{f1} \sim a_{f2} \sim (F^2\lambda)^{1/3} . \tag{4.4.13}$$

For some values of m the caustic indexes are summarized in Table 3.2. At $m = 3$, the first two indexes are identical: $\alpha_1 = \alpha_2 = 2/3$, and the third is half as large: $\alpha_3 = 1/3$. Therefore, the first two indexes should be referred to the transverse dimension of the focal spot, and the third to the longitudinal size:

$$l_{1,2} \sim (F\lambda^2)^{1/3} = l_\perp, \qquad l_3 \sim (F^2\lambda)^{1/3} = l_\parallel .$$

Like for caustics of A_{m+1} series, in this case the focal relations $l_\perp \sim (\lambda l_\parallel)^{1/2}$ and $a_f l_\perp \sim F\lambda$ hold, but for $m > 3$ they break down.

Figure 4.3 shows cross sections of a D_4^- caustic (elliptic umbilic) by planes $\zeta_3 =$ constant. The principal section reveals an angular shape where the field in the π-neighborhood of the vertex is described by the standard integral (4.4.12). On the side branches of the caustic angle the field is described by the Airy integral, and inside the angle there is the range of four-ray propagation. In the section $\zeta_3 > \pi$ the caustic splits into a simple A_2 caustic and an A_3 cusp. In

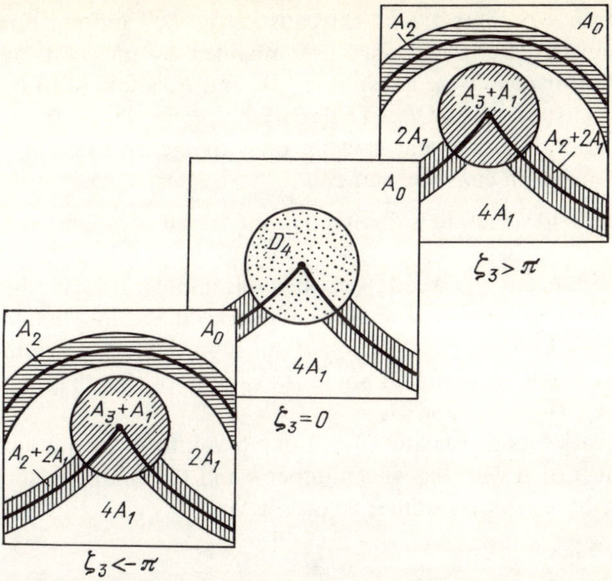

Fig. 4.3. Evolution of the D_4^- caustic near the critical point $\zeta_3 = 0$ accompanied by an indication of subordinate caustics

between there is a two-ray region, and inside the cusp the four-ray region. A similar separation of branches occurs in the section $\zeta_3 < -\pi$.

In dimensional variables, a transition to subasymptotics takes place farther from a critical point than $l_\parallel \sim (F^2\lambda)^{1/3}$ in the longitudinal direction and farther than $l_\perp \sim (F\lambda^2)^{1/3}$ in the transverse direction.

If we compare the integral $I_{D,5}(\zeta)$ for a D_5 caustic (parabolic umbilic) with (4.4.11) we see that the former has t^4 rather than t^3 in the unfolding and has one more external parameter, the exponent acquires one more term, $\zeta_4 t^3$. The metamorphosis diagram of this caustic is given in particular in [4.21]. Since the indexes of this caustic are $\gamma_1 = 3/8$ and $\gamma_2 = 1/4$ the Fresnel scales are also different:

$$a_{f1} \sim F^{5/8}\lambda^{3/8}, \qquad a_{f2} \sim F^{3/4}\lambda^{1/4} .$$

For the D_{m+1} caustic series, the focusing index σ_{foc} is $1/2 - 1/2m$ so that $\sigma_{foc} = 1/3$ for $D_4^\pm$, and $\sigma_{foc} = 3/8$ for D_5.

4.4.5 Caustics with a Large Number of Rays

The caustics A_2, A_5, $D_4^\pm$ and D_5 exhaust the list of seven elementary catastrophes. In the order of increasing complexity, the following caustics are null modal. Besides the A_{m+1} and D_{m+1} series they are extraserial $E_{6,7,8}$ containing more than five rays. For $m = 5$, these are six-ray caustics A_6, D_6, and E_6. For $m = 6$, the seven-ray caustics A_7, D_7, E_7 are accompanied by an eight-ray unimodal caustic D_8, and for $m = 7$, the caustics A_8, D_8 and E_8 are flanked by two unimodal nine-ray caustics X_9 and P_9 (Table 3.4).

All aforementioned and all subsequent caustics obey the general relations illustrated with low-codimension caustics, namely:

(i) The most irregular field structure is observed in the region of interference of many rays.

(ii) When external parameters ζ_p exceed $\pm\pi$, the typical integral $I(\zeta)$ cross over to their subasymptotics in agreement with the adjacency diagrams studied by *Arnold* [4.31], see also [4.1]. The upper portion of this subordinance diagram is shown in Fig. 4.4. We have encountered some transitions of this diagram above, including $A_3 \to A_2$, $D_4 \to A_3$, and $A_4 \to A_3$.

(iii) In dimensional variables a transition from one asymptotic to another occurs at distances about l_p tied with the wavenumber k and the characteristic length F of the problem by a relation similar to (4.4.9), viz.,

$$l_p \sim F/(kF)^{\alpha_p} \sim F^{1-\alpha_p}\lambda^{\alpha_p} ,$$

where α_p are the caustic indexes, and l_p limit the caustic zones around individual elements of the caustic.

(iv) The Fresnel indexes γ_q define the field formation region in the t_q coordinate.

(v) The focusing index σ_{foc} defines the growth of field intensity near a critical point over the level of the typical ray field.

It should be noted that although in general the focusing index σ_{foc} increases with codimension, the rate of growth is rather slow. Table 4.2 summarizes the

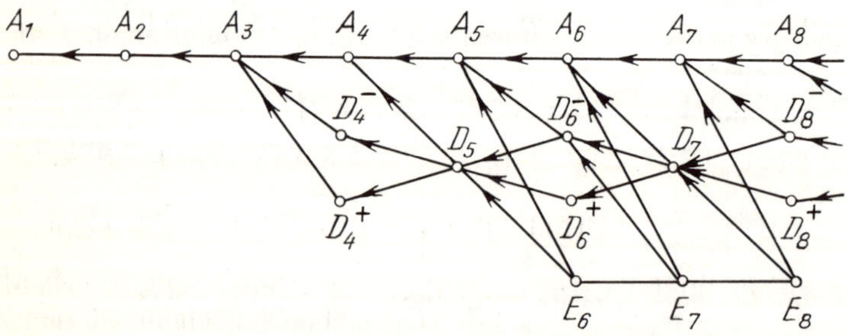

Fig. 4.4. The beginning of the sub-ordinate diagram

Table 4.2. Focusing indexes for complex caustics

Type	P_{m+2}	$R_{p.q}$	$T_{p.q.r}$	Q_{10}	Q_{11}	Q_{12}	Q_{14}	S_{11}	S_{12}	S_{14}	U_{12}	U_{14}	V_{15}	O_{16}
σ_{foc}	$\frac{1}{2}$	$\frac{1}{2}$	$\frac{1}{2}$	$\frac{13}{24}$	$\frac{5}{9}$	$\frac{17}{30}$	$\frac{7}{12}$	$\frac{9}{16}$	$\frac{15}{26}$	$\frac{3}{5}$	$\frac{7}{12}$	$\frac{11}{18}$	$\frac{5}{8}$	$\frac{2}{3}$

values of focusing index for some types of caustic listed in Table 3.4. One may select a caustic which possesses the largest value of σ_{foc} for a given number of rays. Growth of the index of focusing in more complex caustics is associated both with the growth of the numbe of rays convergent at caustics and with the rise of density of the energy fluxes associated with these rays.

In general the values of σ_{foc} listed in Table 4.2 slightly exceed the focusing index for a cylindrical wave, $\sigma_{\text{foc}} = 1/2$, and approach very slowly $\sigma_{\text{foc}} = 1$ characteristic of a perfect spherical wave. Indirectly this points out to a very high (formally infinite) codimension of the spherical wave whose symmetry can be broken down in infinite number of ways.

4.4.6 Calculation of Standard Integrals

Rather efficient techniques have been developed for computation and tabulation of the standard integrals of catastrophe theory. First of all, the exponents in these integrals may be expanded into power series in all arguments. This approach yields good results in the vicinity of a critical point $\zeta = 0$, but at some distance from it convergence of the power series is normally slow [4.21, 32–35].

Fortunately, at large values of arguments the typical integrals allow cross-over to the asymptotic expansions such that the error is under control. Therefore, the values of integrals, calculated at small and moderate values of arguments, can often be matched to one another [4.21, 34].

The direct calculation of integrals of rapidly oscillating exponential functions proved to be a rather inefficient approach because the end effects penalize the accuracy [4.28]. However, a modification of the integration loop in the complex plane transforms the rapidly oscillating exponent into a readily integrable decaying exponent [4.23, 34]. At the present time this is the more popular method for calculating the integrals under consideration, especially because many double integrals may be transformed into single integrals [4.36, 21].

Under some circumstances a competitive technique is found in the evaluation of standard integrals by solving the respective differential equations in partial derivatives. Such equations can be obtained by differentiating the standard integrals with respect to external variables ζ_p and combining the resultant derivatives so that integrable integrands should result. By way of example, for the Pearsey integral $I_p(0, X_2)$, *Connor* et al. [4.37] obtained the equation

$$\frac{\partial^2 I_p}{\partial X_2^2} + \mathrm{i}\frac{X_2 \partial I_p}{2\partial X_2} + \frac{\mathrm{i}}{4} I_p = 0, \quad X_1 = 0 \, .$$

This equation can be solved numerically on the axis $X_1 = 0$ taking as the initial value $I_p(0, 0)$ the one calculated by expanding I_p into a Taylor series. Having determined the Pearsey integral on the X_2 axis, the quantities $I_p(0, X_2)$ may be used as initial values for calculating $I_p(X_1, X_2)$ by integration of the differential equation

$$\frac{\partial^3 I_p}{\partial X_1^3} - \frac{X_2}{2}\frac{\partial^2 I_p}{\partial X_1} - \frac{i}{4} X_1 I_p = 0$$

along the lines X_2 = constant.

The method of differential equations has demonstrated high efficiency in the conditions where one needs values of standard integrals in the entire space of variables $\zeta_1, \ldots, \zeta_m$. It is important to note, that in addition to the integral proper the method yields also its derivatives $\partial I(\zeta)/\partial \zeta_p$ which can be used to determine the uniform asymptotics of the integrals of rapidly oscillating functions, and to construct the uniform asymptotics of the wave field. Early results in computing the derivatives of the Pearsey integral have been obtained by *Dronov* et al. [4.27], and in computing the derivatives of $I_{A,4}$ by *Connor* et al. [4.30].

At the same time the differential equation technique is not free from disadvantages. Its computational speed is markedly inferior to direct integration over a modified contour where the integral is required at a single point. The algorithm loses stability in the caustic shadow for having an exponentially decaying solution as a background, the small errors in finite difference schemes induce a second exponentially growing solution contradicting to the physical setting of the problem.

Practical computations resort to all three basic methods, power series, direct integration, and differential equation, combining them with other techniques, including matching with asymptotic expansions. The latter approach allows the computations of a standard integral to be confined to a neighborhood of the critical point.

Another technique consists in accurate calculation of the integrals at a comparatively sparse set of points with a subsequent interpolation of splines or local power series.

The state of the art is reviewed by *Kryukovskii* et al. [4.21]. This paper gives also computer programs for the standard integrals $I_{A,m}$, $I_{D_4}^{\pm}$, and $I_{D,5}$ and their derivatives. These functions seem to exhaust demands of modern wave theory – catastrophes of higher order are seldom met and hard to identify.

5 Uniform Caustic Asymptotics Derived with Standard Integrals

By representing a wave field in terms of standard caustic integrals one obtains only a local asymptotic solution. Augmenting the standard integral by its derivatives with respect to external parameters (Kravtsov–Ludwig technique) renders the asymptotic uniformly valid, that is, applicable both at short and at long distances from the caustic. We introduce the concepts of transverse and longitudinal caustic scales to derive uniformly-valid applicability conditions for the asymptotic expressions.

5.1 Uniform Airy Asymptotic of a Scalar Field

5.1.1 Heuristic Foundation of the Method of Standard Integrals

For three-dimensional wave problems, the main idea underlying the method of standard integrals is expressed by the principle: *to qualitatively similar configurations of rays and caustics there correspond qualitatively similar structures of the wave field.* This heuristic principle popularized by the authors since the mid-60s [5.1–3] has multiple predecessors in the theory of oscillations and waves. It may be viewed as a particular case of a more general principle of *similarity of solutions in problems with qualitatively similar conditions*, which may be traced out through the entire history of physics and mathematics. The van der Pol method of slow amplitudes in the theory of nonlinear oscillations, the WKBJ method in wave theory, and the Whitham method in the theory of nonlinear wave processes, all of them are applications of the principle of similitude to oscillatory-wave problems.

Geometrical optics is also a typical example of this principle: the method is supported by the assumption that in slowly inhomogeneous media the field has a structure of a plane wave $U(\mathbf{r})\exp[ik\psi(\mathbf{r})]$ with a slowly varying amplitude $U(\mathbf{r})$ and a slowly varying momentum $\mathbf{p}(\mathbf{r}) = \nabla\psi(\mathbf{r})$. Another example from the wave theory may be the uniform Airy asymptotic of a one-dimensional wave function $u(z)$ in the presence of a simple turning point (Sect. 4.2.4). The heuristic principle of similarity serves as a base also for physical optics, the geometrical theory of diffraction due to J.B. Keller, the method of edge waves due to P.Ya. Ufimtsev, and many other techniques of wave theory.

Use of Airy's function to describe the field near a simple caustic seems to be quite natural, indeed Airy introduced this function with this purpose in sight [5.4]. It occurs any time when a local asymptotic of the field near caustics comes into consideration [5.5–7]. However, away from the caustic this local asymptotic of the boundary layer type is no longer valid.

A radical step to the construction of a uniform asymptotic was to add to the Airy function $\mathrm{Ai}(\zeta)$ a term with its first derivative $\mathrm{Ai}'(\zeta)$ [5.8]. The implication was that the combination $c_1\mathrm{Ai}(\zeta) + c_2\mathrm{Ai}'(\zeta)$ could be able to describe the combined contribution of all higher derivatives of Ai (in Sect. 4.2.1 we noted that the latter are expressable via Ai and Ai′), thus taking into account the various deviations from the standard conditions such as the curvature of the caustic, slow inhomogeneity of the medium, etc. An additional argument in support of the term with Ai′ was the work of *Chester* et al. [5.9] which demonstrated that the integral of a rapidly oscillating exponent with two stationary points could be uniformly approximated by the sum $c_1\mathrm{Ai}(\zeta) + c_2\mathrm{Ai}'(\zeta)$.

This approach was taken independently by *Ludwig* [5.10] and *Berry* [5.11]. Ludwig made another important step by constructing for A_{m+1} caustics a uniform asymptotic on the basis of generalized Airy integrals and their derivatives. The uniformity of the asymptotic was ensured by incorporating the derivatives of the standard integrals. Rays played a very important role in such constructions, thus prompting the researchers to characterize the procedure as a whole as "sewing a wave flesh on a geometric skeleton" [5.2, 3].

Without touching upon the generalization of this procedure to diffraction problems, we note that the field of problems allowing for a standard integral solution has had a trend to expand in recent years. Specific investigations in this directions will be mentioned below, thus far we point out to the important conceptual efforts by *Berry* [5.12] and *Mount* [5.13].

Catastrophe theory paved the way for a general theory of uniform asymptotics based on standard integrals and constructed in our work [5.14] to be outlined in Sect. 5.2. As a prelude to this discussion we wish to consider a simpler problem of construction of a uniform Airy asymptotic, which involves essential points of the general theory, at the same time allowing for an explicit solution.

5.1.2 Guessing at a Form of Solution

The exact solution (4.2.17) for a plane wave incident on a linear slab serves as a standard one for all problems with simple caustic A_2. This field describes the interference of two waves in the lit region $z < z_c - \Lambda$, exponential decay in the shadow region $z > z_c + \Lambda$, and yields in a caustic zone $z_c - \Lambda < z < z_c + \Lambda$ a field concentration $(k/\varepsilon_1)^{1/6}$ times higher than the initial field at $z = z^0$. The phase factor $\exp(ikx \sin\theta^0)$ describes the propagation of the wave along the caustic.

Whenever deviations from the standard conditions take place which may be a distortion of the incident wavefront; distortion of the caustic; a change in the $\varepsilon(z)$ profile; or additional inhomogeneities in the medium – one may expect that in addition to the term $A \cdot \tilde{I} \cdot e^{ik\chi}$ containing an Airy integral I, the solution should acquire a term with the derivative of the integral, say $\sim \tilde{I}' e^{ik\chi}$. This reasoning motivates a search for a uniform asymptotic of the wave problem as a sum

$$u(\mathbf{r}) = [A \cdot \tilde{I}(\tilde{\zeta}) + (ik)^{-1} B \cdot \tilde{I}(\tilde{\zeta})] e^{ik\chi} , \tag{5.1.1}$$

where the four functions to be determined are $A(\mathbf{r})$, $B(\mathbf{r})$, $\tilde{\zeta}(\mathbf{r})$ and $\chi(\mathbf{r})$.

The factor $(ik)^{-1}$ in the second term has been introduced to compensate for the factor ik in (4.2.13) giving the derivative $\tilde{I}'(\tilde{\zeta}) = \partial \tilde{I}(\tilde{\zeta})/\partial \tilde{\zeta}$. The choice of the dimensional Airy integral $\tilde{I}(\tilde{\zeta})$ as a principal standard function is explained by a simpler algebra [5.10]. Using this form of solution ensures that the calculations will produce only integer powers of k whereas seeking a solution via the integrals $I(\tilde{\zeta})$, $\mathrm{Ai}(\zeta)$, or $v(\tilde{\zeta})$ has to result in powers of k that are multiples of 2/3 (see the original work [5.8]).

Fractional powers of k are contained in (5.1.1) implicitly and come up in crossing over from $\tilde{I}(\tilde{\zeta})$ to $\mathrm{Ai}(\zeta)$ by the formulas of Sect. 4.2. It can be demonstrated that both approaches are equivalent: (a) through the dimensional integral $\tilde{I}(\tilde{\zeta})$ and integer powers of k and (b) through the dimensionless integral $I(\zeta)$ and fractional powers of k.

For convenience of the following exposition we combine both terms of (5.1.1) in a single integral

$$u(\mathbf{r}) = \left(\frac{k}{2\pi}\right)^{1/2} \int_{-\infty}^{\infty} g(\mathbf{r}, \tilde{t}) \exp[ikf(\mathbf{r}, \tilde{t})] \, d\tilde{t} , \tag{5.1.2}$$

where

$$f(\mathbf{r}, \tilde{t}) = \chi(\mathbf{r}) + \tilde{\zeta}(\mathbf{r})\tilde{t} + \tilde{t}^3/3, \qquad g(\mathbf{r}, \tilde{t}) = A(\mathbf{r}) + \tilde{t}B(\mathbf{r}) . \tag{5.1.3}$$

5.1.3 Equations for Unknown Functions

Differentiating (5.1.2) twice carries the wave equation to the form

$$\begin{aligned} \Delta u + k^2 \varepsilon u = \left(\frac{k}{2\pi}\right)^{1/2} \int_{-\infty}^{\infty} d\tilde{t} \exp(ikf) \{ (ik)^2 g[(\nabla f)^2 - \varepsilon] \\ + ik[2\nabla g \cdot \nabla f + g\Delta f] + \Delta g \} = 0 . \end{aligned} \tag{5.1.4}$$

Here, as usual the addends are grouped in powers of k, but, unlike geometrical optics, the coefficients of equal powers are equated to the integer part of the derivative $\partial f/\partial \tilde{t}$ rather than to zero. This procedure suggested by *Ludwig* [5.8] enables the order of magnitude in k to be reduced by one unit. Indeed, the integral of $T(\tilde{t})\partial f/\partial \tilde{t}$ factored by a rapidly oscillating function $\exp(ikf)$ allows integration by parts

$$\int T\frac{\partial f}{\partial \tilde{t}}\exp(\mathrm{i}kf)\mathrm{d}\tilde{t} = \frac{1}{\mathrm{i}k}\int T\,\mathrm{d}\exp(\mathrm{i}kf)$$

$$= \frac{1}{\mathrm{i}k}\,T\mathrm{e}^{\mathrm{i}kf}\Big|_{-\infty}^{\infty} - \frac{1}{\mathrm{i}k}\int\frac{\partial T}{\partial \tilde{t}}\exp(\mathrm{i}kf)\mathrm{d}\tilde{t}\ ,$$

and if we adopt that $\exp(\mathrm{i}kf)$ can be made to vanish by an appropriate deformation of the integration contour in the complex plane, then

$$\int_{-\infty}^{\infty} T\frac{\partial f}{\partial \tilde{t}}\exp(\mathrm{i}kf)\mathrm{d}\tilde{t} = \frac{1}{\mathrm{i}k}\int\frac{\partial T}{\partial \tilde{t}}\exp(\mathrm{i}kf)\mathrm{d}\tilde{t}\ . \tag{5.1.5}$$

This order-reduction procedure is equivalent to a sequential reduction of the higher derivatives of Airy integral to the integral proper and to its first derivative.

For the first term in (5.1.4) we require that the relation

$$(\nabla f)^2 - \varepsilon = T\frac{\partial f}{\partial \tilde{t}} \tag{5.1.6}$$

should be an identity in $\tilde{t}$. Recognizing that the derivatives of f are polynomials in $\tilde{t}$,

$$\frac{\partial f}{\partial \tilde{t}} = \tilde{\zeta} + \tilde{t}^2, \qquad \nabla f = \nabla\chi + \tilde{t}\nabla\tilde{\zeta}\ ,$$

it would be natural to seek T also as a polynomial, $T = c_0 + c_1\tilde{t} + c_2\tilde{t}^2 + \cdots$. It has been learned that (5.1.5) can be satisfied by meeting the following conditions: first, $c_0 = (\nabla\tilde{\zeta})^2$, $c_1 = c_2 = \cdots = 0$ so that in the final analysis $T = (\nabla\zeta)^2$; secondly, the coefficient of $\tilde{t}$ and the free term in (5.1.6) must vanish,

$$(\nabla\chi)^2 - \tilde{\zeta}(\nabla\tilde{\zeta})^2 - \varepsilon = 0\ , \tag{5.1.7}$$

$$\nabla\chi\cdot\nabla\tilde{\zeta} = 0\ . \tag{5.1.8}$$

The conditions (5.1.7, 8) are the equations to determine the unknown functions $\chi(\mathbf{r})$ and $\tilde{\zeta}(\mathbf{r})$.

Substituting (5.1.6) into (5.1.4) and integrating by parts subject to (5.1.5) yields

$$\Delta u + k^2\varepsilon u = (k/2\pi)^{1/2}\int_{-\infty}^{\infty}\exp(\mathrm{i}kf)\mathrm{d}\tilde{t}$$

$$\times\left[\mathrm{i}k\cdot\left(2\nabla g\nabla f + g\Delta f - \frac{\partial(Tg)}{\partial\tilde{t}}\right) + \Delta g\right] = 0\ . \tag{5.1.9}$$

Expanding the amplitude factor $g = A + B\tilde{t}$ in powers of k

$$g(\mathbf{r},\tilde{t}) = \sum_{n=0}^{\infty}(\mathrm{i}k)^{-n}g_n(\mathbf{r},\tilde{t}) = \sum_{n=0}^{\infty}(\mathrm{i}k)^{-n}[A_n(\mathbf{r}) + \tilde{t}B_n(\tilde{\mathbf{r}})]$$

(this expansion could be done, of course, on an earlier stage) transforms (5.1.9) to the form

$$0=(k/2\pi)^{1/2}\int_{-\infty}^{\infty}\exp(ikf)d\tilde{t}\sum_{n=0}^{\infty}(ik)^{1-n}$$
$$\times\left[2\nabla g_n\cdot\nabla f+g_n\Delta f-\frac{\partial}{\partial\tilde{t}}(g_nT)+\Delta g_{n-1}\right], \qquad (5.1.10)$$

where it is assumed that $g_{-1}=g_{-2}=\cdots=0$.

We can again apply to the leading ($n=0$) term in (5.1.10) the operation of finding the integer about $\partial f/\partial\tilde{t}$ by letting

$$2\nabla g_0\cdot\nabla f+g_0\Delta f-\partial(Tg_0)/\partial\tilde{t}=R_0\partial f/\partial\tilde{t}\ . \qquad (5.1.11)$$

Similar to (5.1.6) this equation should be an identity in $\tilde{t}$; thus, leading to two relations

$$(2\nabla A_0\cdot\nabla\chi+A_0\Delta\chi)+(2\nabla B_0\cdot\tilde{\zeta}\nabla\tilde{\zeta})+B_0\tilde{\zeta}\Delta\tilde{\zeta}+B_0(\nabla\tilde{\zeta})^2=0\ ,$$
$$(2\nabla A_0\cdot\nabla\tilde{\zeta}+A_0\Delta\tilde{\zeta})+(2\nabla B_0\cdot\nabla\chi+B_0\Delta\chi)=0\ , \qquad (5.1.12)$$

[where it has been recognized that $T=(\nabla\tilde{\zeta})^2$], which are the equations for the amplitude factors A_0 and B_0. Simultaneously (5.1.11) yields the coefficient R_0 as

$$R_0=2\nabla B_0\cdot\nabla\tilde{\zeta}+B_0\Delta\tilde{\zeta}\ .$$

A similar reasoning may be applied to transform the terms of $n=1,2$, etc. However, we shall not do so, for we are going to confine ourselves to the fundamental ("zeroth") approximation $g_0=A_0+\tilde{t}B_0$. Accordingly, the subscript 0 will henceforth be dropped.

5.1.4 Relation of the Airy Asymptotic to the Ray Fields

In order to match the caustic field (5.1.1) with the sum of two ray fields

$$u=Ue^{ik\psi_1}+U_2e^{ik\psi_2} \qquad (5.1.13)$$

representing the interference of incident (1) and reflected (2) rays at a distance from the caustic we resort to the asymptotic values of $\tilde{I}$ and its derivative $\tilde{I}'$ outlined in Sect. 4.2.1. Substituting these expressions into (5.1.1) and equating (5.1.1) and (5.1.13), it would be natural to put the smaller exponent $-2k(-\tilde{\zeta})^{3/2}/3+k\chi$ in correspondence to the phase $k\psi_1$ of the incident wave, and the larger exponent $2k(-\tilde{\zeta})^{3/2}/3+k\chi$ in correspondence to $k\psi_2$ to obtain

$$-\tfrac{2}{3}(-\tilde{\zeta})^{3/2}+\chi=\psi_1,\qquad \tfrac{2}{3}(-\tilde{\zeta})^{3/2}+\chi=\psi_2\ . \qquad (5.1.14)$$

These expressions yield the formulas

$$\chi=\tfrac{1}{2}(\psi_1+\psi_2),\qquad \tfrac{2}{3}(-\tilde{\zeta})^{3/2}=\tfrac{1}{2}(\psi_2-\psi_1) \qquad (5.1.15)$$

which allow the sought quantities χ and $\tilde{\zeta}$ to be expressed directly in terms of the eikonals ψ_1 and ψ_2.

Equating the pre-exponential factors yields (in the zeroth order in k)

$$2^{-1/2}[(-\tilde{\zeta})^{-1/4}A + (-\tilde{\zeta})^{1/4}B]e^{i\pi/4} = U_1 \ ,$$
$$2^{-1/2}[(-\tilde{\zeta})^{\ 1/4}A - (-\tilde{\zeta})^{1/4}B]e^{-i\pi/4} = U_2 \ .$$

We use these expressions to represent the amplitude factors A and B in terms of the geometric-optical amplitudes:

$$\begin{aligned} A &= 2^{-1/2}[U_1\exp(-i\pi/4) + U_2\exp(i\pi/4)](-\tilde{\zeta})^{1/4} \\ &= 2^{-1/2}\exp(-i\pi/4)(-\tilde{\zeta})^{1/4}(U_1 + iU_2) \ , \\ B &= 2^{-1/2}[U_1\exp(-i\pi/4) - U_2\exp(i\pi/4)](-\tilde{\zeta})^{-1/4} \\ &= 2^{-1/2}\exp(-i\pi/4)(-\tilde{\zeta})^{1/4}(U_1 - iU_2) \ . \end{aligned} \qquad (5.1.16)$$

A remarkable feature of (5.1.14, 15) obtained for areas separated from the caustic is that actually they are valid in the entire space. To verify this claim it would suffice to check whether or not the quantities χ and $\tilde{\zeta}$ defined by (5.1.14) satisfy (5.1.7, 8), and the quantities A and B defined by (5.1.15) satisfy (5.1.12).

Differentiating (5.1.15) we get

$$\nabla\chi = \tfrac{1}{2}\nabla(\psi_1 + \psi_2), \qquad \sqrt{-\tilde{\zeta}}\nabla\tilde{\zeta} = \tfrac{1}{2}\nabla(\psi_1 - \psi_2)$$

Now, if $\psi_{1,2}$ satisfy the eikonal equation $(\nabla\psi_{1,2})^2 = \varepsilon$, then

$$\begin{aligned} (\nabla\chi)^2 - \tilde{\zeta}(\nabla\tilde{\zeta})^2 &= \tfrac{1}{4}[\nabla(\psi_1 + \psi_2)]^2 + \tfrac{1}{4}[\nabla(\psi_1 - \psi_2)]^2 \\ &= \tfrac{1}{4}[2(\nabla\psi_1)^2 + 2(\nabla\psi_2)^2] = \varepsilon \ , \end{aligned}$$

$$\begin{aligned} \sqrt{-\tilde{\zeta}}\nabla\chi\cdot\nabla\tilde{\zeta} &= \tfrac{1}{4}\nabla(\psi_1 + \psi_2)\cdot\nabla(\psi_1 - \psi_2) \\ &= \tfrac{1}{4}[(\nabla\psi_1)^2 - (\nabla\psi_2)^2] = 0 \ , \end{aligned}$$

which agree with (5.1.7, 8).

In view of (5.1.16) it follows:

$$\nabla A = 2^{-1/2}e^{-i\pi/4}(-\tilde{\zeta})^{1/4}[\nabla U_1 + i\nabla U_2 + \tfrac{1}{4}\nabla\tilde{\zeta}\cdot\tilde{\zeta}^{-1}(U_1 + iU_2)] \ ,$$
$$\nabla B = 2^{-1/2}e^{-i\pi/4}(-\tilde{\zeta})^{-1/4}[\nabla U_1 - i\nabla U_2 - \tfrac{1}{4}\nabla\tilde{\zeta},\tilde{\zeta}^{-1}(U_1 - iU_2)] \ .$$

Substituting these values of ∇A and ∇B along with the amplitudes A and B into (5.1.12) and observing that in view of (5.1.14)

$$\nabla\psi_{1,2} = \nabla\chi \pm (-\tilde{\zeta})^{1/2}\nabla\tilde{\zeta}, \quad \Delta\psi_{1,2} = \Delta\chi \pm (-\tilde{\zeta})^{1/2}\Delta\tilde{\zeta} \mp \tfrac{1}{2}(-\tilde{\zeta})^{-1/2}(\nabla\tilde{\zeta})^2$$

carry the set of equations (5.1.12) to the form

$$\tfrac{1}{2}(M_1 + M_2)(-\zeta)^{1/4} = 0, \qquad \tfrac{1}{2}(M_1 - M_2)(-\zeta)^{-1/4} = 0 \ ,$$

where $M_{1,2} = 2\nabla U_{1,2}\nabla\psi_{1,2} + U_{1,2}\Delta\psi_{1,2}$.

From this system of equations it follows that $M_{1,2} = 0$. Thus, if $U_{1,2}$ satisfy the trans equations $M_{1,2} = 0$, then A and B determined by (5.1.16) obey (5.1.12).

Now the asymptotic solution (5.1.1) may be expressed through the geometric-optical amplitudes $U_{1,2}$ as follows:

$$u = 2^{-1/2}\left[(-\tilde{\zeta})^{1/4}(U_1 + \mathrm{i}U_2)\tilde{I}(\tilde{\zeta}) - \frac{\mathrm{i}}{k}(-\tilde{\zeta})^{1/4}(U_1 - \mathrm{i}U_2)\tilde{I}'(\tilde{\zeta})\right]$$
$$\times \exp(\mathrm{i}k\chi - \mathrm{i}\pi/4) \ . \qquad (5.1.17)$$

This solution may be written directly in terms of the Airy function $\mathrm{Ai}(\zeta)$ or $v(\zeta)$. For example, in agreement with (4.2.9) we have

$$u = \sqrt{\pi}k^{1/6}[(U_1 + \mathrm{i}U_2)(-\tilde{\zeta})^{1/4}\mathrm{Ai}(k^{2/3}\tilde{\zeta})$$
$$- \mathrm{i}k^{-1/3}(U_1 - \mathrm{i}U_2)^{-1/4}(-\tilde{\zeta})^{-1/4}\mathrm{Ai}'(k^{2/3}\tilde{\zeta})]\exp(\mathrm{i}k\chi - \mathrm{i}\pi/4)$$
$$= \sqrt{\pi}[(U_1 + \mathrm{i}U_2)(-\zeta)^{1/4}\mathrm{Ai}(\zeta) - \mathrm{i}(U_1 - \mathrm{i}U_2)(\zeta)^{-1/4}\mathrm{Ai}'(\zeta)]$$
$$\times \exp(\mathrm{i}k\chi - \mathrm{i}\pi/4) \ , \qquad (5.1.18)$$

where $\zeta = k^{2/3}\tilde{\zeta}$, and the quantities $\tilde{\zeta}$ and χ are related to the eikonals ψ_1 and ψ_2 by (5.1.15).

5.1.5 Field in the Caustic Shadow

Now we turn to the region of caustic shadow. Here, we have to deal with the complex eikonals ψ_1 and ψ_2 that are complex conjugates in a nonabsorbing medium,

$$\psi_1 = \psi' + \mathrm{i}\psi'', \qquad \psi_2 = \psi' - \mathrm{i}\psi'' \ . \qquad (5.1.19)$$

Using (5.1.14) we have

$$\chi = \psi', \qquad \tfrac{2}{3}\tilde{\zeta}^{3/2} = \psi'', \qquad \psi'' > 0 \ . \qquad (5.1.20)$$

It follows that in caustic shadow the lines χ = constant and $\tilde{\zeta}$ = constant are the loci of constant real and imaginary parts of ψ_1 and ψ_2.

The general pattern of the isolines χ = constant and $\tilde{\zeta}$ = constant is shown in Fig. 5.1. Positive values of $\tilde{\zeta}$ correspond to the region of shadow, negative $\tilde{\zeta}$ to the lit region. At the caustic, $\tilde{\zeta} = 0$. Dashed lines represent constant eikonals ψ_1 of the incident wave and ψ_2 of the reflected wave, and the arrowed line shows a ray orthogonal to these lines.

The relation of χ and $\tilde{\zeta}$ with the complex eikonals is important in that it helps establish selection rules for complex rays in the shadow region [5.15]. Indeed, making use of the asymptotic formulas (4.2.12) for $\tilde{I}$ and (4.2.13) for $\tilde{I}'$ and the relations (5.1.7, 8) for the field in the shadow region ($\tilde{\zeta} \gg 1$) we have

$$u = U_1 \exp\left(\mathrm{i}k\chi - \frac{2k}{3}\tilde{\zeta}^{3/2}\right) = U_1 \exp(\mathrm{i}k\psi_1) \ . \qquad (5.1.21)$$

While in the lit region the Airy function describes the interference of two ray fields, in the shadow region this function induces only one added decaying away

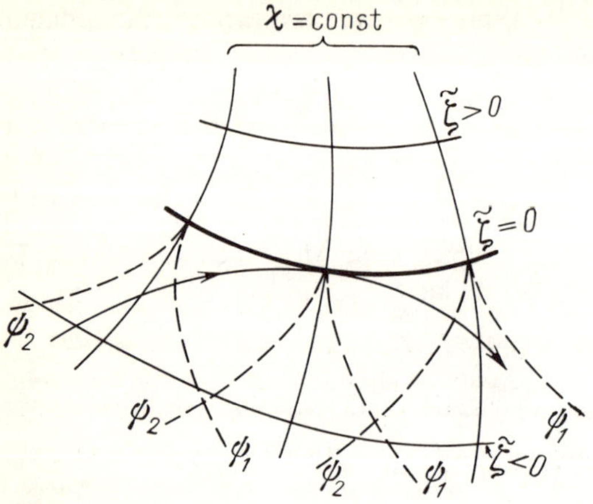

Fig. 5.1. Orthogonality of lines ξ = constant and χ = constant in the vicinity of a caustic

from the caustic as $\exp(\mp k\psi'')$. The asymptotic of (5.1.21) excludes from consideration the second complex wave $U_2 \exp(ik\psi_2)$ which would increase as $\exp(k\psi'')$ inwards the shadow region. Note that in weakly absorbing media with $\operatorname{Im}\varepsilon \equiv \varepsilon'' \ll 1$, the quantities χ and $\tilde{\zeta}$ become complex valued and orthogonality of isolines of χ and $\tilde{\zeta}$ in Fig. 5.1 gives way to a more complex pattern. Accordingly, the wave field acquires a form more complex for analysis, and the argument of the Airy function now involves complex functions. The caustic shifts toward complex space, although the shift is not large, if $\varepsilon'' \ll 1$.

5.1.6 Local Field Asymptotic near a Caustic

A certain advantage of the uniform asymptotic formulas is that they provide a foothold for analysis of the wave field *as a whole*. From the practical side, it is far more convenient to use *local* asymptotics which do not pretend for a global description, but instead lend convenient approximate expressions. One may arrive at local asympotics in a few different ways.

One of them is the method of boundary layer suggested by *Buchal* and *Keller* [5.5] for homogeneous media and generalized to inhomogeneous media by *Gazaryan* [5.6] (for a modern status, see the book of *Babich* and *Kirpichnikova* [5.7]). Our presentation will follow the method of simplification of uniform asymptotics implemented by *Orlov* [5.16]. The idea underlying the method consists in endowing the caustic with coordinates tied to a point M_0 on the caustic where the perpendicular dropped from the point of interest M penetrates the caustic, as shown in Fig. 5.2a. From this it follows that all quantities in (5.1.1) including the argument of the Airy function, ζ, and the amplitudes $u_{1,2}$ may be expressed in terms of fractional powers of the distance to caustic $\nu = MM_0$, where the resultant field is an analytic function of ν. Some details of these calculations follow:

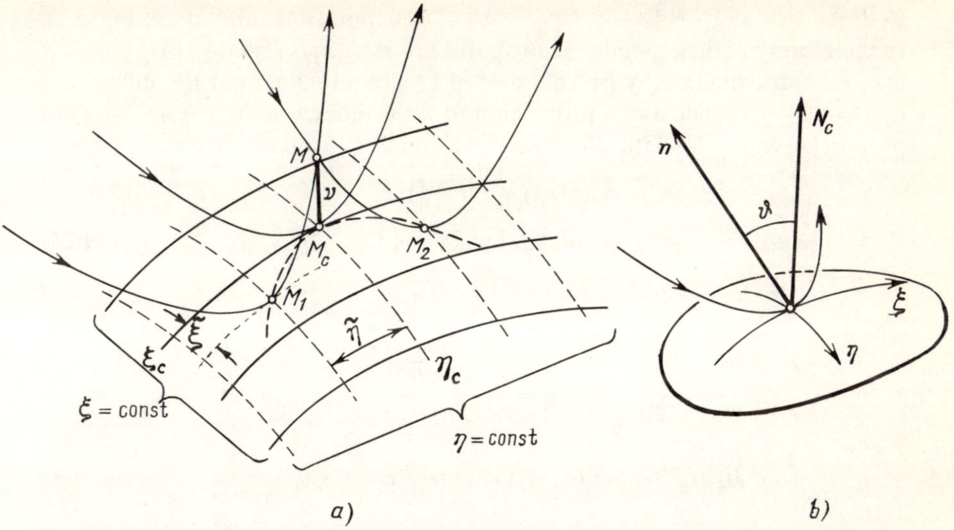

Fig. 5.2. (**a**) Disposition of rays near a simple caustic. (**b**) Position of a ray with respect to a normal to the caustic

Let $\mathbf{r} = \mathbf{R}(\xi, \eta, \psi)$ be the vectorial parameric equation of a ray family forming a simple caustic $\mathbf{r} = \mathbf{r}_c(\xi, \eta)$. The equation of the caustic can be obtained by solving the vanishing Jacobian

$$D(\xi, \eta, \tau) = \frac{\partial(x, y, z)}{\partial(\xi, \eta, \tau)} = \mathbf{r}_\xi \mathbf{r}_\eta \mathbf{r}_\tau = 0 \tag{5.1.22}$$

for τ and substituting this $\tau = \tau_c(\xi, \eta)$ into the ray-family equation

$$\mathbf{r}_c(\xi, \eta) = \mathbf{R}[\xi, \eta, \tau_c(\xi, \eta)] \ . \tag{5.1.23}$$

Note that in (5.2.22) $\mathbf{r}_\xi$, $\mathbf{r}_\eta$, and $\mathbf{r}_\tau$ stand for the partial derivatives of the radius vector $\mathbf{r}$ with respect to the ray parameters ξ, η, and τ.

It is convenient to select the ray coordinates ξ and η such that $\mathbf{r}_\eta = 0$ (yet $\mathbf{r}_\xi \neq 0$) at the caustic (5.1.22). This condition implies that at the caustic rays lying in the plane ξ = constant cross. Accordingly, lines ξ = constant are tangent to rays at the caustic, and the traces left at the caustic by wavefronts ψ = constant are perpendicular to lines ξ = constant and may serve as coordinate lines η = constant (Fig. 5.2). The direction of the ray coordinate on the caustic will be given by the conditions $D_\eta > 0$ and $D_\tau < 0$, the latter implying that the cross section of the ray tube decreases as it approaches the caustic so that near the caustic $D \approx D_\tau(\tau - \tau_0) \approx D_\tau \cdot n^2(\psi - \psi_0)$.

Let M_0 have the ray coordinates ξ_0, η_0, τ_0, the coordinates of the neighboring points being $\tilde{\xi} = \xi - \xi_0$, $\tilde{\eta} = \eta - \eta_0$, and $\tilde{\tau} = \tau - \tau_0$. The point of observation M spaced v from M_0 in the normal to the caustic is hit by two rays tangent to the caustic at $\xi_1 = \xi_0 + \xi_{1,2}, \eta_1 = \eta_0 + \tilde{\eta}_{1,2}$ (points M_1 and M_2 in Fig. 5.2). In the vicinity of these points the ray paths $\mathbf{r} = \mathbf{r}(\xi_1, \eta_1, \tau_1)$ and $\mathbf{r} = \mathbf{r}(\xi_2, \eta_2, \tau_2)$ may be approximated by polynomials in ξ, η, τ revealing that the differences $\tilde{\xi}_{1,2} = \xi_{1,2} - \xi_0$ increase in proportion to v, the differences $\tilde{\eta}_{1,2} = \eta_{1,2} - \eta_0$ in proportion to $v^{1/2}$, and the differences of eikonals in proportion to $v^{3/2}$, namely,

$$\begin{aligned} &\xi_{1,2} - \xi_0 \equiv \tilde{\xi}_{1,2} = -cv/(b|\mathbf{r}_\xi|) + \mathrm{O}(v^{3/2}) \ , \\ &\eta_{1,2} - \eta_0 = \tilde{\eta}_{1,2} = \pm (v/b)^{1/2} - av/(2b^2) + \mathrm{O}(v^{3/2}) \ , \\ &\psi_{1,2} - \psi_0 = \mp ng(v/b)^{3/2} + \mathrm{O}(v^2) \ , \end{aligned} \tag{5.1.24}$$

where

$$b = \frac{n\tilde{D}_\eta}{2|\mathbf{r}_\xi|}, \qquad c = \frac{\mathbf{r}_{\eta\eta}\mathbf{r}_\xi}{2|\mathbf{r}_\xi|}, \qquad g = -\frac{n^3\tilde{D}_\eta\tilde{D}_\tau}{3|\mathbf{r}_\xi|^2} \ ,$$

$$a = \frac{1}{6|\mathbf{r}_\xi|^3}[\tilde{D}_{\eta\eta}|\mathbf{r}_\xi|^2 - 2\tilde{D}_\eta(\mathbf{r}_{\xi\eta}\cdot\mathbf{r}_\xi) - \tilde{D}_\xi(\mathbf{r}_{\eta\eta}\cdot\mathbf{r}_\xi)] \ ,$$

and $\tilde{D} = Dn = \mathrm{d}a/\mathrm{d}\xi\,\mathrm{d}\eta$ is the relative cross section of the ray tube. Subscript 1 indicates the reflected ray after touching the caustic, subscript 2 indicates the incident ray. The upper signs relate to subscript 1. All coefficients are calculated at the reference point ξ_0, η_0.

The last line in (5.1.24) immediately yields

$$\begin{aligned} &k\chi = k(\psi_1 + \psi_2)/2 \approx k\psi_0 + \mathrm{O}(v^{3/2}) \ , \\ &\zeta = -[3k(\psi_2 - \psi_1)/4]^{2/3} \\ &= -\left(\frac{3kng}{2}\right)^{2/3}\frac{v}{b} = -\left(\frac{2k^2n^5D_\tau^2}{|\mathbf{r}_\xi|D_\eta}\right)^{1/3} v = -\left(\frac{2k^2}{\rho_{\mathrm{rel}}}\right)^{1/3} v \ , \end{aligned} \tag{5.1.25}$$

where the quantity

$$\frac{1}{\rho_{\mathrm{rel}}} = \frac{n^3D^2}{|\mathbf{r}_\xi|D_\eta} = \left|\frac{\cos\vartheta}{\rho_{\mathrm{c}}} \pm \frac{1}{\rho_{\mathrm{r}}}\right| \tag{5.1.26}$$

serves as a relative radius of curvature of the ray and caustic, the angle ϑ is between the normal to the caustic, $\mathbf{N}_\mathrm{c}$, and the principal normal to the ray, $\mathbf{n}$, as shown in Fig. 5.2b.

The formula (5.1.26) has been derived by *Gazaryan* [5.6] and reproduced in [5.17] (*Babich* and *Egorov* [5.18] took $\vartheta = 0$ which is true for two-dimensional problems only). The quantities ρ_c and ρ_r should be taken with the same sign if the ray and caustic are at one side from the plane tangent to the caustic, and different signs if on either side. Note that (5.1.25) has already been used in Sect. 2.3.10 in the form $\zeta = -v/\Lambda$ with

$$\Lambda = \left(\frac{\rho_{\mathrm{rel}}}{2k^2}\right)^{1/3} = \frac{1}{k}\left(\frac{k\rho_{\mathrm{rel}}}{2}\right)^{1/3} \tag{5.1.27}$$

playing the role of diffraction scale of the field near the caustic.

An essential feature of (5.1.23) is that it remains valid in the shadow region where $v < 0$ and $\zeta > 0$. Here the function $\zeta(\mathbf{r})$, as well as $\chi(\mathbf{r})$, is analytic at the caustic: although the eikonals ψ_1 and ψ_2 are nonanalytic at the caustic, their combinations $\zeta = [3k(\psi_1 - \psi_2)/4]^{2/3}$ and $\chi = (\psi_1 + \psi_2)/2$ do have regular power expansions there.

Let us look at the amplitude factors. In order to write approximate expressions for the combinations $U_1 \pm \mathrm{i}U_2$ near caustics we resort to (2.1.12) and represent $U_{1,2}$ in the form

$$U_j = \frac{U_{j0}\sqrt{n_0^2 D_0}}{\sqrt{D_j}} \equiv \frac{F(\xi_j, \eta_j)}{\sqrt{D_j}} , \tag{5.1.28}$$

where the numerator and denominator will be expanded into power series in $v^{1/2}$ (v is the distance to the caustic). The leading term in the expansion for F is the constant $F_0 = F(\xi_0, \eta_0)$ associated with point M_0 (Fig. 5.2a),

$$F(\xi_j, \eta_j) = F_0 \pm v^{1/2} F_1 + \mathrm{O}(v) , \tag{5.1.29}$$

whereas the leading term in the expansion for D is proportional to $v^{1/2}$,

$$D_j = \pm\, q_1 v^{1/2} + q_2 v + \mathrm{O}(v^{3/2}) . \tag{5.1.30}$$

The upper signs in these expressions correspond to the incident wave ($j = 1$) and the lower to the reflected wave ($j = 2$). In taking the square root of D_2 we interpret $\sqrt{-1}$ as a factor inducing the caustic phase shift $\exp(-\mathrm{i}\pi/2)$ in U_2, so that

$$U_2 \cong \frac{F(\xi_2, \eta_2)}{\sqrt{n^2(-q_1 v^{1/2})}} \approx \frac{F(\xi_2, \eta_2)\exp(-\mathrm{i}\pi/2)}{\sqrt{n^2 q_1 v^{1/2}}} = -\mathrm{i}|U_2| ,$$

and the combination $\mathrm{i}U_2$ in the lower line of (5.1.18) (we prefer this representation of the field for convenience) may be replaced with $|U_2|$, while U_1 may be always deemed positive. Thus,

$$U_1 + \mathrm{i}U_2 = |U_1| + |U_2| \cong \frac{2F_0}{\sqrt{n^2 q_1}} v^{-1/4} ,$$

$$U_1 - \mathrm{i}U_2 = |U_1| - |U_2| = \frac{2(F_1 - F_0 q_2/2q_1 n^2)v^{1/4}}{\sqrt{n^2 q_1}} ,$$

where

$$F_0 = F(\xi_0, \eta_0), \qquad F_1 = \frac{\partial F}{\partial \eta}\sqrt{2|\mathbf{r}_\xi|/\tilde{D}_\eta} ,$$

$$q_1 = \sqrt{2|\mathbf{r}_\xi| D_\eta / n} ,$$

$$q_2 = \frac{2}{3D_\eta|\mathbf{r}_\xi|}\{D_{\eta\eta}|\mathbf{r}_\xi|^2 - [D_\xi(\mathbf{r}_{\eta\eta}\cdot\mathbf{r}_\xi)] - D_\eta(\mathbf{r}_{\xi\eta}\cdot\mathbf{r}_\xi)]\} .$$

Now equation (5.1.18) becomes

$$u = 2\sqrt{\pi}a_1(k_0b_1)^{1/6}[F_0\mathrm{Ai}(\zeta) - \mathrm{i}(k_0b_1)^{-1/3}(F_\eta + c_1F_0)\mathrm{Ai}'(\zeta)] \times \exp(\mathrm{i}k\chi - \mathrm{i}\pi/4) , \quad (5.1.31)$$

where

$$a_1 = (n^2D_\eta)^{-1/2}, \qquad b_1 = \frac{-n^3D_\eta D_\tau}{2|\mathbf{r}_\xi|^2}, \qquad c_1 = -\frac{q_2}{4n|\mathbf{r}_\xi|} .$$

The coefficient of the derivative of Airy function contains the factor $(kb_1)^{-1/3}$ decreasing at higher frequencies. Therefore, at high frequencies the second term may be neglected and the major role will be played by the first term which can be rewritten as

$$u \approx 2\sqrt{\pi}\frac{U^0\sqrt{D^0}}{\sqrt{D_{\mathrm{eff}}}}\mathrm{Ai}(\zeta)\mathrm{e}^{\mathrm{i}k\chi - \mathrm{i}\pi/4} . \quad (5.1.32)$$

Here, we introduced the effective Jacobian D_{eff} by combining a_1 and $(kb_1)^{1/6}$ as

$$D_{\mathrm{eff}} = [n^2a_1^2(kb_1)^{1/3}]^{-1} = D_\eta(kb_1)^{-1/3} .$$

At this stage, it will be convenient to take the arc length along the line ξ = constant on the caustic as parameter η. Then the quantity $(kb_1)^{-1/3}$ may be thought of as some length $l_\parallel$ that can be transformed to

$$l_\parallel = (b_1)^{-1/3} = \sqrt{2R_{\mathrm{eff}}\Lambda} , \quad (5.1.33)$$

where $\Lambda = (\rho_{\mathrm{eff}}/2k^2)^{1/3}$ is the width of the caustic zone, and $R_{\mathrm{eff}} = |\mathbf{r}_\xi|/nD_\eta$ the effective scale comparable with the radius of curvature of the caustic.

Written in the form $l_\parallel^2/2R_{\mathrm{eff}} = \Lambda$ equation (5.1.33) may be interpreted as a condition for a ray to leave the caustic zone (Fig. 5.3), since the distance from the caustic v increases in the square law $v \sim l^2/2R_{\mathrm{eff}}$ to compare with Λ at $l = l_\parallel$, where l is the distance from M_0 along the caustic. The ray segment from the point of tangency to the point where the ray leaves the caustic zone has the same length accurate to the terms of orders $(\Lambda/R_{\mathrm{eff}})^2$.

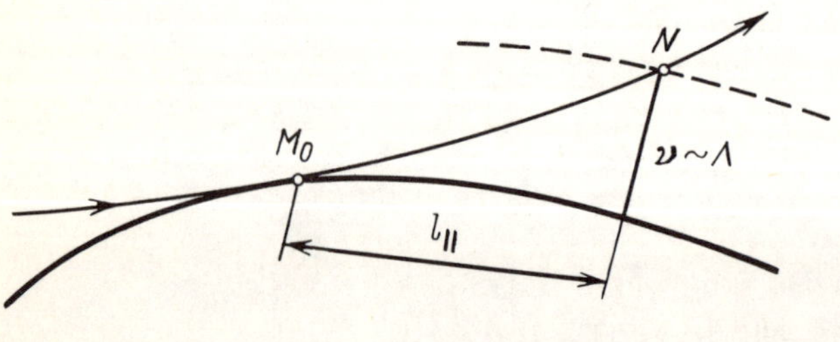

Fig. 5.3. To derive relations between the longitudinal ($l_\parallel$) and transverse (Λ) scales

Another interpretation of the longitudinal scale $l_\parallel$ treats it as a length at which the Fresnel radius $\sqrt{\lambda l_\parallel}$ becomes equal to the transverse scale Λ, i.e., where $\sqrt{\lambda l_\parallel} \sim \Lambda$. Indeed, if we substitute the estimate (5.1.33) we get $\Lambda \sim (\lambda^2 \ R_{\text{eff}})^{1/3} \sim (\rho/k^2)^{1/3}$ which agrees well with the definition of Λ given in (2.3.11).

In two-dimensional problems where $R_{\text{eff}} \approx \rho_{\text{eff}}$, the longitudinal diffraction length $l_\parallel$ amounts to about $(2\rho_{\text{eff}}^2/k_0)^{1/3}$. Precisely this quantity is claimed by *Babich* and *Egorov* [5.18] as the minimum distance (along the ray) beyond which ray theory cannot be used any longer. This agrees well with the fact that geometrical optics gives satisfactory results up to the boundary of the caustic zone.

5.1.7 Interpolation Formula for a Caustic Field

The coefficient 2 in (5.1.31, 32) occurs to reflect the fact of interference of two waves, incident and reflected, in the vicinity of caustics. Conventionally the Airy function $\text{Ai}(\zeta)$ may be represented as the sum $G_{1,2}(\zeta) = \text{Ai}(\zeta) \pm \text{i}\,\text{Bi}(\zeta)$ of two functions that behave as traveling waves in the light region, but lose this physical meaning in the shadow region where $\zeta > 0$ because they involve the exponentially increasing Airy function $\text{Bi}(\zeta)$. By virtue of these functions we may write a peculiar solution [5.16]

$$u = U_1 \exp(-\text{i}k\psi_1) S_1(\zeta) - U_2 \exp(\text{i}k\psi_2) S_2(\zeta) \; ,$$

$$S_1(\zeta) = \sqrt{\pi}(-\zeta)^{1/4} G_2(\zeta) \exp\left[\text{i}\frac{2}{3}(-\zeta)^{3/2} - \frac{\text{i}\pi}{4}\right] , \tag{5.1.34}$$

$$S_2(\zeta) = \sqrt{\pi}(-\zeta)^{1/4} G_1(\zeta) \exp\left[-\text{i}\frac{2}{3}(-\zeta)^{3/2} + \text{i}\frac{\pi}{4}\right] ,$$

built so that in the lit region it transforms into the geometrical sum (5.1.13), for at $|\zeta| \gg 1$ both quantities $S_{1,2}$ tend to unity, and near the caustic it becomes the local asymptotic (5.1.32) since $G_1 + G_2 = 2\text{Ai}$.

This solution, however, should not be viewed as a uniform asymptotic, this is simply an example of a successful interpolation formula which is fit to the uniform asymptotic (5.1.18) near ($|\zeta| < 1$) and far from ($|\zeta| \gg 1$) the caustic. For $|\zeta|$ around -1, the error of (5.1.34) is also insignificant.

5.1.8 Estimating the Coefficient of the Airy Function Derivative

Let us estimate the coefficient of Ai′ in (5.1.31). The sum $F_\eta + c_1 F_0$ can be estimated as F_0/l_η, where l_η is the least of inhomogeneity scales caused by the variation of the amplitude of the incident wave and the curvature of the caustic along the coordinate line $\xi = \text{constant}$. At the same time $(kb_1)^{-1/3}$ is the longitudinal diffraction scale of the caustic zone (5.1.33). Therefore, smallness of the coefficient of $\text{Ai}'(\zeta)$ as compared with the coefficient F_0 of $\text{Ai}(\zeta)$ will be secured provided that

$$l_{\parallel} \ll l_{\eta} \, . \tag{5.1.35}$$

In most familiar problems this condition is satisfied; therefore, the second term in the asymptotic expansion (5.1.1) may be neglected near caustic. However, away from the caustic the relation between the two components changes since the Airy function decreases as $(-\zeta)^{-1/4}$ and the derivative of Airy function increases as $(-\zeta)^{1/4}$. Therefore, for $|\zeta| \gg 1$ one should keep in (5.1.1) both terms that combine in the sum of incident and reflected waves (5.1.13).

5.1.9 The Geometric Backbone and Wave "Flesh"

The fact that the field at a caustic can be expressed in terms of ray field parameters implies that geometrical optics has something more to say than just the information on the ray paths and ray amplitudes, namely it conveys the information about the diffraction character of the field at the caustic. Inside the caustic zone, rays, of course, lose their physical meaning: they cannot be separated by physical instruments any longer. However, rays retain the function of the *geometrical skeleton supporting the wave "flesh"*.

The mehod of uniform asymptotic expansions evaluates the "wave flesh" in two steps. First, for a given initial distribution of the field $u^0 = U^0 \exp(ik\psi^0)$ one finds the rays leaving the initial surface, and the amplitudes and phases of the ray fields along these rays. Then functions $\tilde{\zeta}$ and χ are calculated by (5.1.15) and amplitudes A and B by (5.1.16). While the ray amplitudes U_1 and U_2 are divergent at the caustic, the amplitude factors A and B emerge as finite functions.

The local asymptotic expansion of the fields treats indeterminate forms like 0/0; thus, evaluating the finite amplitudes right on the caustic. Transition to local asymptotics impedes matching of the caustic field with the ray field, but the interpolation formulas (5.1.34) can dramatically facilitate the computations.

It is worth noting that the geometrical framework is important also to determine the field in caustic shadow: it is a helpful vehicle in the evaluation of the eikonals ψ_1 and ψ_2 and then the functions ζ and χ entering the uniform asymptotic (5.1.1). Thus, to find the shadow field at $\zeta \sim 1$ one needs both complex solutions, whereas far inside the shadow region where $\zeta \gg 1$, there survives only one solution (5.1.21) which falls off exponentially with the distance from the caustic. We should emphasize also that comparison of the ray solution (5.1.13) with the uniform solution (5.1.1) is a reliable justification for the incorporation of the caustic phase factor $\exp(-i\pi/2)$.

5.1.10 A Uniform Airy Asymptotic of an EM Field

Following *Kravtsov* [5.19], consider the construction of a uniform Airy asymptotic for Maxwell's equations

$$\operatorname{curl}\mathbf{E} - \mathrm{i}k\mathbf{H} = 0, \qquad \operatorname{curl}\mathbf{H} + \mathrm{i}k\varepsilon\mathbf{E} = 0 \ , \tag{5.1.36}$$

where $\varepsilon = n^2$ is the permittivity of the isotropic medium. It is natural to think that, like with scalar waves, the interference of incident and reflected electromagnetic (EM) waves near a caustic may be described with the aid of Airy functions. Accordingly, we represent the asymptotic solution of Maxwell's equations in the form similar to (5.1.1), viz.,

$$\{\mathbf{E}, \mathbf{H}\} = [\mathbf{A}_{\mathrm{E,H}}\tilde{I}(\tilde{\zeta}) + (\mathrm{i}k)^{-1}\mathbf{B}_{\mathrm{E,H}}\tilde{I}'(\tilde{\zeta})]\mathrm{e}^{\mathrm{i}k\chi} \ , \tag{5.1.37}$$

where $\mathbf{A}_{\mathrm{E,H}}$ and $\mathbf{B}_{\mathrm{E,H}}$ are the vector amplitudes to be determined. They are expandable in powers of inverse wavenumber

$$\mathbf{A}_{\mathrm{E,H}} = \sum_{m=0}^{\infty} (\mathrm{i}k)^{-m}\mathbf{A}^{(m)}_{\mathrm{E,H}}, \quad \mathbf{B}_{\mathrm{E,H}} = \sum_{m=0}^{\infty} (\mathrm{i}k)^{-m}\mathbf{B}^{(m)}_{\mathrm{E,H}} \ . \tag{5.1.38}$$

Substituting (5.1.37, 38) into Maxwell's equations (5.1.36) and putting to zero the coefficients of like powers in k (reduction of the order of terms involving $\partial f/\partial\tilde{t}$ is done by Ludwig's rules outlined in Sect. 5.1.3) leads to the set of recurrence equations for $\mathbf{A}_{\mathrm{E,H}}$ and $\mathbf{B}_{\mathrm{E,H}}$:

$$\begin{aligned}
\nabla\chi\times\mathbf{A}^{(m)}_{\mathrm{E}} + \tilde{\zeta}\nabla\tilde{\zeta}\times\mathbf{B}^{(m)}_{\mathrm{E}} - \mathbf{A}^{(m)}_{\mathrm{H}} &= -\operatorname{curl}\mathbf{A}^{(m-1)}_{\mathrm{E}} \ ,\\
\nabla\chi\times\mathbf{B}^{(m)}_{\mathrm{E}} - \nabla\tilde{\zeta}\times\mathbf{A}^{(m)}_{\mathrm{E}} - \mathbf{B}^{(m)}_{\mathrm{H}} &= -\operatorname{curl}\mathbf{B}^{(m-1)}_{\mathrm{E}} \ ,\\
\nabla\chi\times\mathbf{A}^{(m)}_{\mathrm{H}} + \tilde{\zeta}\nabla\tilde{\zeta}\times\mathbf{B}^{(m)}_{\mathrm{H}} + \varepsilon\mathbf{A}^{(m)}_{\mathrm{E}} &= -\operatorname{curl}\mathbf{A}^{(m-1)}_{\mathrm{H}} \ ,\\
\nabla\chi\times\mathbf{B}^{(m)}_{\mathrm{H}} - \nabla\tilde{\zeta}\times\mathbf{A}^{(m)}_{\mathrm{H}} + \varepsilon\mathbf{B}^{(m)}_{\mathrm{E}} &= -\operatorname{curl}\mathbf{B}^{(m-1)}_{\mathrm{H}} \ ,
\end{aligned} \tag{5.1.39}$$

where it is assumed that $\mathbf{A}^{(m)}_{\mathrm{E,H}} = \mathbf{B}^{(m)}_{\mathrm{E,H}} = 0$ for $m < 0$.

These equations are reducible to equations of geometrical optics. To this end we multiply the first line in (5.1.39) by $(-\tilde{\zeta})^{-1/4}$ and add the result with the second line multiplied by $\pm(\tilde{\zeta})^{1/4}$. Applying the same procedure to the third and fourth lines yields

$$\begin{aligned}
\nabla\psi_{\mp}\times\mathbf{E}^{(m)}_{\pm} - \mathbf{H}^{(m)}_{\pm} &= -\operatorname{curl}\mathbf{E}^{(m)}_{\pm} \pm \frac{1}{4\tilde{\zeta}}\nabla\tilde{\zeta}\times\mathbf{E}^{(m-1)}_{-} \equiv \mathbf{X}^{(m)}_{\pm} \ ,\\
\nabla\psi_{\mp}\times\mathbf{H}^{(m)}_{\pm} + \varepsilon\mathbf{E}^{(m)}_{\pm} &= -\operatorname{curl}\mathbf{H}^{(m-1)}_{\pm} + \frac{1}{4\tilde{\zeta}}\nabla\tilde{\zeta}\times\mathbf{H}^{(m-1)}_{+} \equiv \mathbf{Y}^{(m-1)}_{\pm} \ ,
\end{aligned} \tag{5.1.40}$$

where

$$\begin{aligned}
\{\mathbf{E}^{(m)}_{\pm}, \mathbf{H}^{(m)}_{\pm}\} &= 2^{-1/2}[\mathbf{A}^{(m)}_{\mathrm{E,H}}(-\tilde{\zeta})^{-1/4} \pm \mathbf{B}^{(m)}_{\mathrm{E,H}}(-\tilde{\zeta})^{1/4}]\mathrm{e}^{\pm\mathrm{i}\pi/4} \ ,\\
\psi_{\pm} &= \chi \pm \tfrac{2}{3}(-\tilde{\zeta})^{3/2} \ .
\end{aligned} \tag{5.1.41}$$

These equations remind us of the recurrence equations of geometrical optics for EM fields [5.20] and may be solved in a similar way. To a zero approximation ($m = 0$) we have the system of linear algebraic equations

$$\nabla\psi_{\mp} \times \mathbf{E}_{\pm} - \mathbf{H}_{\pm} = 0, \qquad \nabla\psi_{\mp} \times \mathbf{H}_{\pm} + \varepsilon\mathbf{E}_{\pm} = 0 , \tag{5.1.42}$$

where the subscript "0" was dropped for shortness. For a non-trivial solution to exist, the determinant of this system must be zero, which leads us to a pair of eikonal equations $(\nabla\psi_{\pm})^2 = \varepsilon$ and thereby to equations (5.1.7, 8) for χ and $\tilde{\zeta}$.

Now we use the conditions of consistency of the first order ($m = 1$) equations, which boil down to the requirement of orthogonality of the six-tuple $\{\mathbf{X}^{(0)}_{\pm}, \mathbf{Y}^{(0)}_{\pm}\}$ (vector on the right hand side of Eqs. (5.1.40) at $m = 1$) and the solutions of the transposed system of equations of the zeroth approximation (5.1.42). If $\mathbf{n}_{\mp}$ is the principal normal and $\mathbf{b}_{\mp}$ the binormal to the rays orthogonal to the phase fronts $\psi_{\mp}$ = constant, then the solutions to the transposed system have the form $\{\sqrt{\varepsilon}\mathbf{n}_{\mp}, \mathbf{b}_{\mp}\}$ and $\{\sqrt{\varepsilon}\mathbf{b}_{\mp}, -\mathbf{n}_{\mp}\}$. Therefore, the conditions of consistency for the equations of the first approximation take on the form

$$\sqrt{\varepsilon}(\mathbf{X}^{(0)}_{\pm}, \mathbf{n}_{\mp}) + (\mathbf{Y}^{(0)}_{\pm}, \mathbf{b}_{\mp}) = 0 , \quad \sqrt{\varepsilon}(\mathbf{X}^{(0)}_{\pm}, \mathbf{b}_{\mp}) - (\mathbf{Y}^{(0)}_{\pm}, \mathbf{b}_{\mp}) = 0 . \tag{5.1.43}$$

By virtue of (5.1.42) the fields $\mathbf{E}_{\pm}$ and $\mathbf{H}_{\pm}$ are orthogonal to the phase normals $\mathbf{l}_{\mp} = \nabla\psi_{\mp}/\sqrt{\varepsilon}$ and may be represented as the decompositions in $\mathbf{n}_{\mp}$ and $\mathbf{b}_{\mp}$:

$$\mathbf{E}_{\pm} = \Phi_{\pm}\mathbf{n}_{\mp} + \Psi_{\pm}\mathbf{b}_{\mp}, \qquad \mathbf{H}_{\pm} = \sqrt{\varepsilon}(\Phi_{\pm}\mathbf{b}_{\mp} - \Psi_{\pm}\mathbf{n}_{\mp}) . \tag{5.1.44}$$

Substituting these decompositions in (5.1.43) after a simple algebra yields the transfer equations

$$2\nabla U_{\pm} \cdot \nabla\psi_{\mp} + U_{\pm}\nabla\psi_{\mp} = 0 \tag{5.1.45}$$

for the amplitudes $U_{\pm} = \sqrt{(\Phi^2_{\pm} + \Psi^2_{\pm})}$, and the equation of rotation of the polarization plane (due to S.M. Rytov)

$$\mathrm{d}\gamma_{\mp}/\mathrm{d}\sigma = \kappa_{\mp} \tag{5.1.46}$$

for the angle $\gamma_{\mp} = \arctan(U_{\pm}/\Phi_{\pm})$ between the field vector $\mathbf{E}_{\pm}$ and a normal $\mathbf{n}_{\mp}$, where

$$\kappa_{\mp} = \tfrac{1}{2}(\mathbf{b}_{\mp} \cdot \operatorname{curl} \mathbf{b}_{\mp} - \mathbf{n}_{\mp} \cdot \operatorname{curl} \mathbf{n}_{\mp})$$

is the torsion of the respective rays.

Thereby we have reduced the equations of the method of uniform asymptotics to equations of geometrical optics which enables us to identify the functions ψ_{-}, $\mathbf{E}_{+}$ and $\mathbf{H}_{+}$ with the characteristics of the incident wave ψ_1, $\mathbf{E}_1$, $\mathbf{H}_1$, and the functions ψ_{+}, $\mathbf{E}_{-}$, and $\mathbf{H}_{-}$ with the characteristics of the reflected wave ψ_2, $\mathbf{E}_2$, $\mathbf{H}_2$ [compare (5.1.41) with (5.1.14)]. The immediate implication is that we may use (5.1.41) to express explicitly the vectorial amplitudes $\mathbf{A}_{\mathrm{E,H}}$ and $\mathbf{B}_{\mathrm{E,H}}$ in terms of the ray amlitudes, for example,

$$\mathbf{A}_{\mathrm{E}} = (-\zeta)^{1/4}(\mathbf{E}_1 + \mathrm{i}\mathbf{E}_2)\frac{\mathrm{e}^{-\mathrm{i}\pi/4}}{\sqrt{2}}, \qquad \mathbf{B}_{\mathrm{E}} = (-\zeta)^{-1/4}(\mathbf{E}_1 - \mathrm{i}\mathbf{E}_2)\frac{\mathrm{e}^{-\mathrm{i}\pi/4}}{\sqrt{2}} ,$$

(expressions for $\mathbf{A}_H$ and $\mathbf{B}_H$ are similar). In agreement with (5.1.37) the fields proper can be represented in the form similar to that of the scalar equations (5.1.17, 18), viz.,

$$\begin{aligned}\mathbf{E} &= 2^{-1/2}[(-\tilde{\zeta})^{1/4}(\mathbf{E}_1 + \mathrm{i}\mathbf{E}_2)\tilde{I}(\tilde{\zeta}) - \mathrm{i}(-\tilde{\zeta})^{-1/4}(\mathbf{E}_1 - \mathrm{i}\mathbf{E}_2)\tilde{I}'(\tilde{\zeta})]\mathrm{e}^{\mathrm{i}k\chi - \mathrm{i}\pi/4} \\ &= \sqrt{\pi}[(\mathbf{E}_1 + \mathrm{i}\mathbf{E}_2)(-\zeta)^{1/4}\mathrm{Ai}(\zeta) - \mathrm{i}(\mathbf{E}_1 - \mathrm{i}\mathbf{E}_2)(-\zeta)^{-1/4}\mathrm{Ai}'(\zeta)]\mathrm{e}^{\mathrm{i}k\chi - \mathrm{i}\pi/4} \,.\end{aligned} \tag{5.1.46}$$

In the case of an *anisotropic medium*, a uniform asymptotic of the field is given by the same expressions (5.1.46) which are valid for the isotropic case, but this time the fields $\mathbf{E}_{1,2}$ and $\mathbf{H}_{1,2}$ correspond to a normal wave and can be determined by pertinent formulas different from (5.1.44) e.g., see [5.20]. Equations (5.1.46) are valid for the description of the EM field in the presence of a nondegenerate *space-time caustic*, the fields $\mathbf{E}_{1,2}$ and $\mathbf{H}_{1,2}$ being evaluated by the formulas of space-time geometrical optics [5.21]. Finally, (5.1.46) may be extended to arbitrary vector fields such as elastic waves, waves on water, etc.

5.1.11 Local Asymptotic of an EM Field

Near a caustic where the ray amplitudes $\mathbf{E}_1$ and $\mathbf{E}_2$ become infinite, (5.1.46) may be transformed to a local form [5.16] which involves no indefinite forms of the type $0 \cdot \infty$. Relevant transformations are similar to that involved in the scalar problem; therefore, we will first of all analyze the polarization of the field at caustics.

Write down the ray field (5.1.44) in the form

$$\mathbf{E}_{1,2} = U_{1,2}\mathbf{e}_{1,2} \equiv \mathbf{U}_{1,2}, \quad \mathbf{H}_{1,2} = \sqrt{\varepsilon}U_{1,2}\mathbf{h}_{1,2} \equiv \mathbf{W}_{1,2} \,, \tag{5.1.47}$$

where

$$\mathbf{e}_{1,2} = \mathbf{n}_{1,2}\cos\gamma_{1,2} + \mathbf{b}_{1,2}\sin\gamma_{1,2}$$

and

$$\mathbf{h}_{1,2} = \mathbf{n}_{1,2}\sin\gamma_{1,2} - \mathbf{b}_{1,2}\cos\gamma_{1,2}$$

are the unit vectors of polarization, $\mathbf{e}\cdot\mathbf{h} = 0$, orthogonal to one another, and

$$\gamma_{1,2} = \int_0^{\sigma_{1,2}} \kappa\,\mathrm{d}\sigma$$

are the total angles of rotation of field vectors about the coordinate trihedral **n**, **b**, **l**. With this notation we are in a position to repeat the derivation of Sect. 5.1.6 almost unaltered, treating in (5.1.28) U as vector valued quantity $\mathbf{U} = U\mathbf{e}$ or $\mathbf{W} = W\mathbf{h}$, and in this and subsequent formulas $\mathbf{F}$ as a nonsingular vector valued quantity $\mathbf{F}_U = \mathbf{U}\sqrt{D}$ or $\mathbf{F}_W = \mathbf{W}\sqrt{D}$.

At caustics these vectors are orthogonal, hence, the leading terms of the asymptotic written as a vector analog of (5.1.31), i.e., the coefficients $\mathbf{F}_{U0}$ and $\mathbf{F}_{W0}$ of the Airy functions, are also orthogonal. Clearly, this property is preserved (accurate to small coefficients of Ai′) over the entire caustic zone of width 2Λ, both in light and shadow regions. Deviations from orthogonality can be caused only by such terms as F_η which can give corrections of the order of $l_\parallel/l_\eta$, where $l_\parallel$ is the longitudinal diffraction scale, and l_η is the characteristic length of change of polarization in the incident wave.

Orthogonality of **E** and **H** should also be expected in the region of deep shadow, for the relationships (5.1.42) remain valid also for $\zeta \gg 1$. However, in view of complexity of the vectors $\mathbf{l} = \nabla\psi/\sqrt{\varepsilon_0}$, **n** and **b**, the polarization in the region of shadow may exhibit a different behavior [5.15].

For EM fields, the interpolation formulas can be readily written by replacing in (5.1.34) the scalar functions u, U_1, U_2 with $\mathbf{E}$, $\mathbf{E}_1$, $\mathbf{E}_2$ or $\mathbf{H}$, $\mathbf{H}_1$, $\mathbf{H}_2$ [5.16].

5.1.12 One-Dimensional Problem

In the one-dimensional case when the wave field is described by (4.1.18) there emerges a possibility to write all variables to be found explicitly. First of all, in the one-dimensional problem,

$$\chi = \tfrac{1}{2}(\Psi_1 + \Psi_2) = \text{constant} = \int_{z^0}^{z} \sqrt{\varepsilon}\, dz$$

and $U_2 = -iU_1$, so that $U_1 + iU_2 = 2U_1$ whereas the coefficient $U_1 - iU_2$ of Ai in (5.1.18) vanishes.

Equation (5.1.7) for $\tilde{\zeta}$ takes the form

$$-\tilde{\zeta}\left(\frac{d\tilde{\zeta}}{dz}\right)^2 = \varepsilon$$

from which it follows that $\tilde{\zeta}$ and ε are of opposite sign. Then

$$\frac{2}{3}\frac{d}{dz}(-\tilde{\zeta})^{3/2} = \sqrt{\varepsilon} \quad \text{for } \varepsilon > 0,\ \zeta < 0\ ,$$

and

$$\frac{2}{3}\frac{d}{dz}\tilde{\zeta}^{3/2} = \sqrt{-\varepsilon} \quad \text{for } \varepsilon < 0,\ \zeta > 0$$

whence we get (4.2.21) given in Chap. 4 without derivation. Finally, observing that $U_1(z) = C_1[\varepsilon(z)]^{-1/4}$ the solution (5.1.18) may be written in the form

$$u(z) = 2\sqrt{\pi}C_1(-\zeta/\varepsilon)^{1/4}\mathrm{Ai}(\zeta)$$

agreeing with the solution obtained by Langer (see Sect. 4.2.4). The constant C_1 is selected subject to the initial conditions.

5.1.13 Applicability Conditions for the Airy Asymptotic

The method of standard integrals seems to be promising in deriving a satisfactory uniform approximation provided that the standard solution describes the essential qualitative features of the problem. With this limitation in sight, we will formulate convenient heuristic conditions of applicability for the uniform asymptotics derived with the aid of standard integrals.

Qualitatively an applicability condition will be reduced to a requirement that the key features of the standard problem remain in the problem under consideration [5.3, 22]. Quantitatively, these applicability conditions may be put down as inequalities requiring that the variation of the quantities being constants in the initial problem be small. The demand of small variation of some quantities in these inequalities relates to the characteristic scales of field formation.

By way of example, for applicability of the method of geometrical optics, we require that the ray field $u = U\exp(ik\psi)$ retain the structure of the plane wave within areas measuring at least a wavelength across. Here, the standard problem (plane wave) constants are the amplitude A, momenta $\mathbf{p} = \nabla\psi$, and permittivity of the medium ε. Accordingly, in the method of geometrical optics, these three quantities must be practically constant over lengths comparable with λ in the order of magnitude. This reasoning leads us to the necessary conditions of applicability of the ray method.

At large distances where the radius of a Fresnel zone becomes comparable with the transverse scale of an inhomogeneity of the field, the flow of energy in a ray tube is no longer conserved. If we require conservation of the flow, this will lead us to inequalities (2.2.8) demanding virtually constant parameters of the wave and medium within the Fresnel volume. Keeping in mind these considerations we may formulate the conditions of applicability for the Airy asymptotic approximation. In doing this we shall assume that the applicability conditions of the ray approach to the description of the field beyond the caustic zone are met.

As mentioned above, the Airy function yields an explicit solution to the problem of a plane wave incident on a linear slab (see Sect. 4.2.3). In this case, the standard problem is characterized by (a) the plane form of the caustic surface, $z = z_c$, (b) constancy of the amplitude U^0 of the primary wave and the permittivity of the medium $\varepsilon(z)$ along the caustic, and (c) constancy of the gradient of ε in the normal to the caustic. Proceeding on these lines we require that

(i) the curvature radius ρ_c of the caustic be large compared with the longitudinal scale $l_\parallel$;
(ii) the amplitudes A and B, the curvature radius ρ_c and the permittivity ε vary small along the caustic over the length $l_\parallel$;
(iii) the gradient of ε remain virtually invariable within a layer of width Λ normal to the caustic.

Setting these requirements in the form of inequalities is straightforward,

$$\text{(i)}\quad \rho_c \gg l_\parallel \ ,$$
$$\text{(ii)}\quad l_\parallel|\nabla A| \ll A,\ l_\parallel|\nabla B| \ll B,\ l_\parallel|\nabla_\parallel\varepsilon| \ll \varepsilon,\ l_\parallel|\nabla_\parallel\rho_c| \ll \rho_c \ , \tag{5.1.48}$$
$$\text{(iii)}\quad \Lambda\,|\partial^2\varepsilon/\partial v^2| \ll \partial\varepsilon/\partial v \ .$$

Here, $\mathbf{V}_{\parallel}$ is the derivative along the caustic, $\partial/\partial v$ the derivative along the normal to the caustic, and $l_{\parallel}$ the longitudinal diffraction scale on the caustic.

These conditions refine the inequalities due to *Groshev* and *Kravtsov* [5.22] who formulated the general approach to the evaluation of applicability limits for the method of uniform asymptotic approximations. Condition (iii) has been already mentioned in Sect. 4.2.5 in connection with the applicability conditions of the uniform Airy-asymptotic due to Langer. Inequalities (ii) not only limit the rate of variation of A, B, ε, and ρ_c, but also ensure smallness of the coefficient of the derivative of the Airy function (see Sect. 5.1.6), which seems to be quite natural since the standard problem does not involve a term with Ai′. Examples illustrating how these conditions work will be given in Sect. 5.3.

It would be instructive to derive inequalities of the type (5.1.48) directly from the first principles, i.e., from the general statements leading to uniform asymptotics, say from the theory of path integrals [5.23–25]. In the last case, the applicability conditions (5.1.48) would emerge as the conditions of applicability for the method of stationary phase.

5.2 Uniform Caustic Asymptotics Based on General Standard Integrals

5.2.1 Structure of a Solution

The standard integrals (4.1.1) seem to be natural vehicles for the construction of a uniform asymptotic of the wave field, which could be valid near an arbitrary caustic. Clearly, the form of such solution would be somewhat more involved than in the case of the simplest caustic. While in (5.1.1) there was only one derivative $I'(\tilde{\zeta}) \equiv \partial I(\tilde{\zeta})/\partial\tilde{\zeta}$, now the solution should include M derivatives

$$\frac{\partial \tilde{I}(\tilde{\zeta})}{\partial \tilde{\zeta}_p} \equiv \frac{\partial \tilde{I}(\zeta_1, \zeta_2, \ldots, \zeta_M)}{\partial \tilde{\zeta}_p}, \quad p = 1, 2, \ldots M ,$$

with respect to all external variables to which we refer all free modules. Consequently, the number of unknown arguments ζ_p is equal to $M = m + \mu$, where m is the codimension and μ the modality of the caustic.

Thus, we arrive at the following form of solution:

$$u(\mathbf{r}) = \left[A\tilde{I}(\tilde{\zeta}) + \frac{1}{\mathrm{i}k} \sum_{p=1}^{M} B_p \frac{\partial \tilde{I}(\tilde{\zeta})}{\partial \tilde{\zeta}_p} \right] \mathrm{e}^{\mathrm{i}k\chi} , \tag{5.2.1}$$

where the amplitude factors A and B are representable as series in $1/ik$

$$A = \sum_{n=0}^{\infty} A^{(n)}/(\mathrm{i}k)^n, \qquad B_p = \sum_{n=0}^{\infty} B_p^{(n)}/(\mathrm{i}k)^n .$$

In this form, the solution contains $M + 1$ unknown "phase" arguments $\zeta_p(\mathbf{r})$ and

$\chi(\mathbf{r})$, and $M+1$ amplitude factors $A^{(n)}(\mathbf{r})$ and $B^{(n)}(\mathbf{r})$ in any order of the expansion in $1/ik$.

For practical computations the solution (5.2.1) is convenient to represent in the integral form

$$u(\mathbf{r}) = \left(\frac{k}{2\pi}\right)^{1/2} \sum_{n=0}^{\infty} (ik)^{-n} \int d^l \tilde{t}\, g^{(n)}(\mathbf{r}, \tilde{t}) e^{ikf(\mathbf{r},\tilde{t})} , \tag{5.2.2}$$

where

$$g^{(n)}(\mathbf{r}, \tilde{t}) = A^{(n)}(\mathbf{r}) + \sum_{p=1}^{M} B_p(\mathbf{r}) \frac{\partial \Phi(\tilde{\zeta}, \tilde{t})}{\partial \tilde{\zeta}^p}, \qquad f(\mathbf{r}, t) = \chi(\mathbf{r}) + \Phi(\tilde{\zeta}, \tilde{t}) \tag{5.2.3}$$

Ludwig [5.10] suggested a similar form of solution for caustic of the type A_{m+1}. *Kravtsov* and *Orlov* [5.14] demonstrated that this structure satisfies the wave equation also for general form of caustics if the unknowns are augmented by the free modules. This procedure will be demonstrated below.

5.2.2 Equations for Phase and Amplitude Functions

Equations for phase and amplitude functions may be obtained by substituting (5.2.1) in the Helmholtz equation, namely,

$$\Delta u + k^2 \varepsilon u = \left(\frac{k}{2\pi}\right)^{l/2} \sum_{n=0}^{\infty} \int d^l \tilde{t} \exp(ikf) \{(ik)^2 [(\nabla f)^2 - \varepsilon] g^{(n)} + ik(2\nabla g^{(n)} \cdot \nabla f + g^{(n)} \nabla f) + \Delta g^{(n)}\} = 0 . \tag{5.2.5}$$

Equating the coefficients in like powers of k to the linear combination of derivatives of $f = \Phi + \chi$ with respect to $\tilde{t}_s$

$$\frac{\partial f(\mathbf{r}, \tilde{t})}{\partial \tilde{t}_s} = \frac{\partial \Phi(\mathbf{r}, \tilde{t})}{\partial \tilde{t}_s}, \quad s = 1, 2, \ldots, l , \tag{5.2.6}$$

reduces the order of individual components in $\tilde{t}_s$ and yields in the final analysis equations for all unknowns.

Applying this procedure to the coefficient of $(ik)^2$ we require that the relation

$$[\nabla f(\mathbf{r}, \tilde{t})]^2 - \varepsilon(\mathbf{r}) = \sum_{s=1}^{l} T_s(\mathbf{r}, \tilde{t}) \frac{\partial f(\mathbf{r}, \tilde{t})}{\partial \tilde{t}_s} \tag{5.2.7}$$

be satisfied identically in $t_1, \ldots, t_l$.

It would be natural to seek the function $T_s(\mathbf{r}, \tilde{t})$ in the form of polynomials in internal variables $\tilde{t}$. The requirement of the identical satisfaction in (5.2.7) will yield certain relations which will serve as differential equations for the functions $\chi(\mathbf{r})$ and $\tilde{\zeta}_p(\mathbf{r})$. The thought coefficients of $T_s(\mathbf{r}, \tilde{t})$ will then be expressed through these functions. This done, the terms involving the derivatives $\partial f/\partial \tilde{t}_s$ may be integrated by parts, thus reducing the order in parameter $1/k$.

As a result, in place of (5.2.5) we get

$$\Delta u + k^2 \varepsilon u = \left(\frac{k}{2\pi}\right)^{l/2} \sum_{n=0}^{\infty} (ik)^{1-n} \int d^l \tilde{t} \exp(ikf)$$

$$\times \left\{ 2\nabla g^{(n)} \cdot \nabla f + g^{(n)} \Delta f - \sum_{s=1}^{l} \frac{\partial (g^{(n)} T_s)}{\partial \tilde{t}_s} + \Delta g^{(n-1)} \right\} = 0 \; . \quad (5.2.8)$$

For the leading term ($n = 0$) of the expansion, we also carry out the evaluation of the integer part with respect to $\partial f / \partial \tilde{t}_s$:

$$2\nabla g^{(0)} \cdot \nabla f + g^{(0)} \Delta f - \sum_{s=1}^{l} \frac{\partial (g^{(0)} T_s)}{\partial \tilde{t}_s} = \sum_{s=1}^{l} R_s^{(0)} \frac{\partial f}{\partial \tilde{t}_s} \; . \quad (5.2.9)$$

If we require that this equation be identically satisfied in $\tilde{t}$, we obtain $M + 1$ equations for $M + 1$ unknown quantities $A^{(0)}$ and $B_p^{(0)}$ and also obtain the polynomial functions $R_s^{(0)}$. Repeating this procedure up to the order n leads to the recurrent relations

$$2\nabla g^{(n)} \cdot \nabla f + g^{(n)} \Delta f - \sum_{s=1}^{l} \frac{\partial (T_s g^{(n)})}{\partial \tilde{t}_s} + \Delta g^{(n-1)} - \sum_{s=1}^{l} \frac{\partial R_s^{(n-1)}}{\partial \tilde{t}_s} = \sum_{s=1}^{l} R_s^{(n)} \frac{\partial f}{\partial \tilde{t}_s} \; ,$$

which should be viewed as an implicit form of writing the equations for $A^{(n)}$ and $B_p^{(n)}$.

This computational procedure is a reliable but not shortest way to the goal; indeed, equations to be derived from (5.2.7, 9) will have to be solved. An alternative approach is based, like in the case of the Airy asymptotic, upon reducing the problem to combining the solutions of equations of geometrical optics, i.e., in the final analysis to integration of ordinary differential equations along the rays.

5.2.3 Relation to Geometrical Optics

Substitute the asymptotic expression (4.1.11) for the standard integral calculated by the method of stationary phase into (5.2.2) where the field (to be more precise, its leading term of $n = 0$) is written as the sum

$$u(\mathbf{r}) = \sum_{j=1}^{M+1} W_j(\mathbf{r}) \exp[ikf_j(\mathbf{r})] \; , \quad (5.2.10)$$

where

$$W_j(\mathbf{r}) \equiv g[\mathbf{r}, \tilde{t}^{(j)}] \tilde{h}_j^{-1/2} \exp(i\beta_j \pi/4)$$

$$= \left[A(\mathbf{r}) + \sum_{p=1}^{M} B_p(\mathbf{r}) \frac{\partial \Phi[\tilde{\xi}, \tilde{t}^{(j)}]}{\partial \tilde{\xi}_p} \right] \tilde{h}_j^{-1/2} \exp(i\beta_j \pi/4) \; , \quad (5.2.11)$$

$$f_j(\mathbf{r}) = f[\mathbf{r}, \tilde{t}^{(j)}] = \chi(\mathbf{r}) + \Phi[\tilde{\xi}, \tilde{t}^{(j)}] \; . \quad (5.2.12)$$

(Henceforth, the superscript "0" of g, A and B_p will be dropped.) The roots $t^{(j)}(\mathbf{r})$ will be obtained subject to the conditions of time-invariance ($\partial \Phi/\partial \tilde{t}_s = 0$) or from the equivalent set of equalities

$$\frac{\partial f(\mathbf{r}, \tilde{t})}{\partial \tilde{t}_s} = 0, \quad s = 1, 2, \ldots, l \ , \tag{5.2.13}$$

since $f = \chi + \Phi$ and χ is independent of $\tilde{t}$.

It will be natural to identify the asymptotic representation (5.2.10) with the geometric-optical solution (2.1.15) by letting

$$W_j = U_j \quad \text{and} \quad f_j = \psi_j \ . \tag{5.2.14}$$

However, U_j and ψ_j satisfy, respectively, the equations of eikonal and transport. Does it mean that W_j and f_j possess this property? If so, then there opens up an attractive possibility to find A, B_p, χ, and ζ_p in a purely algebraic manner without having to solve the respective equations.

Let us demonstrate by direct calculations that at a stationary point $\tilde{t}^{(j)} = \tilde{t}^{(j)}(\mathbf{r})$ (5.2.7) boils down to the eikonal equation for f_j, and (5.2.9) to a transport equation for W_j. Letting $\tilde{t} = \tilde{t}^{(j)}(\mathbf{r})$ causes the right hand side in (5.2.7) to vanish in view of the stationarity condition (5.2.13). The left hand side may be transformed on the lines of the chain rule.

Consider a function $F_j = F(\mathbf{r}, \tilde{t}^{(j)}(\mathbf{r}))$. Its gradient is

$$\nabla F_j = \nabla F(\mathbf{r}, \tilde{t})\bigg|_{\tilde{t} = \tilde{t}^j(\mathbf{r})} + \sum_{s=1}^{l} \frac{\partial F(\mathbf{r}, \tilde{t})}{\partial \tilde{t}_s}\bigg|_{\tilde{t} = \tilde{t}^j(\mathbf{r})} \nabla \tilde{t}_s^{(j)} \ . \tag{5.2.15}$$

If we assume that the subscript j flanking a vertical bar implies that $\tilde{t} = \tilde{t}^{(j)}(\mathbf{r})$ is substituted *after* $F(\mathbf{r}, \tilde{t})$ has been differentiated with respect to $\mathbf{r}$ in contrast to an ordinary subscript j indicating that $\tilde{t} = \tilde{t}^j(\mathbf{r})$ has been substituted prior to differentiation, then (5.2.15) rewrites as

$$\nabla F_j = \nabla F|_j + \sum \frac{\partial F}{\partial \tilde{t}_s}\bigg|_j \nabla \tilde{t}_s^{(j)}(\mathbf{r}) \ . \tag{5.2.16}$$

For $F = f$ the second term in (5.2.15) vanishes in view of the stationary condition (5.2.13), and from (5.2.7) we obtain the eikonal equation

$$(\nabla f_j)^2 = \varepsilon \tag{5.2.17}$$

identical to the equation $(\nabla \psi_j)^2 = \varepsilon$.

Let us now turn to (5.2.9) which at $\tilde{t} = \tilde{t}^{(j)}$ takes the form (here, as agreed above, $g = g^{(0)}$)

$$2\nabla f_j \nabla g_j + g_j \Delta f_j - g_j \sum_{s=1}^{l} \left(\frac{\partial T_s}{\partial \tilde{t}_s}\right)_j - \sum_{s=1}^{l} T_{sj} \frac{\partial g_j}{\partial \tilde{t}_s} = 0 \ . \tag{5.2.18}$$

First of all we make use of (5.2.16) to represent Δf_j and ∇g_j in terms of $\Delta f|_j$ and

$\nabla g|_j$:

$$\Delta f_j = \nabla(\nabla f_j) = \nabla\cdot(\nabla f|_j) = \Delta f|_j + \sum \left.\frac{\partial \nabla f}{\partial \tilde{t}_s}\right|_j \cdot \nabla \tilde{t}_s^{(j)} ,$$

$$\nabla g_j = \nabla g|_j + \left.\frac{\partial g}{\partial \tilde{t}_s}\right|_j \nabla \tilde{t}_s^{(j)} .$$

With reference to (5.2.11) we express g_j through W_j as $g_j = W_j \tilde{h}_j^{1/2}$ to obtain the equation

$$\begin{aligned}\tilde{h}_j^{1/2}\Bigg[(2\nabla W_j \nabla f_j + W_j \Delta f_j) - \sum_{s=1}^{l} \left.\frac{\partial W}{\partial \tilde{t}_s}\right|_j (2\nabla f \nabla \tilde{t}_s + T_s)_j \\ - W_j \sum_{s=1}^{l}\left(\left.\frac{\partial T_s}{\partial \tilde{t}_s}\right|_j + \left.\frac{\partial \nabla f}{\partial \tilde{t}_s}\right|_j \cdot \nabla \tilde{t}_s^{(j)}\right) - W_j \nabla f_j \cdot \nabla \ln \tilde{h}_j\Bigg] = 0\end{aligned} \qquad (5.2.19)$$

which assumes the desired form of a transfer equation

$$2\nabla W_j \cdot \nabla f_j + W_j \Delta f_j = 0 \qquad (5.2.20)$$

provided that

$$(T_q + 2\nabla f \cdot \nabla \tilde{t}_q)_j = 0 , \qquad (5.2.21)$$

$$\sum_{s=1}^{l}\left(\frac{\partial T_s}{\partial \tilde{t}_s} + \frac{\partial \nabla f}{\partial \tilde{t}_s}\cdot \nabla \tilde{t}_s\right)_j + \nabla f_j \cdot \nabla \ln \tilde{h}_j = 0 . \qquad (5.2.22)$$

In order to obtain (5.2.21) we differentiate (5.2.7) with respect to $\tilde{t}_q$ [(5.2.7) is identical in this variable]

$$2\nabla f \cdot \frac{\partial \nabla f}{\partial \tilde{t}_s} = \sum_{q=1}^{l} T_q \frac{\partial^2 f}{\partial \tilde{t}_s \partial \tilde{t}_q} + \sum_{q=1}^{l} \frac{\partial T_q}{\partial \tilde{t}_s}\frac{\partial f}{\partial \tilde{t}_q} . \qquad (5.2.23)$$

Substituting $\tilde{t} = \tilde{t}^j$ causes the second term on the right side to vanish and we have

$$\left(2\nabla f \cdot \frac{\partial \nabla f}{\partial \tilde{t}_s}\right)_j - \left(\sum_{q=1}^{l} T_q \frac{\partial^2 f}{\partial \tilde{t}_s \partial \tilde{t}_q}\right)_j = 0 . \qquad (5.2.24)$$

Now, we take the gradient of equation (5.2.13) that serves to determine the roots $\tilde{t}^{(j)}(\mathbf{r})$ and, hence, is an identity in $\mathbf{r}$:

$$\left(\frac{\partial \nabla f}{\partial \tilde{t}_s}\right)_j + \left(\sum_{q=1}^{l} \frac{\partial^2 f}{\partial \tilde{t}_s \partial \tilde{t}_q} \nabla \tilde{t}_q\right)_j = 0 . \qquad (5.2.25)$$

Premultiplying this relation scalarly by $2\nabla f_j$ gives

$$2\left(\nabla f \cdot \frac{\partial \nabla f}{\partial t_s}\right)_j + 2\sum_{q=1}^{l}\left(\frac{\partial^2 f}{\partial \tilde{t}_s \partial \tilde{t}_q} \nabla \tilde{t}_q \cdot \nabla f\right)_j = 0 . \qquad (5.2.26)$$

This relation differs from (5.2.24) only in the substitution of $(2\nabla t_q \nabla f)_j$ for $-(T_q)_j$, whence it follows (5.2.21).

The manipulations necessary to derive (5.2.22) in a general form, i.e., for an arbitrary multiplicity l, are rather unwieldy. Therefore, we will just outline the procedure for the general case and display specific calculations for $l = 1$ only.

Differentiating (5.2.23) with respect to $\tilde{t}_p$, i.e., taking the second derivative of (5.2.7) at $\tilde{t} = \tilde{t}^j(\mathbf{r})$ we obtain the set of l^2 relationships ($p, s = 1, \ldots, l$):

$$\left(2\frac{\partial \nabla f}{\partial \tilde{t}_s}\cdot\frac{\partial \nabla f}{\partial \tilde{t}_p} + 2\nabla f \frac{\partial^2 \nabla f}{\partial \tilde{t}_p \partial \tilde{t}_s}\right)_j = \sum_q \left(\frac{\partial T_q}{\partial \tilde{t}_p}\frac{\partial^2 f}{\partial \tilde{t}_s \partial \tilde{t}_q} + \frac{\partial T_q}{\partial \tilde{t}_s}\frac{\partial^2 f}{\partial \tilde{t}_q \tilde{t}_p} + T_q \frac{\partial^3 f}{\partial \tilde{t}_s \partial \tilde{t}_q \partial \tilde{t}_p}\right)_j, \tag{5.2.27}$$

which is used to determine the set of l^2 derivatives $\partial T_q/\partial \tilde{t}_p$. This is the most complicated part of these calculations.

In the one-dimensional case ($l = 1$) the system (5.2.27) degenerates into a single equation for $\partial T/\partial \tilde{t}$:

$$2\left(\frac{\partial \nabla f}{\partial \tilde{t}}\right)^2_j + 2\left(\nabla f \cdot \frac{\partial^2 \nabla f}{\partial \tilde{t}^2}\right)_j = 2\left(\frac{\partial T}{\partial \tilde{t}}\frac{\partial^2 f}{\partial \tilde{t}^2}\right)_j + \left(T\frac{\partial^3 f}{\partial \tilde{t}^3}\right)_j .$$

Substituting into this expression $T_j = -2(\nabla f \cdot \nabla \tilde{t})_j$ from (5.2.21) we have

$$\left(\frac{\partial T}{\partial \tilde{t}}\right)_j = \left(\frac{\partial^2 f}{\partial \tilde{t}^2}\right)^{-1}_j \left[\left(\frac{\partial \nabla f}{\partial \tilde{t}}\right)^2 + \nabla f \cdot \left(\frac{\partial^2 \nabla f}{\partial \tilde{t}^2} + \frac{\partial^3 f}{\partial \tilde{t}^3}\nabla \tilde{t}\right)\right]_j .$$

The coefficient of ∇f may be recognized as the gradient of $\partial^2 f_j/\partial \tilde{t}^2$, and if we let $l = 1$ in (5.2.25) we obtain that $\partial \nabla f/\partial \tilde{t}^2$ is equal to $\nabla \tilde{t}\, \partial^2 f/\partial \tilde{t}^2$. Recognizing that in the one-dimensional case $\tilde{h} = \partial^2 f/\partial \tilde{t}^2$ we arrive at the expression

$$\left(\frac{\partial T}{\partial \tilde{t}} + \nabla \tilde{t} \cdot \frac{\partial \nabla f}{\partial \tilde{t}}\right)_j - \nabla f_j \cdot \nabla \ln \tilde{h}_j = 0$$

which proves (5.2.22) for $l = 1$.

Thus, we have verified that the quantity W_j defined by (5.2.11) satisfies (at least for $l = 1$ and $l = 2$) the transfer equation (5.2.20), and that the quantity f_j given by (5.2.12) obeys the eikonal equation (5.2.17).

5.2.4 General Scheme to Compute Caustic Fields

The efforts spent in proving (5.2.17, 20) will pay off by presenting us with a convenient way to find the quantities χ, ζ_p, A, and B_p in a purely algebraic manner without having to solve the differential equations that follow from (5.2.7, 9). Indeed, we have verified that outside the caustic zone the quantities f_j and W_j can be identified with the parameters ψ_j and U_j of geometrical optics. At the same time, equations (5.2.17, 20) for these quantities formally are valid everywhere [for the initial equations (5.2.7, 9) are also valid everywhere] including the points inside the caustic zone where the ray description is no longer valid. Therefore, $M + 1$ relations $f_j = \psi_j$ of (5.2.14) should be viewed as $M + 1$ algebraic equations

$$\chi(\mathbf{r}) + \Phi[\tilde{\zeta}(\mathbf{r}), \tilde{t}^{(j)}(\mathbf{r})] = \psi_j(\mathbf{r}), \quad j = 1, 2, \ldots, M + 1 \tag{5.2.28}$$

to find $M + 1$ quantities χ and ζ_p both inside and outside caustic zones.

Likewise, $M + 1$ quantities $A(\mathbf{r})$ and $B_p(\mathbf{r})$ are found from $M + 1$ equations $W_j = u_j$ which may be written in the form

$$A(\mathbf{r}) + \sum_{p=1}^{M} B_p(\mathbf{r}) \frac{\partial \Phi[\tilde{\zeta}(\mathbf{r}), \tilde{t}^{(j)}(\mathbf{r})]}{\partial \tilde{\zeta}_p} \equiv U_j(\mathbf{r}) \tilde{h}_j^{1/2} \exp(-\mathrm{i}\beta_j \pi/4) \ . \tag{5.2.29}$$

Here the values of $\tilde{t}^{(j)}$ are determined from the stationarity condition (5.2.13).

Summarizing the steps involved in the evaluation of a uniform asymptotic approximation for the caustic field on the basis of standard functions we see them as follows:

(a) Solve the equations of geometrical optics to define the ray field $u = \sum_{j=1}^{M+1} U_j \exp(\mathrm{i}k\,\psi_j)$. The number of rays $M + 1$, the type of caustic and thereby the type of standard integral are determined simultaneously.

(b) Use equations (5.2.28, 13) to find the "phase" functions $\chi(\mathbf{r})$ and $\tilde{\zeta}_p(\mathbf{r})$, and equations (5.2.29, 13) to find the amplitude factors $A(\mathbf{r})$ and $B_p(\mathbf{r})$.

This procedure carries over the center of gravity in the construction of a uniform asymptotic to the solution of the problem in the ray approximation, which implies that the amplitudes and eikonals at the point of observation and then the unknown functions χ, ζ_p, A, and B_p will be determined from the algebraic equations (5.2.28, 29).

5.2.5 Uniform Caustic Asymptotic of an EM Field

By analogy with the scalar problem, an asymptotic solution of Maxwell's equations valid near an arbitrary caustic can be obtained with the aid of type (5.2.1) of caustic expansions yet with vectorial coefficients,

$$\{\mathbf{E}, \mathbf{H}\} = \left[\mathbf{A}^{\mathrm{E,H}} \tilde{I}(\zeta) + (\mathrm{i}k)^{-1} \sum_{p=1}^{M} \mathbf{B}_p^{\mathrm{E,H}} \frac{\partial}{\partial \tilde{\zeta}_p} \tilde{I}'(\tilde{\zeta})\right] \mathrm{e}^{\mathrm{i}k\chi} \ , \tag{5.2.30}$$

where the coefficients $A^{\mathrm{E,H}}$ and $\mathbf{B}_p^{\mathrm{E,H}}$ are asymptotic series in powers of $1/\mathrm{i}k$.

Substituting (5.2.30) in Maxwell's equations one may obtain equations for the phase functions χ and ζ_p and for the zero approximation vectorial amplitudes $\mathbf{A}^{\mathrm{E,H}}$ and $\mathbf{B}_p^{\mathrm{E,H}}$. However, like in the scalar problem, the sought functions can be expressed algebraically through the geometrical characteristics. In particular, χ and ζ_p are determined from (5.2.28) and $\mathbf{A}^{\mathrm{E},H}$ and $\mathbf{B}_p^{\mathrm{E,H}}$ from the vectorial analogs of (5.2.29). For example, for the amplitudes of the electric field

$\mathbf{A}^{\mathrm{E}}$ and $\mathbf{B}_p^{\mathrm{E}}$ we obtain the equations

$$\mathbf{A}^{\mathrm{E}} + \sum \mathbf{B}_p^{\mathrm{E}} \left(\frac{\partial \Phi}{\partial \zeta_p}\right)_j = E_j(\mathbf{r})\,\tilde{h}_j^{1/2} \exp(\mathrm{i}\beta_j \pi/4) \ , \tag{5.2.31}$$

where $\mathbf{E}_j$ is the vector amplitude associated with the jth ray.

These equations may be used to analyze the polarization of an EM field near a caustic. Clearly, similar relations must exist for vectorial fields of other physical nature including elastic waves.

5.2.6 The Ray Skeleton and Uniform Caustic Asymptotics

Though rays lose their physical meaning near caustics, they continue to play the part of a framework supporting the wave field. Worth mentioning are the following properties of uniform caustic asymptotics, which are associated with the behavior of rays.

(a) Although the ray amplitudes U_j diverge at a caustic, all amplitudes A and B_p remain finite. This may be attributed to the fact that in (5.2.29) the singularity of U_j is offset by the factor $\tilde{h}_j^{1/2}$, which in combination with divergence $\mathscr{J}^{-1/2}$ gives finite values, indeed, even for $\mathscr{J} \to 0$, $(\tilde{h}/\mathscr{J})^{1/2} < \infty$. This compensation may be thought of as a manifestation of some diffraction content concealed in the ray parameters.

(b) The phase factors $\exp(\mathrm{i}\beta_j\pi/4)$ emerge in (5.2.10) when we come from the exact values of caustic integrals to their asymptotics by the method of stationary phase. These factors are associated with a change of the inertia index β in the Hessian $\{\tilde{h}_{\alpha\beta}\} = \{\partial^2 f/\partial\tilde{t}_\alpha \partial\tilde{t}_\beta\}$, and for general form of caustics, they may justify the incorporation of caustic phase shifts. If the factor $\exp(\mathrm{i}\beta_j\pi/4)$ is incorporated in U_j in advance, then in the algebraic equations (5.2.29) it will be compensated by the factor $\exp(-\mathrm{i}\beta_j\pi/4)$ so that in the light region we shall actually have positive quantities.

(c) In caustic shadow, some real-valued stationary points disappear (usually in pairs) so that a pair of oscillating fields gives way to an exponentially decaying field. Thus, standard integrals help formulate the selection rules for complex rays in typical situations.

(d) Farther from the central point of a caustic singularity the standard integral gradually acquires a simpler structure because the stationary points decrease in number. In the vicinity of separate branches of the caustic the integral is expressed in terms of simple standard integrals. Thus, the caustics seem to be subordinated not only at the level of geometric forms but also at the level of typical caustic integrals relying on these forms.

5.2.7 Some Specific Situations

For caustics of the series A_{m+1}, standard integrals are given by (4.4.1) and the uniform asymptotic written in terms of the dimensionless integral $I(\zeta)$ and its dimensional analog $\tilde{I}(\tilde{\zeta})$ has the form

$$u = \left[A\tilde{I}(\tilde{\zeta}) + \frac{1}{\mathrm{i}k}\sum B_p \frac{\partial \tilde{I}(\tilde{\zeta})}{\partial \tilde{\zeta}_p}\right]\mathrm{e}^{\mathrm{i}k\chi}$$

$$= k^{\sigma_{\mathrm{foc}}}\left[AI(\zeta) + \frac{1}{\mathrm{i}}\sum B_p \frac{\partial I(\zeta)}{\partial \zeta_p}\right]\mathrm{e}^{\mathrm{i}k\chi} , \quad (5.2.32)$$

where $\sigma_{\mathrm{foc}} = 1/2 - 1/(m+2)$ is the focusing index.

By virtue of the explicit expression for the generating function $\Phi(\zeta, t)$ in (4.4.1), (5.2.28, 29) may be written as

$$\chi + \sum_{p=1}^{m} \tilde{\zeta}_p\left(\frac{1}{p} - \frac{1}{m+2}\right)\tilde{t}_j^p = \psi_j ,$$

$$A + \sum_{p=1}^{m} B_p \tilde{t}_j^p / p = U_j\left[\sum_{p=1}^{m}(p - m - 2)\tilde{\zeta}_p \tilde{t}_j^{p-2}\right]^{1/2} \exp(-\mathrm{i}\beta_j \pi/4) , \quad (5.2.33)$$

where $\tilde{t}_j$ are the roots of the algebraic equation

$$\frac{\partial \Phi}{\partial \tilde{t}} = \tilde{t}^{m+1} + \sum \tilde{\zeta}_p \tilde{t}_j^{p-1} = 0 . \quad (5.2.34)$$

For (5.2.33), the highest power $\tilde{t}^{m+1}$ is expressed through smaller powers according to (5.2.34).

The particular case of $m = 1$ (A_2, fold) has been analyzed already in Sect. 5.1. At $m = 2$ (A_3, cusp) the use is made of the standard function (4.3.1) and the uniform asymptotic is given by

$$u(\mathbf{r}) = k^{1/4}\left(A\tilde{I} - \mathrm{i}k^{-1/4}B_1\frac{\partial \tilde{I}}{\partial \tilde{\zeta}_1} - \mathrm{i}k^{-1/2}B_2\frac{\partial \tilde{I}}{\partial \tilde{\zeta}_2}\right)\mathrm{e}^{\mathrm{i}k\chi}$$

$$= \left(A\tilde{I} + \frac{1}{\mathrm{i}k}B_1\frac{\partial \tilde{I}}{\partial \tilde{\zeta}_1} + \frac{1}{\mathrm{i}k}B_2\frac{\partial \tilde{I}}{\partial \tilde{\zeta}_2}\right)\mathrm{e}^{\mathrm{i}k\chi} . \quad (5.2.35)$$

The quantities χ, $\tilde{\zeta}_1$, and $\tilde{\zeta}_2$ are defined from the equations

$$\chi + \tfrac{3}{4}\tilde{\zeta}_1\tilde{t}_j + \tfrac{1}{4}\tilde{\zeta}_2\tilde{t}_j^2 = \psi_j, \quad j = 1,2,3 , \quad (5.2.36)$$

where $\tilde{t}_j$ are the roots of the cubic equation

$$\tilde{t}^3 + \tilde{\zeta}_2\tilde{t} + \tilde{\zeta}_1 = 0 .$$

The amplitudes A, B_1, and B_2 are determined from the system of equations

$$A + \tilde{t}_j B_1 + \tfrac{1}{2}\tilde{t}_j^2 B_2 = U_j(2\tilde{\zeta}_2\tilde{t}_j + 3\tilde{t}_j^2)^{1/2}\exp(-\mathrm{i}\beta_j\pi/4) .$$

For $m = 2$, the roots $\tilde{t}_j$ may be expressed via $\tilde{\zeta}_1$ and $\tilde{\zeta}_2$ using Cardan's solution of the cubic; however, explicit expressions of χ, $\tilde{\zeta}_1$ and $\tilde{\zeta}_2$ through ψ_j defy evaluation.

The equations for χ, $\tilde{\zeta}_1$ and $\tilde{\zeta}_2$ obtained from (5.2.7), after the integer part of $\partial f/\partial \tilde{t}$ has been singled out, are as follows:

$$(\nabla\chi)^2 - \tilde{\zeta}_1(\nabla\tilde{\zeta}_1 \cdot \nabla\tilde{\zeta}_2) = n^2(\mathbf{r}) \ ,$$
$$(\nabla\tilde{\zeta}_1)^2 + (\nabla\chi \cdot \nabla\tilde{\zeta}_2) - \tfrac{1}{4}\tilde{\zeta}_2(\nabla\tilde{\zeta}_2)^2 = 0 \ , \qquad (5.2.37)$$
$$\tfrac{1}{4}\zeta_1(\nabla\zeta_2)^2 + \tilde{\zeta}_2(\nabla\tilde{\zeta}_1 \cdot \nabla\tilde{\zeta}_2) - 2(\nabla\chi \cdot \nabla\tilde{\zeta}_1) = 0 \ .$$

The equations for the amplitudes A, B_1 and B_2 following from (5.2.8) are

$$\mathscr{L}_0(A) + \tfrac{1}{2}\tilde{\zeta}_1[\mathscr{L}_2(B_1) - \mathscr{L}_1(B_2)] + \tfrac{1}{4}(\nabla\tilde{\zeta}_2)^2 A + (\nabla\tilde{\zeta}_1 \cdot \nabla\tilde{\zeta}_2)(B_1 + B_2) = 0 \ ,$$
$$\tfrac{1}{2}\mathscr{L}_2(A) - \mathscr{L}_1(B_1) - \tfrac{1}{2}\mathscr{L}_0(B_2) + \tfrac{1}{2}\tilde{\zeta}_2\mathscr{L}_2(B_2) + \tfrac{1}{8}(\nabla\tilde{\zeta}_2)^2 B_2 = 0 \ ,$$
$$\mathscr{L}_1(A) - \mathscr{L}_0(B_1) + \tfrac{1}{2}\tilde{\zeta}_2[\mathscr{L}_2(B_1) + \mathscr{L}_1(B_2)] + \tfrac{1}{4}\tilde{\zeta}_1\mathscr{L}_2(B_2)$$
$$+ \tfrac{1}{4}(\nabla\tilde{\zeta}_2)^2(2B_1 + B_2) = 0 \ ,$$

where $L_0(F) = 2\nabla F \cdot \nabla\chi + F\Delta\chi$, and $\mathscr{L}_{1,2}(F) = 2\nabla F \cdot \nabla\tilde{\zeta}_{1,2} + F\Delta\tilde{\zeta}_{1,2}$.

For a four-ray swallowtail ($m = 3$), the roots $\tilde{t}_j$ satisfy already an equation of fourth order, and for a butterfly ($m = 4$), an equation of fifth order. These equations defy analytical solutions; therefore, all hopes of solution will be placed on numerical methods.

By way of example, we also give an equation for the phase and amplitude functions of umbilic series D_{m+1} ($m \gg 3$). Equations for $\tilde{t}_1$ and $\tilde{t}_2$ may be obtained from the conditions (4.1.8) and have the form

$$2\tilde{t}_2\tilde{t}_1 + \tilde{\zeta}_1 = 0, \qquad \pm m\tilde{t}_2^{m-1} + \tilde{t}_1^2 + \sum_{p=2}^{m} \tilde{\zeta}_p(p-1)\tilde{t}_2^{p-2} = 0 \ . \qquad (5.2.38)$$

In particular, at $m = 3$ (hyperbolic and elliptic umbilics, $D_4^{\pm}$) the second of these equations is only of the second order in $\tilde{t}_2$, viz.,

$$\pm 3\tilde{t}_2^2 + \tilde{t}_1^2 + \tilde{\zeta}_2 + 2\tilde{\zeta}_3\tilde{t}_2 = 0 \ .$$

However, if we substitute $\tilde{t}_1 = -\tilde{\zeta}_1/2\tilde{t}_2$ from the first line of (5.2.38) in this equation, its order rises to four.

Thus, an analytical solution of the system (5.2.38) is unattainable already at $m = 3$ to say nothing of other multiray caustics with $m \gg 3$. Nonetheless, the prospects for construction of uniform asymptotics for $m \gg 3$ are far from being hopeless if numerical methods are invoked to determine the desired quantities. Success may also be achieved in the lines of matching local asymptotics which taken in combination may estimate the fields to satisfactory accuracy. Available possibilities will be discussed below.

5.2.8 Local Asymptotics

One of the ways to overcome hindrances emerging in the construction of uniform asymptotic expansions involves crossing over to local expansions matched to one another in some transition regions.

Consider a field near a critical point $\tilde{\zeta} = 0$ of a catastrophe. At the critical point all $M + 1$ rays coalesce rendering all the eikonals ψ_j and ray amplitudes U_j identical. The system (5.2.29) degenerates, for all lines in this system coincide at $\tilde{\zeta} = 0$.

At the caustic, all amplitude factors B_p enter (5.2.29) with coefficients $\partial\Phi/\partial\tilde{\zeta}_p$ which vanish at the critical point $\tilde{\zeta} = 0$ since all stationary values $\tilde{t}_j$ are also zero, so that from the entire system (5.2.29) we may retain only the relation

$$A = \lim_{v \to 0} U |\tilde{h}|^{1/2} \, e^{i\beta\pi/4} \qquad (5.2.39)$$

which defines the leading term of the asymptotic (5.2.1). The limit for $v \to 0$ describes symbolically an approach to the critical point where it is convenient to treat v as the distance from this point in the coordinate space. For example, in the case of the Airy asymptotic, the quantity v has been represented by the distance in a normal to the caustic.

Representing the amplitude U like in (5.2.28) through the initial value U^0

$$U = F_0/\sqrt{D}, \qquad F_0 = U^0 \sqrt{D^0}$$

enables (5.2.39) to be rewritten as

$$A = F_0 K, \qquad K_{\text{foc}} = \lim_{v \to 0} |\tilde{h}/D|^{1/2} , \qquad (5.2.40)$$

where the coefficient K_{foc} is a function of wavenumber k

$$K_{\text{foc}} \sim k^{\sigma_{\text{foc}}} , \qquad (5.2.41)$$

and σ_{foc} is the focusing index. The quantity K_{foc}, may be thought of as a focusing factor, for it characterizes the rise of field amplitude at a critical point over the amplitude of the primary wave.

It will be not unexpected if the quantity A defined by (5.2.39) or (5.2.40) is nearly constant within some neighborhood of the critical point. The condition for A to be nearly constant may be represented by

$$\Lambda_p |\nabla_p A| \ll A , \qquad (5.2.43)$$

where Λ_p is the size of the caustic zone in the direction of $\tilde{\zeta}_p$ (gradient of the amplitude is also taken in this direction). This inequality provides one of the condition of applicability of the asymptotic representation (5.2.1); to be more specific, it requires that the amplitude A be constant within the caustic zone because in the standard problem this quantity is strictly constant.

In the neighborhood of the critical point $\tilde{\zeta} = 0$ all parameters may be expanded into power series in the increments of coordinates x, y, z and parameters $\alpha_1, \ldots, \alpha_m$ about the critical value. The leading terms in these expansions may be expected to be linear, while the subsequent terms will play the role of corrections. With such expansions at hand we hope to determine the *analytical* behavior of χ, $\tilde{\zeta}_p$, $\tilde{t}_j$ and $\tilde{h}_j$ near the critical point and thereby an approximation to the amplitude factors B_p and a refinement of the "principal" amplitude A.

If expansions constructed in this manner cover all the caustic with a sufficient accuracy, then it is admissible to proceed to matching these expansions with the caustic asymptotics of lower codimension, which describe the field in the adjacent areas of the caustic separated from the critical point farther than Λ_p.

For the A_2 caustic, this plan has been realized by *Orlov* [5.16] and proved to be effective. However, for caustics of higher codimension, this is yet to be implemented. This scheme seems to combine the principal advantage of uniform asymptotics – possibility to describe the field over the entire space at all values of external variables – with the advantages of local description which allows one to obtain numerical values at the end of the calculation.

Embedded in this "localized" version, the uniform theory may be expected to be nowhere less accurate and computationally slower than other asymptotic techniques since it relies on a ready field structure and seeks only for details of this structure. It should be clear that all the advantages of this approach can be exploited only if the field structure is already known – this is the penalty which has to be paid for the convenience and tractability of the analysis.

5.3 Illustrative Examples

5.3.1 The Circular Caustic

This specific problem has been frequently reiterated in the literature (e.g., [5.26, 27] and also [5.15]) since it allows for an exact solution and all available types of asymptotic. Let us write down the equation of caustic $r = \sqrt{x^2 + y^2} = a$ in the parametric form $x_c = a\cos\xi$, $y_c = a\sin\xi$. Rays tangent to the caustic are straight lines:

$$x = a\cos\xi - \tau\sin\xi \ , \qquad y = a\sin\xi + \tau\cos\xi \ , \tag{5.3.1}$$

which form a line ray surface (hyperboloid of revolution) in the space x, y, ξ, as shown in Fig. 5.4. Projecting this ray surface onto the coordinate plane (x, y) gives rise to a circular caustic $r = a$.

Let r, α be the cylindrical coordinates such that $x = r\cos\alpha$, $y = r\sin\alpha$. Parameters ξ and τ corresponding to the incident and reflected rays, respectively, follow from (5.3.1) as

$$\xi = \alpha \pm \arccos(a/r), \qquad \tau = \pm\sqrt{r^2 - a^2} \ ,$$

the respective eikonal $\psi = a\xi + \tau$ being

$$\psi_{1,2} = a\alpha \pm [a\arccos(a/r) - \sqrt{r^2 - a^2}] \ . \tag{5.3.2}$$

If the amplitude A of the ray field depends only upon r, but is independent of α (cylindrical symmetry), then by virtue of (2.1.12) we have

$$U_1 = c(r^2 - a^2)^{-1/4}, \qquad U_2 = -\mathrm{i}(r^2 - a^2)^{-1/4} \ . \tag{5.3.3}$$

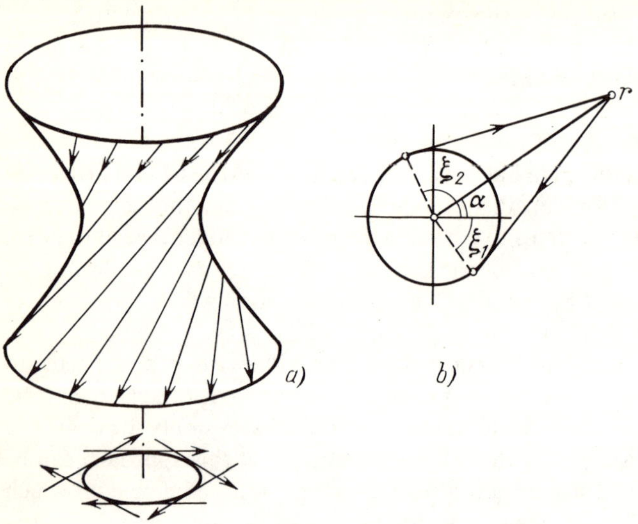

Fig. 5.4. (**a**) Ray surface inducing a circular caustic. (**b**) Angular parameters characterizing the position of rays

These data are quite sufficient to construct a uniform field asymptotic by the formula (5.1.18) as

$$u = C\left(\frac{-\zeta}{r^2 - a^2}\right)^{1/4} \mathrm{Ai}(\zeta)\exp(ika\alpha + \mathrm{i}\pi/4) , \tag{5.3.4}$$

where the argument of Airy function is given by the equation

$$\tfrac{2}{3}(-\zeta)^{3/2} = k[(r^2 - a^2)^{1/2} - a\arccos(a/r)] \tag{5.3.5}$$

in which $\zeta > 0$ for $r < a$ and $\zeta < 0$ for $r > a$.

In the illuminated region ($r > a, \zeta < 0$), for $|\zeta| \gg 1$, we get the sum of two ray fields

$$u = C(r^2 - a^2)^{-1/4}[\exp(ik\psi_1) + \exp(ik\psi_2 - \mathrm{i}\pi/2)] , \tag{5.3.6a}$$

and in the shadow region ($r < a, \zeta > 0$) we have the decaying field

$$u = C(a^2 - r^2)^{-1/4}\exp\{-k[\cosh^{-1}(a/r) - \sqrt{a^2 - r^2}]\} . \tag{5.3.6b}$$

In the vicinity of the caustic, where $|\zeta| \leq 1$, an approximate dependence for ζ may be obtained from (5.3.5) as

$$\zeta(r) \approx (2k^2/a)^{1/3}(a - r) . \tag{5.3.7}$$

Then (5.3.4) yields the local asymptotic

$$u \approx C(k/2a^2)^{1/6}\,\mathrm{Ai}[2^{1/3}k^{2/3}a^{-2/3}(a - r)] . \tag{5.3.7}$$

All approximate formulas given above agree with the exact solution

$$u(r) = C_1 J_{ka}(kr)\mathrm{e}^{ika\alpha} , \tag{5.3.8}$$

where $J_{ka}(kr)$ is the Bessel function of order ka:

$$J_{ka}(kr) = \frac{1}{2\pi} \int_{-\pi}^{\pi} \exp[i(kat - kr \sin t)] \, dt$$

(in the approximate formulas, the constant C should be put equal to $(2/\pi)^{1/2} C_1 k^{1/2} e^{-i\pi/4}$). The ray solution (5.3.6) corresponds to the Debye asymptotic of Bessel functions. This asymptotic may be obtained by computing the integral that represents the Bessel function by the stationary phase method. A shadow asymptotic may be obtained by deforming the path of integration in the plane of complex variable t. The uniform asymptotic (5.3.4) corresponds to the uniform formulas of *Olver* [5.28] (it is also deducible on the lines outlined in [5.9]), and the local asymptotic (5.3.7) corresponds to a linearized dependence of $\zeta(r)$ in the vicinity of caustic $r = a$.

In this case the caustic scale is equal to $\Lambda = (2k^2/a)^{-1/3}$. If in agreement with the applicability conditions (Sect. 5.1) we require that the caustic radius a be large compared to Λ, we obtain the inequality $(2k^2/a)^{-1/3} \ll a$ equivalent to $ka \gg 1/\sqrt{2}$. With this condition in mind, one might expect that the uniform Airy asymptotic (5.3.4) would ensure a satisfactory approximation of the exact solution (5.3.8) at sufficiently large radii of the caustic.

On the other hand, an accuracy of asymptotic expansions sufficient for applications is achieved already at $ka \sim 1$. This can be seen from Fig. 5.5 showing the wave function versus radius at $ka = 1$. At $ka = 1$ the radial dependence for the uniform asymptotic (5.3.4) coincides with the plot of the exact dependence $J_1(kr)$, with the graphic accuracy. The plot of the ray field (5.3.6) having a singularity at the caustic (at $kr = 1$) is also given.

Decreasing the parameter $ka = 2\pi a/\lambda$ which is the ratio of the perimeter of the caustic to the wavelength, one can verify that at the caustic where $r = a$ the

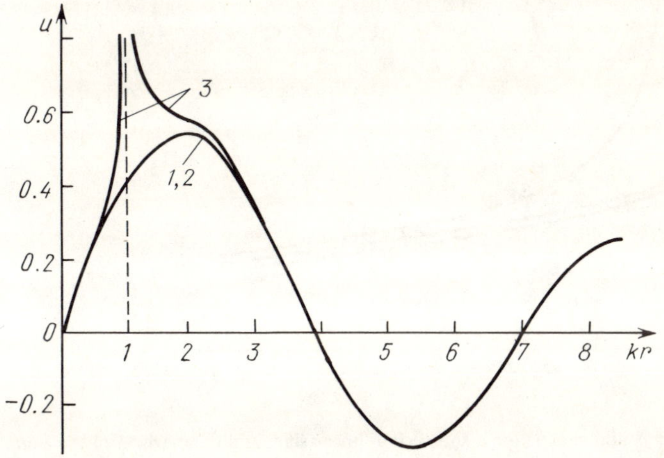

Fig. 5.5. Behavior of (1) exact solution $J_{ka}(kr)$, (2) Airy asymptotic, and (3) WKB asymptotic in the case of a circular caustic at $ka = 1$

deviation between the exact solution and the uniform Airy asymptotic (5.3.4) rises to 10% for $ka \sim 1/2$ [5.22]. Hence, the Airy asymptotic may be used in this problem up to $a \sim 1/2k \sim \lambda/12$, i.e., up to the caustic radius about 1/12 as large as the wavelength. At the limit of applicability, the "large" parameter $M = a/\Lambda = (2k^2a^2)^{1/3}$ of the problem amounts to only 0.8.

5.3.2 Point Source in a Linear Slab

Let a linear source be embedded at x^0, z^0 in a linear layer of permittivity $\varepsilon(z) = \varepsilon_0 - \varepsilon_1 z$. To simplify the solution we consider the two-dimensional case. Rays emanating from x^0, z^0 form in this layer a parabolic caustic (Fig. 5.6a) $4(n^0)^2(\varepsilon_0 - \varepsilon_1 z) = \varepsilon_1^2(x - x^0)^2$, where $n^0 = (\varepsilon_0 - \varepsilon_1 z^0)^{1/2}$ is the refractive index near the source. The solutions of the equations of geometrical optics for this problem have the form [5.15]

$$\psi_{1,2} = \frac{2}{3\varepsilon_1}(f_1^{3/2} \mp f_2^{3/2}) ,$$
$$U_1 = \sqrt{n^0/(\tau_1\sqrt{f_1 f_2})}, \qquad U_2 = -\mathrm{i}\sqrt{n^0/(\tau_1\sqrt{f_1 f_2})} \tag{5.3.9}$$

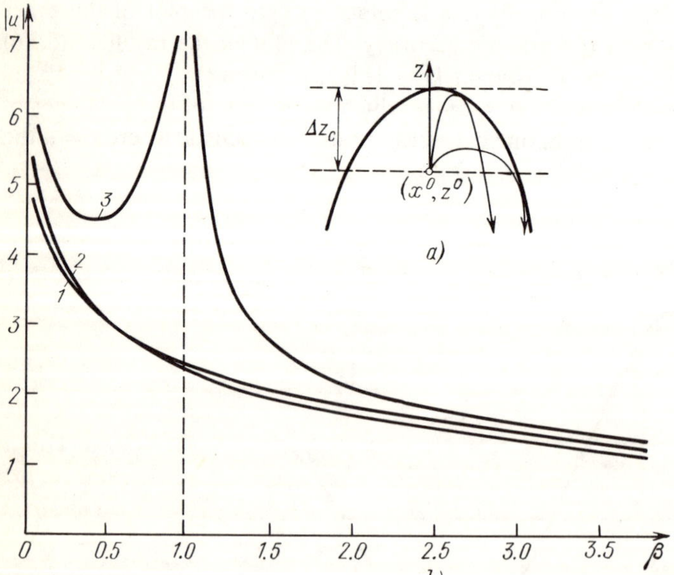

Fig. 5.6. Geometry of rays and caustic for a source in a linear slab, and comparison of the exact solution (1) with the Airy asymptotic (2) and WKB asymptotic (3) for a horizontal section of the field at the level of the source. Here $\beta = \varepsilon_1|x - x^0|/(\varepsilon_0 - \varepsilon_1 z^0)$ is a dimensionless distance from the source

where

$$f_{1,2} = \tfrac{1}{2}[2\varepsilon_0 - \varepsilon_1(z + z^0) \pm \varepsilon_1 R] \ ,$$

$$\tau_{1,2} = \frac{2}{\varepsilon_1}(f_1^{1/2} \mp f_2^{1/2}), \qquad R = \sqrt{(x - x^0)^2 + (z - z^0)^2} \ .$$

For $R \to 0$, the amplitude U_1 of the primary wave has a cylindrical singularity of the type $R^{-1/2}$. In addition, both amplitudes U_1 and U_2 diverge at the caustic where f_2 vanishes.

Using (5.3.9) in (5.1.18) we get for the uniform asymptotic

$$\tilde{\zeta} = \varepsilon_1^{-2/3} f_2 \ , \qquad \chi = \frac{2}{3\varepsilon_1} f_1^{3/2} \ ,$$

$$A = -\mathrm{i}\sqrt{n^0}\,\varepsilon_1^{1/6}(1/\sqrt{\tau_1} + 1/\sqrt{\tau_2})(2f_2)^{-1/2} f_1^{-1/4} \ ,$$
$$B = -\mathrm{i}\sqrt{n^0}\,\varepsilon_1^{1/6}(1/\sqrt{\tau_1} - 1/\sqrt{\tau_2})(2f_2)^{-1/2} f_1^{-1/4} \ .$$

It is an easy matter to verify that the amplitude factor B remains finite at the caustic, though f_2 in the denominator vanishes at the caustic. To demonstrate, we observe that

$$\begin{aligned} 1/\sqrt{\tau_1} - 1/\sqrt{\tau_2} &= (\tau_2 - \tau_1)/[\sqrt{\tau_1\tau_2}(\sqrt{\tau_1} + \sqrt{\tau_2})] \\ &= 4\sqrt{f_2}/[\varepsilon_1\sqrt{\tau_1\tau_2}(\sqrt{\tau_1} + \sqrt{\tau_2})] \end{aligned}$$

and rewrite B in the form

$$B = \frac{4\mathrm{i}\sqrt{n^0}\,\varepsilon_1^{-5/6}}{\sqrt{2}f_1^{1/4}\sqrt{\tau_1\tau_2}(\sqrt{\tau_1} + \sqrt{\tau_2})} \ ,$$

which has no singularities.

As a result, we arrive at a uniform asymptotic expression for the wave field

$$\begin{aligned} u = \sqrt{\pi n^0} f_1^{-1/4}(1/\sqrt{\tau_1} + 1/\sqrt{\tau_2})(k/\varepsilon_1)^{1/6} \Big\{ &\mathrm{Ai}[-(k/\varepsilon_1)^{2/3} f_2] \\ &- \mathrm{i}\left(\frac{k}{\varepsilon_1}\right)^{-1/3} \frac{1}{\sqrt{f_1} + \sqrt{\varepsilon_1 R}} \mathrm{Ai}'[-(k/\varepsilon_1)^{2/3} f_2] \Big\} \\ &\times \exp[\mathrm{i}2kf_1/(3\varepsilon_1) + \mathrm{i}\pi/4] \ . \end{aligned} \tag{5.3.10}$$

This same expression may be obtained by considering the asymptotic for the exact solution to the point-source problem. Such a solution has been obtained by *Kormilitsyn* [5.29].

Near the caustic $f_2 = 0$ and we have

$$f_1 \approx \varepsilon_1 R \ , \qquad \tau_{1,2} \approx 2f_1^{1/2}/\varepsilon_1 \approx 2\sqrt{R/\varepsilon_1} \ , \qquad f_2 \approx -\nu/\Lambda \ ,$$

where ν is the normal distance to the caustic, $\Lambda = (k^2\varepsilon_1)^{-1/3}/\cos\gamma$ is the characteristic caustic length, γ is the angle between the z axis and the normal to the caustic, and $\cos\gamma = [1 + \varepsilon_1^2(x - x^0)^2/(2n^0)^2]^{-1/2}$. Substituting these quantities in (5.3.10) we arrive at the local a symptotic

$$u \approx \frac{\sqrt{2\pi n^0}\,\varepsilon_1^{1/4}}{R^{3/4}}\left(\frac{k}{\varepsilon_1}\right)^{1/6}\left\{\mathrm{Ai}\left(-\frac{v}{\Lambda}\right) - \frac{\mathrm{i}(k/\varepsilon_1)^{-1/3}}{2(\varepsilon_1 R)^{1/2}}\mathrm{Ai}'\left(-\frac{v}{\Lambda}\right)\right\}$$

$$\times\exp[\mathrm{i}2kf_1/(3\varepsilon_1) + \mathrm{i}\pi/4]$$

which is valid within the Λ-neighborhood of the caustic.

Now, we write the condition for applicability of the uniform asymptotic expression (5.3.10) as $\rho_c \gg \Lambda$, e.g., see (5.1.48). In the problem on hand the caustic curvature radius is

$$\rho_c = \rho_0/\cos^3\gamma = \rho_0[1 + \varepsilon_1^2(x - x^0)^2/(2n^0)^2]^{3/2} \ ,$$

where $\rho_0 = 2(n^0)^2/\varepsilon_1$ is the curvature radius at the vertex of the caustic. It can be demonstrated that the condition $\rho_c \gg \Lambda$ will be satisfied everywhere so long as

$$\rho_0/\Lambda_0 \approx 2\Delta z_c/\Lambda_0 \gg 1 \ , \qquad (5.3.11)$$

where $\Lambda_0 = (k^2\varepsilon_1)^{-1/3}$ is the width of the caustic zone near the vertex, and $\Delta z_c = z_{c0} - z^0$ is the distance from the source to the vertex of the caustic. This condition dictates that the source must be separated from the caustic by a distance markedly exceeding the caustic field scale Λ_0.

In this case, the heuristic criterion (5.3.11) allows a validity check because the problem of a point source in a linear slab has an exact solution in the form of a single integral [5.29]. We compare the uniform asymptotic (5.3.10) with the exact solution for a rather unfavorable situation of $M = \rho_0/\Lambda_0 = 1$ signifying that the source is only $\Lambda_0/2$ distant from the vertex of the caustic. Figure 5.6b shows the behavior of the modulus of the exact and approximate solutions in the horizontal plane $z = z^0$ containing the source. The discrepancy between these solutions never exceeds 12%. For comparison there is the plot of the magnitude of the geometric-optical solution which runs quite satisfactory near the source, $\beta \equiv \varepsilon_1(x - x^0)/(\varepsilon_0 - \varepsilon_1 z^0) \to 0$, and in the shadow region, $\beta \gtrsim 2$, but is absolutely inapplicable in the vicinity of the caustic on which $\beta = 1$.

5.3.3 Swallowtail Caustics in a Linear Layer Bordering upon a Homogeneous Halfspace

Notwithstanding frequent occurrence of caustics in a host of applications of wave theory, the literature is lacking in substantial analyses demonstrating how the shape of the caustic surface changes when some or other parameters vary under real conditions. To illustrate how the shape of caustics and pattern of the wave field evolve in response to the variation of problem parameters we quote here the results of detailed computations performed by *Orlov* [5.30–33] primarily before the advent of catastrophe theory (these results are partially reflected in Vol. 6 of this series [5.20]).

Let us consider the reflection of rays from a halfspace $z > 0$ with linear permittivity law $\varepsilon(z) = 1 - \varepsilon_1 z$ assuming that for $z \leq 0$ the medium is homogeneous, $\varepsilon = 1$, as shown in Fig. 5.7a. This problem is of practical significance for the ionospheric propagation of short radio waves. It has been learned that if the point source is not far from the linear slab, $\zeta = \varepsilon_1 z^0 \geq -2/3$, and not very deep in the slab, $\zeta \leq 1/9$, then the caustic curve has the form of a loop with two cusps. Each point within the loop is hit by four rays (Fig. 5.7b).

Figure 5.8 shows the evolution of a caustic stimulated by a variation of the parameter $\zeta = \varepsilon_1 z^0$. These diagrams lead us to an assumption that the caustic loop is a cross section of a swallowtail existent in a space of large dimensionality.

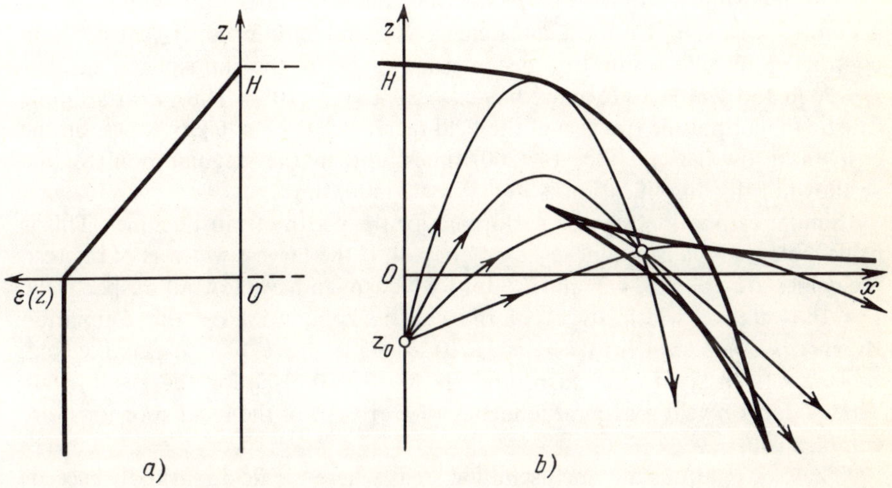

Fig. 5.7. Electromagnetic waves reflected from a plasma layer with linear permittivity profile (**a**) form a caustic loop (**b**) (trace of a swallowtail) when the dimensionless height of the source $\zeta = \varepsilon_1 z^0$ varies

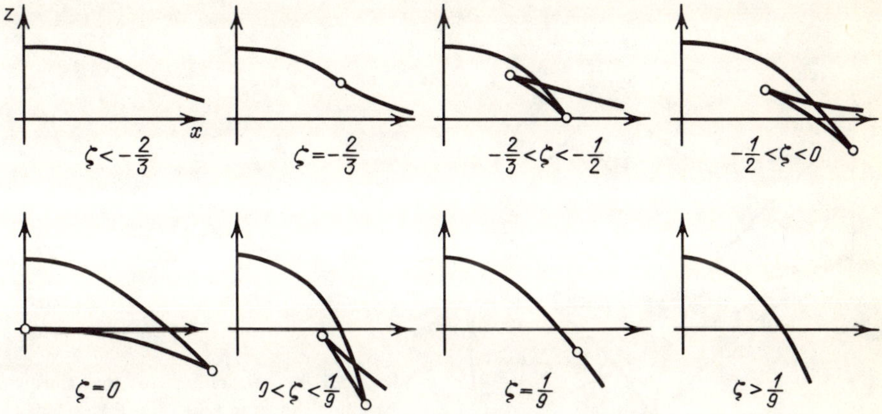

Fig. 5.8. Evolution of a caustic shape in a linear plasmic layer

Indeed, if we unfold these series of sections in ζ, we obtain a typical shape of a swallowtail in the extended space (x, ζ, z). Figure 5.9 shows a part of the resultant shape for $\zeta > 0$.

Orlov [5.34] has also analyzed the field structure as a function of relation between problem parameters. In what follows we estimate only the amplification of the field at the caustic over the field of a spherical wave in a homogeneous space. The characteristic scales of these problems are the wavelength λ, the scale of the slab $1/\varepsilon_1$, and the distance of the source from the boundary of the slab, z^0. Caustic loops are formed for $-2/(3\varepsilon_1) < z^0 < 1/(9\varepsilon_1)$, so that z^0 may be deemed of the order of $1/\varepsilon_1$, hence, the problem keeps essentially only one small dimensionless parameter $\mu = \varepsilon_1/k$.

The focusing factor is $F \sim \mu^{-\sigma_{\text{foc}}}$, the focusing index σ_{foc} being 1/6 for a simple A_2 caustic, 1/4 for a caustic cusp A_3, and 3/10 for an A_4 caustic loop contracted into a point. For decametric waves in the ionosphere, i.e., for $\lambda = 20$ m and $1/\varepsilon_1 = 100$ km, we have $\mu = \varepsilon_1/k = 3 \times 10^{-5}$. Thus, near a simple caustic the amplitude (power) of the field increases $\mu^{-1/6} \approx 6$ (36) times, on the cusp the growth is $\mu^{-1/4} \approx 14$ (200) times, and in the singular point of the swallowtail the amplification is $\mu^{-3/10} \approx 22$ (500) times.

Similar estimation may be performed for the width of caustic zones. This is achievable with caustic indexes, e.g., Chap. 4. If the largest width is of interest, the values are $\alpha = 2/3$, 1/2, and 2/5 for fold, cusp, and swallowtail, respectively. For the relative width $\Delta v_c \varepsilon_1$ of the caustic zone we have the estimation $\Delta v_c \varepsilon_1 \approx (k/\varepsilon_1)^{-\alpha}$, which at $k/\varepsilon_1 \approx 3 \times 10^{-5}$ yields $\Delta v_c \approx 10^{-3}/\varepsilon_1$ for the fold, $\Delta v_c \approx 5 \times 10^{-3}/\varepsilon_1$ for the cusp, and $\Delta v_c \approx 15 \times 10^{-3}/\varepsilon_1$ for the swallowtail. These figures reveal a general tendency of a growth of the focal spot for more complex caustics.

These estimations may be useful for areas where the field markedly exceeds the values typical for the particular locality. Specifically these estimates are suitable to reveal the range where the nonlinear processes developing in strong electromagnetic (or caustic) fields become essential and may lead to appreciable distortions of the field.

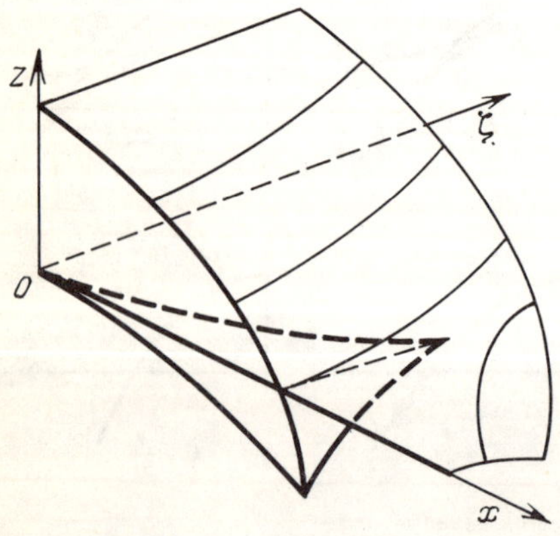

Fig. 5.9. Representation of the caustic evolution shown in Fig. 5.8 in the extended space of parameters $\{x, z, \zeta\}$

5.3.4 Butterfly in a Parabolic Plasma Layer

Let us look at the behavior of rays and fields in a parabolic plasma layer of permittivity

$$\varepsilon(z) = \begin{cases} 1 - \dfrac{\omega_p^2}{\omega^2}\left[1 - \left(\dfrac{2z - z_1}{z_1}\right)^2\right], & \text{for } 0 < z < z_1 , \\ 1, & \text{for } z < 0 \text{ and } z > z_1 . \end{cases} \tag{5.3.12}$$

Here, $\omega_p^2 = 4\pi e^2 N/m$ is the square of the plasma frequency at the middle of the layer, i.e., at $z = z_1/2$.

For a point source placed at a height z^0, different caustic configurations in the parameter plane $\{|z^0/z_1|, \omega/\omega_p\}$ are depicted in Fig. 5.10 [5.32, 33]. A swallowtail forms in a comparatively narrow interval $1.00 \leq \omega/\omega_p \leq 1.08$; however, the maximum amplification is around $\mu^{-1/3}$, corresponding to the butterfly degenerated into a point, is achieved precisely in this interval. An additional

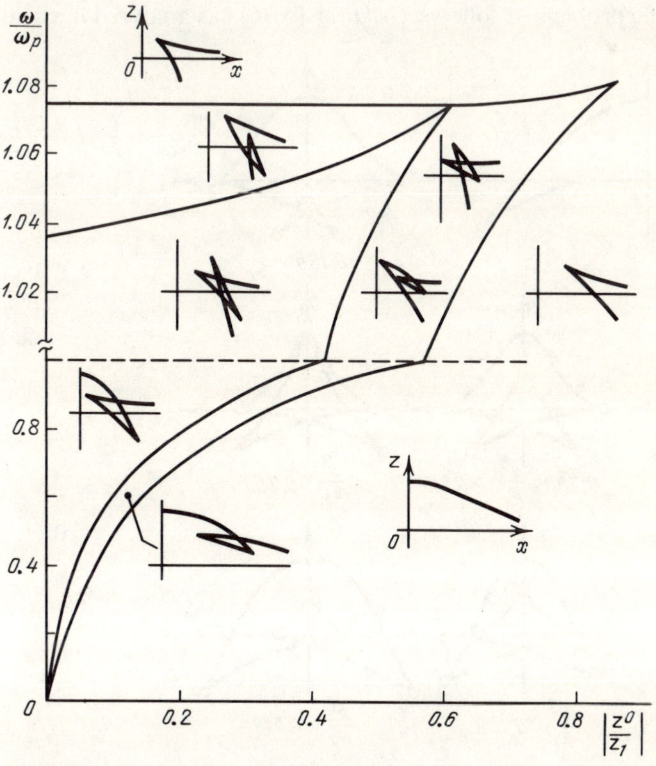

Fig. 5.10. Metamorphosis of a caustic pattern (cross sections of a butterfly) in the plane of ω/ω_p versus $|z^0/z_1|$ for a parabolic plasma layer

focusing, i.e., over that of the swallowtail, is insignificant in magnitude, though the complication of the field structure is quite sensible.

5.3.5 Elliptic Umbilic Formed by an Antenna in a Plasma Layer

Let a focused antenna with a phase across it obeying the square law $S_0(x) = -k\beta x^2$ be embedded in a plasma layer of permittivity $\varepsilon(z) = 1 - \varepsilon_1 z$. Refraction causes a whimsical caustic pattern to occur in the medium, as shown in Fig. 5.11. In the upper part, $z > 0$, this pattern has the structure of a D_4^+ caustic (elliptic umbilic) while in the lower part, $z < 0$, one can recognize elements of a butterfly [5.35]. The evolution of the caustic presented in the figure corresponds to variation of the dimensionless parameter β/ε_1 from negative values (diverging wave) to a value of $\beta/\varepsilon_1 \gtrsim 0.419$.

5.3.6 Elliptic Umbilics in Underwater Acoustics

The utility of the general classification of standard integrals can be demonstrated with a specific problem as follows: *Grikurov* [5.36] has analyzed the field

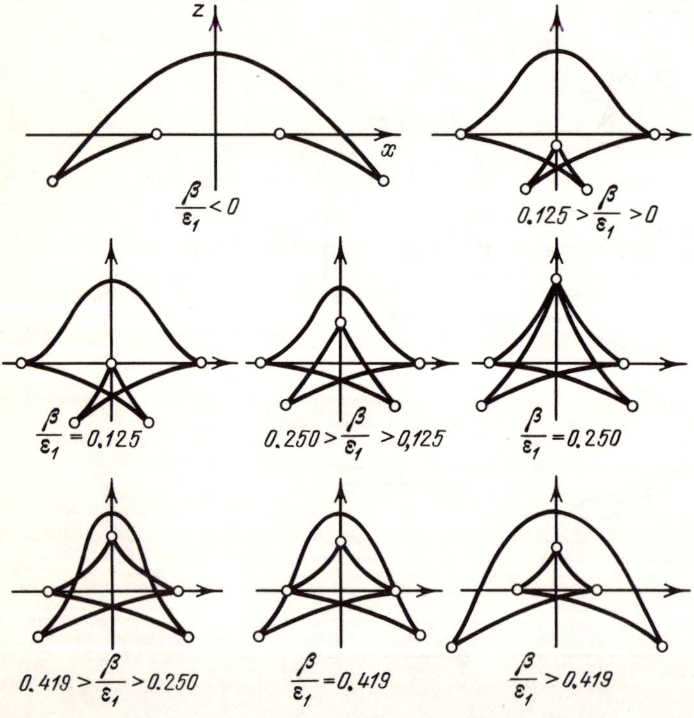

Fig. 5.11. Evolution of a caustic which occurs in a linear plasma layer irradiated by a wave with a square variation of phase in the plane $z = 0$

asymptotic in an underwater acoustic channel for a simple caustic branch passing near a caustic cusp. To this end he has introduced a new function

$$G(\alpha,\beta,\gamma) = \int_0^\infty \frac{dt}{\sqrt{t}} \exp\left(-\frac{\gamma t^2}{2}\right) \cos\left(\frac{t^3}{12} - \alpha t - \frac{\beta^2}{t} + \frac{\pi}{4}\right) .$$

Kryukovskii et al. [5.37] have noted that the above caustic is classified as a hyperbolic umbilic D_4^+ and, if so, then the field should be described by the respective standard integral with a phase function from Table 3.1 (accurate to a factor)

$$\Phi_+(\zeta_1,\zeta_2,\zeta_3) = \iint_{-\infty}^{\infty} dt_1\, dt_2 \exp[i(\tau_2^2\tau_1 + \tau_1^3 + \zeta_1\tau_1^2 + \zeta_2\tau_1 + \zeta_3\tau_2)] .$$

One and the same field could not be described by two different functions G and Φ_+. Indeed, it has been established [5.37] that these functions correspond to each other one-to-one, viz.,

$$\Phi_+(\zeta_1,\zeta_2,\zeta_3) = 2\sqrt{\pi}(12)^{-1/6} G[-\zeta_2(12)^{-1/3}, \tfrac{1}{2}\zeta_1(12)^{1/6}, \zeta_3(18)^{-1/3}] .$$

This achievement is another proof of utility of the universal classification of caustics. Indeed, what has been a set of alien exotic phenomena acquires an internal logic. With some experience one may restore missing details by separate fragments and predict evolution of caustic surfaces and the interference structure of the field as a whole.

The literature abounds with other examples of caustics that occur in actual wave problems. We wish to expose some problems of practical application of catastrophe theory to wave field computation and do so in the form of questions and answers.

5.3.7 How Far Can We Advance in Constructing Caustic Asymptotics?

Although the standard integrals of catastrophe theory solve the problem of construction of local and uniform asymptotics for caustics of arbitrary dimensionality, the actual evaluation of fields presents formidable difficulties.

First of all, the geometrical part of the problem is rather involved as it requires that all rays associated with the caustic should be evaluated and the type of caustic should be identified. Analysis of experimental data presents special difficulties as the available information is as a rule incomplete. By way of example, the reader is suggested to identify the structure of caustics brought about by a light beam passed through a turbulent medium, as shown in Fig. 1.3. The number of rays passing through each point of the image plane varies here from 3 to 6. If no other sections of the wave field are available, it is rather hard to restore the full structure of the caustics and adequately identify them. The information about the longitudinal behavior of the field may be incomplete too – one may need to know how the elements of a caustic traverse and rearrange when medium parameters vary. An insight in the problem complexity can be obtained by considering the data obtained by *Martin* and *Flatté* [5.38] in computer simulation of light beam passage through a turbulent medium.

If the problem of classification is solved, complications occur in the tabulation of the respective standard integrals. The limit attained today is the caustics of codimension 4 and 5, see Chap. 4. To master higher codimensions calls for considerable efforts. Some simplification occurs, conversely, where the number of rays is very large, the resulting field may be deemed random and analyzed with the methods developed in the statistical physics of radio wave propagation [5.39, 40], specifically, with the methods of description of speckle-inhomogeneous fields. It seems that a conditional boundary between the deterministic and statistical descriptions of caustic fields should be drawn where the proportion of area occupied by caustic zones amounts to 10–20%.

5.3.8 Do Swallowtails Exist in Two Dimensions?

Thus far we have avoided statements of exact mathematical evidence for the types of caustics admissible in a space having a given codimension. One of relevant theorems formulated by H. Whitney claims that projecting of an arbitrary smooth surface onto a plane produces only bifurcation sets of the fold catastrophe (A_2) and of the cusp catastrophe (A_3), i.e., only lines and cusp points (points of return). Another theorem claims that in three dimensions there exist only singularities A_2, A_3, A_4 and $D_4^{\pm}$ listed in Table 3.1.

Recognizing importance of this type of theorems for classification of caustics, we would like to draw attention to certain difficulties that occur in using these theorems in evaluating wave fields. As an example, consider two close cusps whose cusp points are separated by less than the width of the caustic zone (in Fig. 4.2 this condition is met by $\zeta_2 = -\pi$). If in agreement with Whitney's theorem we would treat the caustic loop as a set of two cusps A_3, then the field near these singularities should be described by the Pearsey integral (4.3.1). On the other hand, this approach is invalidated by the coalescence of the caustic zones of the cusps, forcing us to resort to a more intricate integral, $I_{A,4}$, corresponding to a swallowtail. However, by Whitney's theorem a swallowtail cannot exist in two dimensions!

This discrepancy owes its existence, of course, to terminological formulations. In the circumstances, the caustic cusp should be viewed as a trace of a swallowtail in an extended space of parameters: two geometric coordinates in the plane should be augmented by one more parameter unfolding the loop into a swallowtail. An example of such an unfolding is presented in Fig. 5.8b, the developing parameter being the coordinate of the source, $\zeta = \varepsilon_1 z_0$. In exactly the same way the butterflies in Fig. 5.9 and elliptic umbilics in Fig. 5.10 should be treated as the traces of these singularities produced by the unfolding in the respective parameters.

Addition of an unfolding parameter is especially significant when there are some suspicions that the system lies near a point of bifurcation. For example, if there are grounds to expect a birth of a caustic loop (as in the diagrams of Fig. 5.8a) it would be advantageous to perform unfolding in an additional parameter, orienting foresightedly to an A_4 singularity (swallowtail) and the respective standard integral, rather than the Airy function corresponding to a fold. With this foresight we can be on a safe side in evaluating field structures and will be safe from errors associated with too rigorous use of theorems of catastrophe theory.

6 Maslov's Method of the Canonical Operator

This method evolved from a very simple idea of describing the asymptotic of the field first in a mixed coordinate–momentum space and performing the Fourier transformation next in the configuration space. This idea proved to be extremely fruitful and effective, and leads to the formulation of heuristic applicability conditions of Maslov's method.

6.1 Principal Relationships

6.1.1 The Wave Equation in the Coordinate–Momentum Representation

The main idea of Maslov's method is that the asymptotic series in powers of inverse large parameter should be constructed not for the wave function $u(\mathbf{r})$ but rather for its Fourier transform in one or two space variables (for intricate waveforms, a Fourier transform with respect to time may be performed). In quantum mechanical terms, it means the evaluation of a wave function in a mixed coordinate–momentum representation.

This beautiful and fruitful idea was formulated by *Maslov* [6.1] in 1965. It has been the subject matter of numerous papers, several review papers [6.2–5] and monographs [6.6, 7], see also [6.8]. The following short recapitulation of the method will be along the lines of one of these reviews [6.2] which has been addressed to researchers and engineers interested in the new mathematical methods of wave theory.

In this chapter we denote the radius vector $\mathbf{r} = (x, y, z)$ by $\mathbf{x} = (x_1, x_2, x_3)$, momentum by $\mathbf{p} = (p_1, p_2, p_3)$, and the coordinates in the coordinate–momentum space by $\mathbf{y} = (p_1, x_2, x_3)$ or $\mathbf{y} = (p_1, p_2, x_3)$. Consider first the space $\mathbf{y} = (p_1, x_2, x_3)$. In this space the wave function $\hat{u}(\mathbf{y})$ is related to $u(\mathbf{x})$ by the Fourier transformation

$$u(\mathbf{x}) = C\int \hat{u}(\mathbf{y})\exp(\mathrm{i}kp_1x_1)\mathrm{d}p_1 \tag{6.1.1}$$

and satisfies, as can be readily verified, the equation

$$\left[-p_1^2 + \frac{\partial^2}{\partial x_1^2} + \frac{\partial^2}{\partial x_2^2} + \varepsilon\left(-\frac{1}{\mathrm{i}k}\frac{\partial}{\partial p_1}, x_2, x_3\right)\right]\hat{u}(\mathbf{y}) = 0\ . \tag{6.1.2}$$

6.1.2 Asymptotic Solution of the Wave Equation

We will seek solution of this equation in the form of an asymptotic expansion in powers of the inverse wavenumber, similar to the development of geometrical optics,

$$\hat{u}(\mathbf{y}) = \left[\hat{U}^{0}(\mathbf{y}) + \frac{1}{ik}\hat{U}^{(1)}(\mathbf{y}) + \cdots \right] e^{ik\psi(\mathbf{y})} , \tag{6.1.3}$$

where the cap implies that the quantity relates to the space $\mathbf{y} = (p_1, x_2, x_3)$. Substituting (6.1.3) into (6.1.2) and using the Debye procedure of setting to zero the coefficients of the like powers of k, we obtain the following equations for $\hat{\psi}(\mathbf{y})$ and $\hat{U}(\mathbf{y}) \equiv \hat{U}^{(0)}(\mathbf{y})$

$$p_1^2 + (\nabla_\perp \hat{\psi}) - \hat{\varepsilon} = 0 , \tag{6.1.4}$$

$$\frac{\partial \hat{U}}{\partial p_1}\frac{\partial \hat{\varepsilon}}{\partial x_1} + 2\nabla_\perp \hat{U} \cdot \nabla_\perp \hat{\psi} - \frac{\hat{U}}{2}\frac{\partial}{\partial p}\left(\frac{\partial \hat{\varepsilon}}{\partial x_1}\right) + \hat{U}\Delta_\perp \hat{\psi} = 0 , \tag{6.1.5}$$

where

$$\hat{\varepsilon} = \varepsilon\left(\frac{\partial \hat{\psi}}{\partial p_1}, x_2, x_3\right), \qquad \nabla_\perp = \left(0, \frac{\partial}{\partial x_2}, \frac{\partial}{\partial x_3}\right), \qquad \Delta_\perp = \frac{\partial^2}{\partial x_2^2} + \frac{\partial^2}{\partial x_3^2} .$$

Equation (6.1.4) is the equation of eikonal in the space $\mathbf{y} = (p_1, x_2, x_3)$. An easier way to a solution of this equation would be by analogy with analytical mechanics which establishes a relation between the action $S(\mathbf{x})$ in the coordinate space and the action $S(\mathbf{y})$ in the mixed representation $\mathbf{y} = (p_1, x_2, x_3)$ [6.1]. Because the eikonal in wave theory is similar to the action in mechanics, we have

$$\hat{\psi}(\mathbf{y}) = \psi(\mathbf{x}) - x_1 p_1 = \psi_0(\xi, \eta) + \int_0^\tau \varepsilon \, d\tau - p_1 x_1 , \tag{6.1.6}$$

where x_1, ξ, η, and τ should be expressed in terms of $\mathbf{y}$ using the equations of rays in space $(\mathbf{x}, \mathbf{p})$,

$$x_k = X_k(\xi, \eta, \tau), \quad p_k = P_k(\xi, \eta, \tau) .$$

The amplitude $\hat{u}(\mathbf{y})$ and constant C in (6.1.1) will be found subject to the condition that the leading term in (6.1.1) should become the geometric-optical approximation $u(\mathbf{x}) = U(\mathbf{x}) \cdot \exp[ik\psi(\mathbf{x})]$ in the region where the divergence $\mathcal{J}$ is other than zero. Substituting $\hat{u}(\mathbf{y}) = \hat{U}(\mathbf{y}) \cdot \exp[ik\psi(\mathbf{y})]$ in (6.1.1) we calculate the integral by the method of stationary phase. A stationary point, $p_{1\mathrm{st}}$, is the one where

$$\frac{\partial \psi}{\partial p_1} + x_1 = 0 . \tag{6.1.7}$$

By the rules of transformation of the eikonal in coming from one space of variables to another, $\partial\hat{\psi}(\mathbf{y})/\partial p_1 = -x(\mathbf{y})$. Therefore at the stationary point

$$x_1(\mathbf{y}_{\mathrm{st}}) = x_1, \qquad \mathbf{y}_{\mathrm{st}} = (p_{1\mathrm{st}}, x_2, x_3) . \tag{6.1.8}$$

Since

$$\frac{\partial\hat{\psi}}{\partial p_1} = \frac{\partial\psi}{\partial x_1}\frac{\partial x_1(\mathbf{y})}{\partial p_1} - p_1\frac{\partial x_1(\mathbf{y})}{\partial p_1} - x_1(\mathbf{y}) \ ,$$

observing (6.1.8) in (6.1.7) we get

$$p_{1\mathrm{st}} = \frac{\partial}{\partial x_1}\psi[x_1(\mathbf{y}_{\mathrm{st}}), x_2, x_3] \ .$$

In other words, the quantity $p_{1\mathrm{st}}$ satisfies the equations of rays and by (6.1.6)

$$\hat{\psi}(\mathbf{y}) + x_1 p_1|_{p_{1\mathrm{st}}} = \psi(\mathbf{x}) \ .$$

Recognizing (6.1.7) we have

$$\frac{\partial^2}{\partial p_1^2}(\hat{\psi} + x_1 p_1)|_{p_1 = p_{1\mathrm{st}}} = -\frac{\partial x_1(\mathbf{y}_{\mathrm{st}})}{\partial p_1} \ . \tag{6.1.9}$$

Standard calculations along the lines of the stationary phase method yields

$$u(\mathbf{x}) \sim C\hat{U}(\mathbf{y}_{\mathrm{st}})\sqrt{\frac{2\pi}{\mathrm{i}k}}\left(\frac{\partial x_1(\mathbf{y}_{\mathrm{st}})}{\partial p_1}\right)^{-1/2} \mathrm{e}^{\mathrm{i}k\psi(\mathbf{x})} \ . \tag{6.1.10}$$

Comparing this expression with the ray solution (2.1.12) we get

$$\hat{U}(\mathbf{y}) = \frac{U^0(\xi,\eta)}{\sqrt{\mathcal{J}}}\sqrt{\frac{\partial x_1(\mathbf{y})}{\partial p_1}}, \qquad C = \sqrt{\frac{\mathrm{i}k}{2\pi}} \ , \tag{6.1.11}$$

where ξ, η, τ, and x_1 should be expressed through $\mathbf{y}$ in agreement with the equations of rays. This result may be obtained in a rigorous derivation, by solving (6.1.5) with the use of Liouville's theorem. Finally, for the leading term of the asymptotic series we obtain

$$u(\mathbf{x}) = \sqrt{\frac{\mathrm{i}k}{2\pi}}\int\frac{U^0(\xi,\eta)}{\sqrt{\hat{\mathcal{J}}(\mathbf{y})}}\exp\left[\mathrm{i}k\int_0^\tau\hat{\varepsilon}\,\mathrm{d}\tau - x_1(\mathbf{y})p_1 + x_1 p_1\right]\mathrm{d}p_1 \ , \tag{6.1.12}$$

where

$$\hat{\mathcal{J}}(\mathbf{y}) = \mathcal{J}(\mathbf{x})\frac{\partial p_1}{\partial x_1}\bigg|_{x_1 = x_1(\mathbf{y})} \tag{6.1.13}$$

is the divergence in the space $\mathbf{y} = (p_1, x_2, x_3)$.

In Maslov's pioneering work [6.1], the divergences $\mathcal{J}$ and $\hat{\mathcal{J}}$ are presented as the Jacobians of transition from the initial problem parameters $(\alpha_1, \alpha_2, \alpha_3)$ to $\mathbf{x}$ and $\mathbf{y}$

$$\begin{aligned} \mathcal{J}(\mathbf{x}) &= \frac{\partial(x_1, x_2, x_3)}{\partial(\alpha_1, \alpha_2, \alpha_3)} = \det\left[\frac{\partial x_i}{\partial \alpha_j}\right] , \\ \hat{\mathcal{J}}(\mathbf{y}) &= \frac{\partial(p_1, x_2, x_3)}{\partial(\alpha_1, \alpha_2, \alpha_3)} = \det\left[\frac{\partial y_i}{\partial \alpha_j}\right] . \end{aligned} \tag{6.1.14}$$

To establish relation of these expressions with (2.1.13) based on the ray parameters ξ, η, τ, we observe that Maslov has specified the initial wave function $u^0(\alpha) = U^0(\alpha)\exp[ik\psi^0(\alpha)]$ in a volume, whereas in our setting the initial field $u^0(\xi, \eta) = U^0(\xi, \eta)\exp[ik\psi^0(\xi, \eta)]$ is given on the surface Q whose equation $\mathbf{r} = \mathbf{r}^0(\xi, \eta)$ may be written in the form $\alpha_i = \alpha_i(\xi, \eta)$ if $\alpha_1, \alpha_2, \alpha_3$ are thought of as initial coordinates x_1^0, x_2^0, x_3^0. Transition from two ray parameters ξ, η to the three initial parameters $\alpha_1, \alpha_2, \alpha_3$ needs some artificial technique which is essentially a transfer of the boundary conditions on u^0 from surface Q onto some spatial neighborhood.

Because the right sides of the ray equations (2.1.5) do not depend explicitly on τ we may introduce an additive constant τ_0 in the solutions $\mathbf{x} = \mathbf{X}(\xi, \eta, \tau)$ and $\mathbf{p} = \mathbf{P}(\xi, \eta, \tau)$ and rewrite (2.1.9) as $x_i(\tau) = X_i(\xi, \eta, \tau + \tau_0)$. Then the initial coordinates α_i may be defined as $\alpha_i = X_i(\xi, \eta, \tau_0)$ assuming that to $\tau_0 = 0$ there corresponds the initial surface Q. In view of $\partial/\partial\tau = \partial/\partial\tau_0$

$$\frac{\partial(x_1, x_2, x_3)}{\partial(\xi, \eta, \tau)} = \frac{\partial(x_1, x_2, x_3)}{\partial(\xi, \eta, \tau_0)}, \qquad \left.\frac{\partial(x_1, x_2, x_3)}{\partial(\xi, \eta, \tau)}\right|_{\tau=0} = \frac{\partial(\alpha_1, \alpha_2, \alpha_3)}{\partial(\xi, \eta, \tau_0)}.$$

Coupling these relations with (2.1.13) yields the first of expressions in (6.1.14). The trajectory $x_i(\alpha, \tau)$ is determined by eliminating ξ, η, and τ between $x_i = X_i(\xi, \eta, \tau + \tau_0)$ and $\alpha_i = X_i(\xi, \eta, \tau_0)$.

The second expression in (6.1.14) follows from the first and (6.1.13).

In a similar manner we may demonstrate that for a double Fourier transform, when the problem boils down to seeking a wave function in the mixed space $\mathbf{y} = (p_1, p_2, x_3)$, the asymptotic solution to Helmholtz's equation is given by

$$u(\mathbf{x}) = (ik/2\pi) \int U^0(\xi, \eta)/\sqrt{\hat{\hat{\mathscr{J}}}} \times \exp\left[ik\left(\int_0^\tau \varepsilon\, d\tau - x_1(y)p_1 - x_2(y)p_2 + x_1p_1 + x_2p_2\right)\right] dp_1\, dp_2\,, \tag{6.1.15}$$

where the divergence is given by

$$\hat{\hat{\mathscr{J}}}(\mathbf{y}) = \mathscr{J}(\mathbf{x})\frac{\partial(p_1, p_2)}{\partial(x_1, x_2)} = \frac{\partial(p_1, p_2, x_3)}{\partial(\alpha_1, \alpha_2, \alpha_3)} = \det\left[\frac{\partial y_i}{\partial \alpha_j}\right]. \tag{6.1.16}$$

6.1.3 Elimination of Field Divergence at Caustics

An important feature of expressions (6.1.12, 15) is that a suitable choice of coordinates in $\mathbf{y}$ can always eliminate the singularity of amplitudes $\hat{U}$ and $\hat{\hat{U}}$ at caustics, thus securing that the field at the caustic $u(\mathbf{x})$ there is finite. A choice of a mixed space in which $\hat{\mathscr{J}}(\mathbf{y}) \neq 0$ or $\hat{\hat{\mathscr{J}}}(\mathbf{y}) \neq 0$ though $\mathscr{J}(\mathbf{x}) = 0$ relies upon a certain topological properties of the ray surface $\mathbf{x} = \mathbf{X}(\xi, \eta, \tau)$, $\mathbf{p} = \mathbf{P}(\xi, \eta, \tau)$ in the phase space, i.e., upon the properties of the Lagrangian manifold [6.1] which is understood as the congruence of the paths $\mathbf{x} = \mathbf{X}(\xi, \eta, \tau)$ and $\mathbf{p} = \mathbf{P}(\xi, \eta, \tau)$.

In the phase space, rays (trajectories) do not intersect: in view of uniqueness of the solution to the dynamic equation only one trajectory passes through a phase point having coordinate $\mathbf{x}$ and momentum $\mathbf{p}$. Therefore, in this space we may always choose a plane $\hat{\mathbf{y}} = (\hat{y}_1, \hat{y}_2, \hat{y}_3)$ to which trajectories will be projected locally one-to-one though the projections of trajectories on the configurational space intersect. Thus the divergence $\hat{\mathscr{J}}(\mathbf{y})$ will be nonzero in the $\hat{\mathbf{y}}$ space nonwithstanding $\mathscr{J}(\mathbf{x}) = 0$.

A plane $\hat{\mathbf{y}}$ satisfying the aforementioned requirements may be found in the lines of the following procedure. Let at a caustic the functional determinant $\mathscr{J}$ have a first order zero, i.e., the rank of $[\partial x_i/\partial \alpha_j]$ decreases by one unit at the caustic. We perform an orthogonal rotation of the axes $x_i' = a_{ij}x_j$ such that one row of $[\partial x_i'/\partial x_j]$, say the first, vanishes, i.e.,

$$\frac{\partial x_1'}{\partial \alpha_j} = 0, \quad j = 1, 2, 3 \ .$$

This condition implies that the coordinate x_1' is orthogonal to the caustic surface. If we take the local coordinates $\mathbf{y}' = (p_1', \hat{x}_2, \hat{x}_3)$ with $p_i' = a_{ij}p_j$ as new variables, then the field $u(\mathbf{x})$ will be represented by a Maslov integral (6.1.12). When the rank of $[\partial x_i/\partial \alpha_j]$ is decreased by two units, it would be natural to represent the field as a double integral (6.1.15).

6.1.4 Canonical Operator

From the derivations of (6.1.12, 16) it is clear that these expressions describe the local asymptotic of the field, which is valid where the divergence $\mathscr{J}(\mathbf{y}) = \det[\partial y_i/\partial \alpha_j]$ is other than zero. It is impossible to meet this condition simultaneously over the entire space; therefore, to arrive at a global asymptotic we have to break down all the space into domains of applicability for one or other asymptotic expression. At the boundaries of these domains and in some neighborhood of the boundaries two (or more) solutions of the type (6.1.12) or (6.1.15) are valid which are asymptotically equivalent for k tending to infinity. This ensures the uniformity of an asymptotic approximation over the entire space.

The uniform asymptotic for the Helmholtz equation may be written as a single formula with the aid of Maslov's canonical operator [6.1] which will be described below in very general terms.

The canonical operator essentially unites local formulas, i.e., expressions for the ray field $u = A\exp(ik\psi)$ and integral representations like (6.1.12, 15) each of which is valid in a limited domain. Let $u_j(\mathbf{x})$ be an asymptotic expression for the field in domain V_j. Such a field is represented by a Fourier transform whose multiplicity depends on the degree of degeneracy of the functional determinant $\mathscr{J}(\mathbf{x}) = [\partial x_i/\partial \alpha_j]$, namely, the ray field corresponds to the Fourier transform of zero multiplicity, the field (6.1.12) to a single Fourier transform, and the field (6.1.15) to a double Fourier transform.

We assign to each domain V_j an infinitely differentiable function $e_j(\mathbf{x})$ which is unity within V_j and gradually falls off outside. The sum of all $e_j(\mathbf{x})$ must be unity,

$$\sum e_j(\mathbf{x}) = 1 \; .$$

Now, the full field $u(\mathbf{x})$ in the entire space can be represented as the sum

$$u(\mathbf{x}) = \sum e_j(\mathbf{x}) u_j(\mathbf{x})$$

which is equal to $u_j(\mathbf{x})$ within V_j and ensures a gradual transition from one asymptotic expression to another in crossing the boundary between two adjacent domains.

The field $u_j(\mathbf{x})$ may be viewed as the result of mapping of the initial data $u^0(\xi, \eta)$ onto domain V_j. Symbolically, this may be written as the result of action of an operator T_j, namely, $u_j(\mathbf{x}) = T_j u^0(\xi, \eta)$, where T_j is expressed through Fourier transforms of multiplicity zero, one, or two. The resultant formula

$$u(\mathbf{x}) = \left(\sum e_j(\mathbf{x}) T_j\right) u^0(\xi, \eta) \equiv K u^0(\xi, \eta) \tag{6.1.17}$$

provides a definition of Maslov's canonical operator K [6.1]. This operator is advantageous in that it provides a uniform asymptotic of the field in the circumstances where each of the particular expressions for $u_j(\mathbf{x})$ ensures only a local asymptotic within V_j. On the boundary of two domains, two (or more) expressions for $u_j(\mathbf{x})$ are valid; they are *asymptotically equivalent* as $k \to \infty$. A proof of such equivalence is given in Maslov's work [6.1].

Using Maslov's formalism, one does not necessarily give the initial wave function $u^0(\xi, \eta)$ in the ray form $U^0(\xi, \eta) \cdot \exp(\mathrm{i} k \psi^0(\xi, \eta)]$. The canonical operator enables one to employ initial conditions of more general form, namely, such as that described by (6.1.12, 15). To be more specific, in this case the initial conditions for the field $\hat{u}(\mathbf{y})$ are given in the coordinate-momentum space $\mathbf{y}$, and $\hat{u}(\mathbf{y})$ may be found by taking the inverse of the integrals (6.1.12, 15).

This canonical operator may be used to describe asymptotic solutions for a wide class of equations of the "wave" type. In addition to the Helmholtz equation this class includes equations of quantum mechanics, both relativistic and non-relativistic, and vector equations of electrodynamics and theory of elasticity. Maslov's method is applicable to solving nonstationary problems such as the propagation of field breaks. Finally, the asymptotic calculation of eigenvalues [6.1] may be of considerable interest for applied wave theory. Possibilities of the method will be illustrated with specific problems in Sect. 6.2.

6.1.5 Remarks on Applicability Conditions

Operating with the amplitudes and phases of wave functions in the coordinate–momentum representation it would be natural to impose on them some conditions similar to the applicability conditions of geometrical optics and the method of uniform caustic asymptotics. Constraints on the variation rate of parameters of the wave and medium in the configuration space are evident – these parameters should not experience jumps within the width a_f of the Fresnel volume. As for the variation rate in the momentum variables p_x, p_y, and p_z, it deserves appropriate Fresnel scales which would enable one to speak of a fast or slow variation; however, this problem is yet to be solved.

6.2 Specific Problems

6.2.1 Plane Wave in a Linear Layer

This problem has an exact solution which has been delineated in Sect. 2.4.2. We reiterate this problem to demonstrate how fields are described in the mixed coordinate–momentum representation.

In the plane $x_1 = x$, $x_2 = z$ the ray paths are given by equations (2.4.1), the eikonal $\psi(\tau)$ is given by

$$\psi(\tau) = n^0 \xi \sin\theta^0 + n^{02} - \tfrac{1}{2}\varepsilon_1 n^0 \tau^2 \cos\theta^0 + \varepsilon_1^2 \tau^3/12 \ , \tag{6.2.1}$$

and the divergence is

$$\mathscr{J} = \frac{n^0 \cos\theta^0 - \varepsilon_1 \tau/2}{n^0 \cos\theta^0} \ . \tag{6.2.2}$$

If we substitute here (2.4.2) and (2.4.3) for $\tau = \tau(x, z)$ and $\xi = \xi(x, z)$, then below the caustic, i.e., for $z < z_c = z^0 + (n^0 \cos\theta^0)^2/\varepsilon_1$, we obtain two values for the eikonal (2.4.4) and two values of divergence

$$\begin{aligned} \mathscr{J}_{1,2} &= \pm [(n^0 \cos\theta^0)^2 - \varepsilon_1 (z - z^0)]^{1/2}/n^0 \cos\theta^0 \\ &= \pm [\varepsilon_1 (z_c - z)]^{1/2}/n^0 \cos\theta^0 \ , \end{aligned} \tag{6.2.3}$$

which aid in describing the field as a sum of incident and reflected waves (2.4.6). Near a caustic and above it, the ray expression (2.4.6) is no longer valid, but for $z \geq z_c$ the field can be represented by Maslov's method.

Let the integration variable be the z component of momentum, p_z. By (2.4.1) the quantities ξ, τ, and z can be expressed via x and p_z as follows:

$$\begin{aligned} \xi &= x - 2n^0 \sin\theta^0 (n^0 \cos\theta^0 - p_z)/\varepsilon_1 \ , \\ \tau &= 2(n^0 \cos\theta^0 - p_z)/\varepsilon_1 \ , \\ z &= z^0 + [(n^0 \cos\theta^0)^2 - p_z^2]/\varepsilon_1 \ . \end{aligned} \tag{6.2.4}$$

It can be readily demonstrated that

$$\partial p_z/\partial z = -\varepsilon_1/2p_z \ , \qquad \hat{\mathscr{J}} = p_z/n^0 \cos\theta^0 \ ,$$

$$\hat{\psi}(x, p_z) = s(x) - z_c p_z + p_z^3/3\varepsilon_1 \ , \qquad s(x) = x n^0 \cos\theta^0 + 2(n^0 \cos\theta^0)^3/3\varepsilon_1 \ .$$

Thus, in the space (x, p_z) the divergence $\hat{\mathscr{J}}(x, p_z) = \mathscr{J}(x, z).\ \partial p_z/\partial z = -\varepsilon_1/2n^0 \cos\theta^0$ is nonzero everywhere, whereas, in the configuration space $\mathscr{J}(x, z)$ vanishes at the caustic surface $z = z_c$.

Substituting these expressions in (6.1.12) we have

$$u(x, z) = \sqrt{\frac{ik}{2\pi}\frac{2n^0 \cos\theta}{-\varepsilon_1}}\, U^0 \exp[iks(x)] \int \exp\{ik[(z - z_c)p_z + p_z^3/3\varepsilon_1]\}\, dp_z\ . \tag{6.2.5}$$

This integral is expressed in terms of the Airy function, so that the field can be rewritten as

$$u(x, z) = 2\sqrt{\pi}\, U_0 e^{-i\pi/4}(k/\varepsilon_1)^{1/6}(n^0 \cos\theta^0)^{1/2} e^{iks(x)} \times \mathrm{Ai}[k^{2/3}\varepsilon_1^{1/3}(z - z_c)]\ . \tag{6.2.6}$$

This expression coincides with the exact solution of the problem (4.2.17). Away from the caustic, in the lit region ($z < z_c$) it becomes the sum of two ray fields (2.4.6) and in the shadow region ($z > z_c$) it decays by the exponential law (2.4.7), thus, justifying the selection rules of complex rays.

Finite values of the field (6.2.5) at caustics stem from the absence of singularities in projection of the Lagrange's manifold (ray surface) on the coordinate–momentum plane (x, p_z), despite a fold singularity (A_2) in the physical space (x, z). This situation is illustrated in Fig. 6.1 showing a ray surface in the form of a parabolic cylinder in the three space (x, z, p_z). While the rays (2.4.1) form a caustic in the x, z plane, their projections do not intersect in the x, p_z plane.

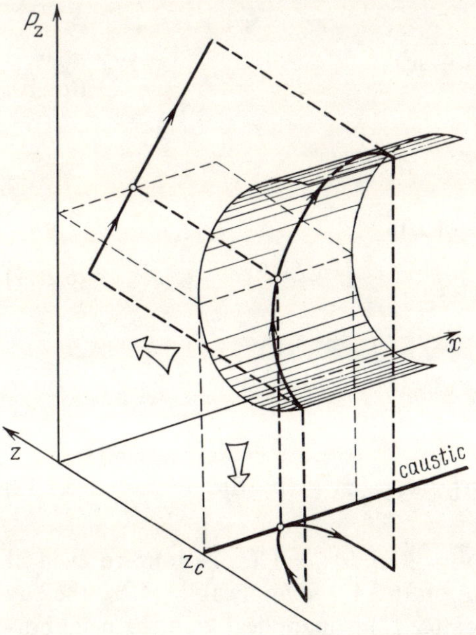

Fig. 6.1. Ray surface in the mixed coordinate–momentum space (x, z, p_z) forms a caustic in projection onto the configuration space (x, z), yet has no singularities in projecting onto a "mixed" plane (x, p_z)

6.2.2 Diffraction on a Phase Screen

Consider the diffraction of a wave passing through a phase-amplitude screen in free space. For simplicity we limit our consideration to the case of a one-dimensional screen. Let at $z = 0$ the initial field passing through the screen be $u^0(\xi) = U^0(\xi)\exp[ik\psi^0(\xi)]$ with $U^0(\xi)$ and $\psi^0(\xi)$ being the amplitude and the eikonal of the wave. The ray equations in free space can be integrated directly to give the following solutions:

$$x(\tau) = \xi + \tau f'(\xi), \qquad z(\tau) = \tau\sqrt{1 - f'^2(\xi)} \; ,$$
$$p_x(\tau) = f'(\xi), \qquad p_z(\tau) = \sqrt{1 - f'^2(\xi)} \; , \tag{6.2.7}$$

where $f(\xi) = \psi^0(\xi)$.

By the formulas of Chap. 2 we find

$$\psi = \psi^0(\xi) + \tau, \qquad \mathscr{J}(z, \xi) = 1 + zf''(\xi)[1 - f'^2(\xi)]^{-3/2} \; , \tag{6.2.8}$$

so that in the region of applicability of geometrical optics

$$u(x, z) = U^0 \mathscr{J}^{-1/2}_{(z,\xi)} \exp\{ik[\psi^0(\xi) + \tau]\} \; , \tag{6.2.9}$$

where ξ and τ are expressed through x and z with the aid of equations following from (6.2.7), viz.,

$$(x - \xi)/z = f(\xi)/\sqrt{1 - f'^2(\xi)}, \qquad \tau = \sqrt{[x - \xi(x, z)^2 + z^2} \; .$$

To define the field near caustics where $\mathscr{J} = 0$ we make use of (6.1.12) letting at the caustic $x_1 = x$ and $p_1 = p_x$. The dependence of ξ, τ, and x on z and p_x is given by the equations

$$f'[\xi(p_x)] = p_x, \qquad \tau(p_x, z) = z/\sqrt{1 - p_x^2} \; ,$$
$$x(p_x, z) = \xi(p_x) + zp_x/\sqrt{1 - p_x^2} \; . \tag{6.2.10}$$

From (6.2.2) and (6.2.10) it follows:

$$\hat{\psi}(p_x, z) = f[\xi(p_x)] + z\sqrt{1 - p_x^2} - \xi(p_x)p_x \; ,$$
$$\mathscr{J} = 1 + zf''[\xi(p_x)]\cdot(1 - p_x^2)^{-3/2} \; , \tag{6.2.11}$$
$$\frac{\partial p_x}{\partial x} = \frac{f''[\xi(p_x)]}{\mathscr{J}} \; .$$

Now the field near caustics is

$$u(x, z) = \sqrt{\frac{k}{2\pi i}} \int \frac{U^0[\xi(p_x)]}{\sqrt{-f''[\xi(p_x)]}} \exp[ik\hat{\psi}(p_x, z) + xp_x]\,dp_x \; . \tag{6.2.12}$$

In this expression, the Jacobian $\hat{\mathscr{J}} = \mathscr{J}\partial p_x/\partial x = f''(\xi)$ is nonzero even at $\mathscr{J} = 0$; therefore, the asymptotic solution (6.2.12) is applicable where the ray approximation (6.2.9) is not. However, (6.2.12) is inapplicable in the neighborhood of those values of p_x for which $\hat{\mathscr{J}} = f''[\xi(p_x)] = 0$. It should be natural to assign to these p_x caustics in the p_x, z space.

For $f'' = 0$ the Jacobian $\mathscr{J}$ has to be other than zero, as by (6.2.8) it is unity; therefore, along the rays where it vanishes one may use the ray formula (6.2.9). Thus, for an arbitrary point of observation (x, z), one of the expressions (6.2.9) or (6.2.12) is valid. Where both $\mathscr{J}$ and $\hat{\mathscr{J}}$ are nonzero, either of these formulas is valid (of course, asymptotically, i.e., as $k \to \infty$). This is an easy matter to verify by taking ξ as integration variable in (6.2.12). Then $dp_x = f''(\xi)d\xi$ and the integral

$$u(x, z) = (k/2\pi i)^{1/2} \int U^0(\xi)\sqrt{-f''(\xi)}\exp(ik\{\hat{\psi}[p_x(\xi), z] + xp_x(\xi)\})d\xi \qquad (6.2.13)$$

is calculated by the method of stationary phase. Note that here in view of (6.2.10) $p_x = f'(\xi)$.

It is instructive to compare the asymptotic formula (6.2.11) with what follows from the Huygens–Kirchhoff principle, namely,

$$u(x, z) = \sqrt{\frac{k}{2\pi i}} \int U^0(\xi)\frac{\exp\{ik[f(\xi) + R]\}}{\sqrt{R}}d\xi\ , \qquad (6.2.14a)$$

where $R = \sqrt{(x - \xi)^2 + z^2}$. This formula yields a good description of the diffraction provided that $kR \gg 1$, because when this condition is met, the Bessel function $H_0(kR)$ in terms of which the two-dimensional Green function is expressed, may be replaced with its asymptotic:

$$H_0(kR) \to \exp(ikR - i\pi/4)\sqrt{2/\pi kR}\ .$$

It would be natural to ask if the "weaker" asymptotic formula (6.2.13) can describe the same diffraction phenomena as the exact formula (6.214a), and if so, then under what conditions. Comparing these formulas we can conclude that they yield close descriptions, if within the region where $U^0(\xi) \neq 0$ $\sqrt{-f''(\xi)} \approx 1/\sqrt{R}$ and $z\sqrt{1 - f'^2(\xi)} + (x - \xi)f'(\xi) \cong R$. The first of these conditions secures the equality of the amplitudes and the second the equality of the phases in the integrands. These conditions will be met approximately if $|f'| \ll 1$ (this corresponds to paraxial wave fields) and the observation point (x, z) lies near the caustic. The condition $-f'' \sim 1/R$, for example, immediately follows by letting the Jacobian $\mathscr{J}$ be approximately zero with $|f'| \ll 1$ and $z \sim R$.

We illustrate this reasoning with a specific problem of diffraction of a focused wave $u_0(\xi) = U^0(\xi)\exp(-ik\xi^2/2F)$ at a $2a$ wide slit so that we may safely let $U^0(\xi) = 0$ for $|\xi| > a$. We assume that the focal length F is larger compared to a, represent R approximately as $z + (x - \xi)^2/z$ in the exponent and as z in the denominator of (6.2.14a), i.e., use the Fresnel approximation, and transform the Kirchhoff integral (6.2.14a) to the form

$$u(x, z) \cong \sqrt{\frac{k}{2\pi iz}}\exp\left(ikz + \frac{ikx^2}{2z}\right)\int_{-a}^{a} U^0(\xi)\exp\left[ik\left(-\frac{x\xi}{z} + \frac{\xi^2}{2F}\frac{F - z}{z}\right)\right]d\xi\ . \qquad (6.2.14b)$$

In the paraxial approximation $|f'| \ll 1$ the Maslov integral (6.2.13) takes the form

$$u(x,z) \cong \sqrt{\frac{k}{2\pi \mathrm{i} F}} \exp(\mathrm{i}kF) \int_{-a}^{a} U^0(\xi) \exp\left[\mathrm{i}k\left(-\frac{x\xi}{F} + \frac{\xi^2}{2F}\frac{F-z}{F}\right)\right] \mathrm{d}\xi \ . \tag{6.2.15}$$

Comparing these expressions we see that they are practically coincident in a certain neighborhood of focal point $(0, F)$, to be more specific, within the rectangle $|x| \lesssim \sqrt{F/k}$, $|z-F| \lesssim \sqrt{F^3/ka^2}$. This rectangle is about $\sqrt{ka^2/F}$ times the size of the focal spot. Indeed, if $l_\perp \sim F/ka^2$ and $l_\| \sim F^2/ka^2$ are the transverse and longitudinal dimensions of the spot, then

$$\frac{x}{l_\perp} \lesssim \left(\frac{ka^2}{F}\right)^{1/2}, \qquad \frac{|z-F|}{l_\|} \lesssim \left(\frac{ka^2}{F}\right)^{1/2} .$$

As a rule in optical instruments $ka^2/F \gg 1$; therefore, the dimensions of the domain of applicability of (6.2.15) are many times the size of the focal spot.

6.2.3 Asymptotic Solution of the Parabolic Equation

Suppose that the field $u^0(x, y) = U^0(x, y) \exp[\mathrm{i}k\psi^0(x, y)]$ given in the plane $z = 0$ produces a weakly divergent wave (paraxial field) in the half-space $z > 0$. Seeking a solution to the Helmholtz equation (2.1.1) as a plane wave $\exp(\mathrm{i}kz)$ with variable amplitude $w(x, y, z)$ we obtain for w the equation

$$\Delta w + 2\mathrm{i}k\frac{\partial w}{\partial z} + k^2(\varepsilon - 1)w = 0 \ . \tag{6.2.16}$$

Dropping in this equation $\partial^2 w/\partial z^2$ we arrive at the parabolic equation

$$\Delta w + 2\mathrm{i}k\frac{\partial w}{\partial z} + k^2(\varepsilon - 1)w = 0 \tag{6.2.17}$$

which is known in diffraction theory. This equation is similar to the Schrödinger equation and therefore, allows for a Maslov's asymptotic solution.

In this case the ray trajectories $x(z)$ and $y(z)$ are determined from the set of Hamilton's equations

$$\frac{\mathrm{d}x}{\mathrm{d}z} = p_x, \qquad \frac{\mathrm{d}p_x}{\mathrm{d}z} = \frac{1}{2}\frac{\partial \varepsilon}{\partial x}, \qquad \frac{\mathrm{d}y}{\mathrm{d}z} = p_y, \qquad \frac{\mathrm{d}p_y}{\mathrm{d}z} = \frac{1}{2}\frac{\partial \varepsilon}{\partial y} , \tag{6.2.18}$$

which should be solved subject to the initial conditions:

$$x^0 = \alpha_1, \qquad y^0 = \alpha_2, \qquad p_x^0 = \frac{\partial \psi^0(\alpha_1, \alpha_2)}{\partial x}, \qquad p_y^0 = \frac{\partial \psi^0(\alpha_1, \alpha_2)}{\partial y} .$$

The phase Ψ along a trajectory is found from the eikonal equation

$$\frac{d\psi}{dz} = \frac{1}{2}(p_x^2 + p_y^2 + \varepsilon - 1), \qquad p_x = \frac{\partial\psi}{\partial x}, \qquad p_y = \frac{\partial\psi}{\partial y},$$

corresponding to the parabolic equation (6.2.17) with initial conditions given by the equation $\psi^0 = f(\alpha_1, \alpha_2)$. The amplitude $w(x, y, z)$ equals $U^0(\alpha_1, \alpha_2)[\partial(x, y)/\partial(\alpha_1, \alpha_2)]^{-1/2}$.

By way of example, consider a plane diffraction problem for a wave passing through a one-dimensional screen $z = 0$, $x = \xi$ into the free half-space $z > 0$. Dropping the derivation we give only the final results borrowed from [6.2].

Let $\alpha_1 \equiv \xi$ be a point on the screen. It can be demonstrated that in free space $\mathscr{J} \equiv \partial x/\partial\xi = 1 + zf''(\xi)$. for $\mathscr{J} \neq 0$ the field is representable by the ray expression

$$w(x, y) = \frac{U^0(\xi)\exp\{ik[f(\xi) + (x - \xi)^2/2z]\}}{\sqrt{1 + zf''(\xi)}}, \tag{6.2.19}$$

where $\xi = \xi(x, z)$ is determined from the equation $x = \xi + zf'(\xi)$, and in the neighborhood of caustics ($\mathscr{J} \approx 0$) the field may be described by the Maslov integral

$$w(x, z) = \sqrt{\frac{ik}{2\pi}}\int\frac{U^0(\xi)}{f''(\xi)}\exp\left\{ik\left[f(\xi) - \frac{1}{2}zp_x^2 + (x - \xi)p_x\right]\right\}dp_x, \tag{6.2.20}$$

where $\xi = \xi(p_x)$ is given by the equation $f'(\xi) = p_x$.

Integrating (6.2.20) with respect to ξ we obtain

$$w(x, z) = \sqrt{\frac{ik}{2\pi}}\int U^0(\xi)\sqrt{f''(\xi)}\exp\left\{ik\left[f(\xi) - \frac{1}{2}zf'^2(\xi) + (x - \xi)f'(\xi)\right]\right\}d\xi. \tag{6.2.21}$$

It is not hard to trace out a relation between (6.2.12, 13) and the formulas (6.2.20, 21) for the field $w = u\exp(kz)$. The latter follow from the former for $|f'| \ll 1$, i.e., under conditions of paraxiality.

In a homogeneous medium the parabolic equation has an exact solution:

$$w(x, z) = \sqrt{\frac{k}{2\pi iz}}\int U^0(\xi)\exp\{ik[f(\xi) + (x - \xi)^2/2z]\}\,d\xi. \tag{6.2.22}$$

Comparing the approximate formulas (6.2.20, 21) with the exact solution (6.2.22) one may draw conclusions similar to that quoted at the end of Sect. 6.2.2.

6.2.4 Miscellaneous Problems

Arnold [6.3] and *Ziolkowski* and *Deschamps* [6.4] used the Maslov approach to solve a number of problems. Specifically, Arnold considered the behavior of the field near a caustic cusp. Results of this analysis were, of course, the same as in the analysis based on the Kirchhoff integral; therefore, we avoid giving them in this text. For analyses of the cusp and more complex catastrophes, the reader is referred to [6.9, 10].

Quite useful evidence for ionospheric propagation problems was obtained by *Lukin* and *Palkin* [6.7, 11]. Numerical methods of field evaluation by Maslov's formalism deserve special mention. The monographs of *Lukin* and *Palkin* [6.7] and of *Kryukovsky* et al. [6.12] are completely devoted to such numerical schemes.

7 Method of Interference Integrals

The method of interference integrals devised by Yu.I. Orlov represents the wave field as a superposition of partial wavelets, each of which satisfies the equation of geometrical optics while the sum describes the caustic field. In the particular case when the momentum is chosen as a parameter of the family of wavelets, Orlov's method becomes Maslov's method. This chapter also considers Orlov's interference integrals in which partial wavelets are caustic fields (Airy asymptotics or more complex expansions).

7.1 Ray Type Integrals

7.1.1 Wide and Narrow Sense Interpretations

In wide sense any integral superposition of wave fields may be called interference integral. Most popular field representations are those in the form of Kirchhoff integral (superposition of spherical waves), in the form of a Fourier integral (superposition of plane waves), and in the form of cylindrical and spherical harmonics. Specific interference integrals are Maslov's integral representations (summation is over the projections of momentum), superpositions of Gaussian beams, and integral superpositions of quasi-plane WKB waves or Airy asymptotics in layered media.

The narrow-sense interpretation suggested by Yu.I. Orlov relates to superpositions of an infinite number of ray (eikonal) partial waves of the form

$$u_\alpha(\mathbf{r}) \equiv u(\mathbf{r}, \alpha) = U(\mathbf{r}, \alpha) e^{ikL(\mathbf{r}, \alpha)} , \tag{7.1.1}$$

where $L(\mathbf{r}, \alpha)$ is the full integral of the eikonal equation

$$(\nabla L)^2 = \varepsilon(\mathbf{r}) , \tag{7.1.2}$$

that is the family of solutions to equation (7.1.2) depending on a parameter α, and $U(\mathbf{r}, \alpha)$ is the solution of the transport equation

$$2\nabla U \cdot \nabla L + U \Delta L = 0 . \tag{7.1.3}$$

Since the partial wave (7.1.1) approximately satisfies the wave equation

$$\nabla^2 u_\alpha + k^2 \varepsilon(\mathbf{r}) u_\alpha = 0 , \tag{7.1.4}$$

the superposition

$$u(\mathbf{r}) = \int u_\alpha(\mathbf{r})\,\mathrm{d}\alpha = \int U(\mathbf{r},\alpha)\mathrm{e}^{ikL(\mathbf{r},\alpha)}\,\mathrm{d}\alpha \tag{7.1.5}$$

of the partial waves (7.1.1) also will be an approximate solution of (7.1.4). This superposition of waves is the interference integral due to *Orlov* [7.1]. To be more specific, equation (7.1.5) is an interference integral of the ray type, unlike the caustic type of interference integral [7.1] which uses in place of the eikonal solutions (7.1.1) the superpositions of caustic fields. Different facets of the theory of interference integrals are highlighted in [7.2–7].

Following the early works of *Orlov* [7.1, 8], *Vainstein* et al. developed the method of virtual rays [7.9–11] whose principal idea – the representation of a wave field as a superposition of quasi-ray fields – is close to that of the method of interference integrals. The difference between the two methods consists in the choice of rays and the associated partial waves.

Very close uniform representations of the field were reported by *Arnold* [7.12, 13] who exploited the spectral approach and operated with the term "oscillatory integral".

In this chapter, relaying upon the materials of a review paper [7.7], we wish to expound on the key ideas of the method of interference integrals and expose the basic evidence gathered with problems of diffraction and propagation of waves in inhomogeneous media.[1]

7.1.2 Eikonals and Amplitudes of Partial Waves

To make a neat derivation we limit ourselves to the two-dimensional problem with a single parameter in the solution (7.1.5). A three-dimensional problem would call for two parameters, α_1 and α_2. A choice of parameter α is decided by the problem statement. Consider a particular case of two dimensions when the initial field

$$u(\mathbf{r})|_\Gamma = u^0(\xi) = U^0(\xi)\exp[ikL^0(\xi)] \tag{7.1.6}$$

is specified on a curve Γ described by the equation $\mathbf{r} = \mathbf{r}^0(\xi)$. If in the expansion (7.1.5) we use ξ as parameter α, then the solutions of the eikonal equations over which the decomposition is effected must satisfy on Γ the conditions

$$L|_{\mathbf{r}\in\Gamma} = L^0(\xi)\ , \tag{7.1.7}$$

$$\left.\frac{\partial L}{\partial \xi}\right|_{\mathbf{r}\in\Gamma} = 0\ , \tag{7.1.8}$$

which ensure the asymptotic (for $k \to \infty$) transition to the initial field (7.1.6) of the interference integral (7.1.5).

---

[1] One of the authors of this book (Yu.A. Kravtsov) expresses his gratitude to the authors – V.B. Avdeyev, A.V. Demin, A.P. Yarygin, and M.V. Tinin – for permission to use in this text a considerable portion of their paper [7.7].

It is worth noting that the conditions (7.1.7, 8) represent a more general problem statement in eikonal equation solving than the Cauchy problem which corresponds to the traditional method of geometrical optics. Recognizing that all solutions of the type (7.1.2) are contained in the full and general integrals of the eikonal equation, we look in turn at these two integrals. Let

$$L = L_1(\mathbf{r}, p) + b \tag{7.1.9}$$

be one of the infinite set of full integrals of equation (7.1.2) with b being a constant, and p a parameter of the family of solutions. In the case of separable variables, p may be the separation parameter.

The class of solutions (7.1.9) satisfying the conditions (7.1.7, 8) is given by

$$L(\mathbf{r}, \xi) = L_1[\mathbf{r}, p(\xi)] - L_1[\mathbf{r}^0(\xi), p(\xi)] + L^0(\xi) \ , \tag{7.1.10}$$

where $p(\xi)$ is defined from the stationarity condition

$$\frac{\partial L^0}{\partial \xi} - \frac{\partial L_1(\mathbf{r}^0(\xi), p)}{\partial \xi} = 0 \ . \tag{7.1.11}$$

The spectral density $U(\mathbf{r}, \alpha)$ may be obtained from (7.1.3) if we invoke the representation

$$\nabla L = \sqrt{\varepsilon}\mathbf{q}/q$$

and require that vector $\mathbf{q}$ should satisfy the condition $\operatorname{div}\mathbf{q} = 0$. Then the solution of the transfer equation (7.1.3) will be

$$U(\mathbf{r}, \xi) = C_0(\xi)\sqrt{q/\sqrt{\varepsilon}} \ , \tag{7.1.12}$$

where the constant C_0 depends in the general case on the integration variable in the decomposition (7.1.5). The vector $\mathbf{q}$ has been defined by Orlov [7.1, 8] and for the two-dimensional case is representable in the form

$$\mathbf{q} = \nabla\frac{\partial L}{\partial \xi} \times \boldsymbol{\eta} \ , \tag{7.1.13}$$

where $\boldsymbol{\eta}$ is a unit vector perpendicular to the plane of the problem.

In view of (7.1.10, 12, 13) the interference integral (7.1.5) takes on the form

$$u(\mathbf{r}) = \varepsilon^{-1/4}\int_\Gamma C_0(\xi)\sqrt{\partial\nabla L/\partial\xi} \times \exp[ik\{L[\mathbf{r}, p(\xi)] - L[\mathbf{r}^0(\xi), p(\xi)] + L^0(\xi)\}]\,d\xi \ . \tag{7.1.14}$$

The constant $C_0(\xi)$ will be determined from the boundary condition (7.1.6) using the asymptotic expansion of the integral (7.1.14) at a stationary point which exists by the condition (7.1.11). If we assume that the evaluated family of solutions (7.1.10) has a stationary point of first order on the initial curve Γ, i.e., $\partial^2 L/\partial\xi^2 \neq 0$ on Γ then

$$C_0(\xi) = U^0(\xi)\left(\frac{k\varepsilon^{1/2}}{2\pi i}\frac{\partial^2 L}{\partial \xi^2}\right)_\Gamma^{1/2} \left|\frac{\partial \nabla L}{\partial \xi}\right|_\Gamma^{-1/2} . \tag{7.1.15}$$

When the stationary point is of a higher order than the first, then $C_0(\xi)$ should be determined with the aid of respective formulas of asymptotic expansion of the integral (7.1.5).

Let us look now at the possibility of representing the interference integral (7.1.5) on the basis of the solutions contained in the general integral of the eikonal equation. By definition of the general integral [7.14] it can be derived from the complete integral by imposition of the constraint

$$\frac{\partial L}{\partial p} + \frac{\partial L}{\partial b}\frac{db}{dp} = 0 , \tag{7.1.16}$$

where b is an arbitrary function of p, and equation (7.1.16) defines parameter p as a function of the point $\mathbf{r}$ under investigation. Choosing

$$b(p) = L^0(\xi) - L_1[\mathbf{r}^0(\xi), p]$$

and substituting it in (7.1.16) yields

$$\frac{\partial L_1(\mathbf{r}, p)}{\partial p} - \frac{\partial L_1(\mathbf{r}^0(\xi), p)}{\partial p} = 0 . \tag{7.1.17}$$

This is the equation of a family of geometric-optical rays emanating from points $\mathbf{r}^0(\xi)$ of curve Γ with different values of parameter p.

If we express from (7.1.17) parameter p as a function $p[\mathbf{r}, \mathbf{r}^0(\xi)]$ and substitute it in the complete integral (7.1.9), then we obtain solutions entering the general integral of the eikonal equation (7.1.2)

$$L = L_1\{\mathbf{r}, p[\mathbf{r}, \mathbf{r}^0(\xi)]\} - L_1\{\mathbf{r}(\xi), p[\mathbf{r}, \mathbf{r}^0(\xi)]\} + L^0(\xi) . \tag{7.1.18}$$

This expression describes a family of integral curves depending on a parameter. It may be also viewed as another complete integral of the eikonal equation. Therefore, it may be used as a basis to construct a different type of interference integral in which expansion will be effected over another system of eikonal waves.

As an example, we take up the problem of a linear source in a homogeneous medium. If we take a circle of radius a as an initial curve Γ, and x^0 as parameter ξ (i.e., the x component of the initial vector $\mathbf{r}^0$) then to the family of plane waves there corresponds the eikonal

$$L(x, y, \xi) = \frac{x^0}{a}(x - x^0) + \frac{y^0(x^0)}{a}[y - y^0(x^0)] + a , \tag{7.1.19}$$

and to the family of cylindrical waves leaving different points of Γ there corresponds the eikonal

$$L(x, y, \xi) = \sqrt{(x - x^0)^2 + [y - y^0(x^0)]^2} + a , \tag{7.1.20}$$

where $y^0(x^0) = \sqrt{a^2 - (x^0)^2}$. The corresponding phase fronts of the partial waves are shown in Fig. 7.1a,b.

Clearly there exist infinitely many interference integrals of the ray type like an infinite number of complete integrals of the eikonal equation. From this

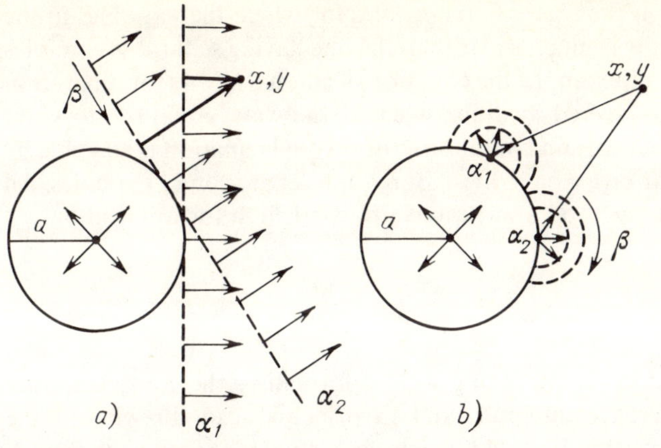

Fig. 7.1. Two-parameter family of virtual rays corresponding to wavelets with (**a**) plane and (**b**) cylindrical phase fronts

standpoint the ray type of interference integral should be treated as integral expansions of the desired field over different systems of partial waves whose phase fronts differ appreciably from the familiar plane, spherical or cylindrical phase surfaces.

Examples of complete integrals may be the integrals of the eikonal equation obtained by separation of variables in curvilinear systems of coordinates. In such circumstances the integration variable in (7.1.5) should be the parameter p of separation of variables in the equation of eikonal rather than parameter ξ of the initial curve Γ. The family of solutions to the eikonal equations over which expansion is effected will be determined from the conditions:

$$L = L^0(\xi), \quad \text{for } \mathbf{r} \text{ in } \Gamma , \tag{7.1.21}$$

$$\frac{\partial L}{\partial p} = 0, \quad \text{for } \mathbf{r} \text{ in } \Gamma , \tag{7.1.22}$$

which are similar to the conditions (7.1.7, 8). The interference integral obtained in this way will be equivalent to the expansion (7.1.14) if we change variables in the latter using the relation of the separation parameter p with the parameter ξ of curve Γ, $\xi = \xi(p)$, which follows from (7.1.11).

By way of example, we note that *Vainshtein* and *Ufimtsev* [7.11] have written the integral expansion of the Green function of a two-dimensional source (filament) in a homogeneous medium on the base of a complete integral of the eikonal equation in polar coordinates, whereas *Orlov* [7.8] has made use of another complete integral obtained by separation of variables in a Cartesian system of coordinates, and both expansions differ only in the type of partial waves.

A significant advantage of the method of interference integrals is that it operates with the equation of eikonal rather than directly with the wave equation. As it is known the class of $\varepsilon(\mathbf{r})$ profiles for which the variables in the eikonal equation separate is much wider than the one having separable variables in the wave equation. For example, the equation of eikonal allows for separation of variables in a wedge-layered medium, whereas the wave equation does not [7.15]. In addition, under certain conditions, the eikonal equation is solvable by the method of perturbations, which does not require separation of variables. All these features make the method an attractive poroposition for applications.

7.1.3 Virtual Rays

The algorithm outlined in the preceding section formalizes the construction of the ray type of interference integrals, and in principle, can do without the concept of virtual rays used in [7.9–11]. However, virtual rays may be incorporated in this case as well as the rays corresponding to all feasible phase fronts of partial waves in the expansion (7.1.5). The whole set of virtual rays may be labeled by two indexes, say α and β, the former showing the number of a partial wave (wave front) and the latter numbering rays in the phase front of a given partial wave.

Some properties of the two-parametric family of virtual rays are as follows:

(i) From the continuum of virtual rays only a finite number are "real rays", i.e., the ones leaving the source and coming into the point of observation. To these rays there correspond stationary points defined by the condition $\partial L/\partial\alpha = 0$. Thus, the point of observation is hit by infinitely many virtual rays (one or a few rays for each phase front from the continuum of the phase fronts corresponding to the superposition of partial waves), but only a finite number originates from the source, i.e., belongs to the class of "real rays".

(ii) Given a one-parametric family of rays corresponding to a fixed value of α, the point of observation is hit by a finite number of rays none of which may turn out to be a "real ray" emitted by the source.

(iii) The source emits an infinite set of real rays each of which belongs to some or other subfamily of virtual rays, i.e., to one or another partial wave.

It should be noted that *Vainshtein* et al. [7.9, 11] have introduced virtual rays in a different way than in the method of interference integrals, namely, these rays satisfy the ray equations over the entire ray path, and the corresponding phases depending on a parameter need not necessarily be solutions of the eikonal equation.

The concept of virtual rays possesses a tractability fundamental to geometrical optics and allows for a rather simple account of boundary conditions by using for these rays the common laws of reflection of geometrical optics. This approach has been employed, for example, in solving a problem of diffraction on a wedge [7.11]. Also, virtual rays take into account the radiation pattern of transmitting and receiving antennas by means of an "instrument function" [7.9, 10].

By way of illustration Fig. 7.1a shows a two-parameteric family of virtual rays corresponding to partial waves with plane phase front (7.1.9), and the diagram of Fig. 7.1b shows a family for the cylindrical partial waves (7.1.20).

7.1.4 Specific Problems

By the time of this writing a number of problems of diffraction and propagation of waves in inhomogeneous media have been solved by the method of interference integrals [7.2–7, 9–11].

(a) Plane-stratified Medium

An example of interference integral may be the function of a point source embedded at $(0, z^0)$ in a medium with monotonously falling permittivity $\varepsilon(z)$, viz.,

$$G(x, z) = \frac{1}{4\pi\mathrm{i}} \frac{\exp(\mathrm{i}kL') + \exp(\mathrm{i}kL'' + \mathrm{i}\pi/2)}{[(\varepsilon - \alpha^2)(\varepsilon^0 - \alpha^2)]^{1/4}} \, d\alpha \,, \tag{7.1.23}$$

where

$$L' = -\alpha x - \int_{z^0}^{z_<} \sqrt{\varepsilon - \alpha^2} \, dz, \qquad L'' = -\alpha x - \left(\int_{z^0}^{\bar{z}} + \int_{z_>}^{\bar{z}} \right) \sqrt{\varepsilon - \alpha^2} \, dz,$$

$z_< = \max(z^0, z)$, $z_> = \min(z^0, z)$, $\varepsilon^0 = \varepsilon(z^0)$, and $\bar{z}$ is the turning point defined by $\varepsilon(\bar{z}) = \alpha^2$. The parameter of expansion α is the sine of the angle of incidence, $\alpha = \sin\theta$.

(b) Reflection from a Layer with Tunneling

If $\varepsilon(z)$ reaches a minimum $\varepsilon_m > 0$ at $z = z_m$, then the point source function is representable as a sum $G = G_1 + G_2$ with

$$G_1 = \frac{1}{4\pi\mathrm{i}} \int_{-\sqrt{\varepsilon_m}}^{\sqrt{\varepsilon_m}} \frac{\exp[\mathrm{i}kL'(z)]}{[(\varepsilon - \alpha^2)(\varepsilon^0 - \alpha^2)]^{1/4}} \, d\alpha \tag{7.1.24}$$

describing the contribution of virtual rays which have not been turned ($|\alpha| \leq \sqrt{\varepsilon_m}$), and

$$G_2 = \frac{1}{4\pi\mathrm{i}} \left(\int_{-\infty}^{-\sqrt{\varepsilon_m}} + \int_{\sqrt{\varepsilon_m}}^{\infty} \right) \frac{\exp(\mathrm{i}kL') + \exp(\mathrm{i}kL'' + \mathrm{i}\pi/2)}{[(\varepsilon - \alpha^2)(\varepsilon^0 - \alpha^2)]^{1/4}} \, d\alpha \tag{7.1.25}$$

corresponding to rays reflected from the layer. The notation for L' and L'' is essentially the same as in (7.1.23).

(c) Radially Inhomogeneous Medium

The interference integral for the point source function has the form

$$G = \frac{1}{\pi} \sum_{m=-\infty}^{\infty} \int_0^R \cos\left(k \int_{\bar{r}}^{r_<} \frac{\sqrt{R^2 - \alpha^2}}{r} dr\right) \{(R^2 - \alpha^2)[(R^0)^2 - \alpha^2]\}^{-1/4}$$
$$\times \exp\left(-ik \int_{\bar{r}}^{r_>} \frac{\sqrt{R^2 - \alpha^2}}{r} dr - \frac{i\pi}{4}\right) \cos[\alpha(\phi - \phi^0) + 2\pi m] d\alpha \ , \quad (7.1.26)$$

where $R = r\sqrt{\varepsilon(r)}$, $R^0 = r^0\sqrt{\varepsilon(r^0)}$, and (r^0, ϕ^0) are the coordinates of the point source, $r_> = \max(r, r^0)$, $r_< = \min(r, r^0)$. The expansion parameter is taken to be $r \sin\phi$. A turning point $\bar{r}$ is determined from the condition $\bar{r}^2 \varepsilon(\bar{r}) - \alpha^2 = 0$. Calculations with these formulas have been reported by *Orlov* and *Demin* [7.2, 3].

The phase fronts and corresponding virtual rays for one of the partial waves with fixed α are shown in Fig. 7.2. This family of phase fronts is associated with a circular caustic of radius $\bar{r}(\alpha)$. Thus, the field (7.1.26) is a superposition of waves having circular caustics.

(d) Two-Dimensionally Inhomogeneous Medium

As will be recalled such media are hard to analyze analytically. *Avdeyev* et al. [7.16, 17] have constructed an interference integral for a 2-D inhomogeneity with the profile $\varepsilon = 1 - b^2 f(\theta)/r^2$. When this inhomogeneity is irradiated by a plane wave propagating along the symmetry axis of the inhomogeneity, the full

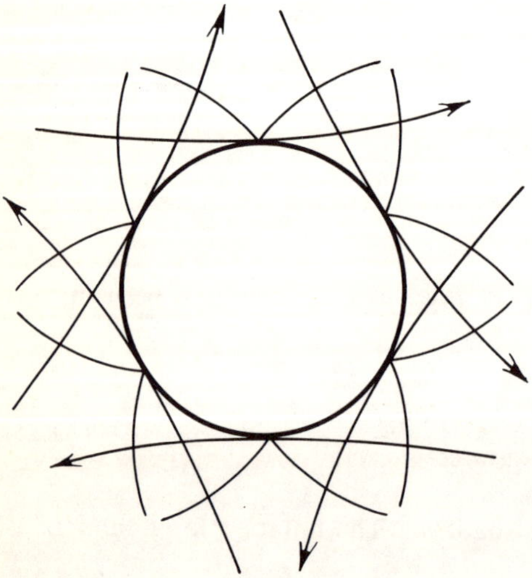

Fig. 7.2. Rays and phase fronts corresponding to a circular caustic in an axially symmetric inhomogeneous medium

3-D field is given by the integral

$$U(r) = \frac{2\exp(i\pi/4)}{\sqrt{kr\sin\theta}} \int_0^\infty \nu\mu^{-1/2} e^{-i\pi\mu/2} B(\mu, kr)$$

$$\times \cos\left[\mu(\theta-\theta_p) - \pi/4 + \int_0^{\theta_p} \sqrt{\mu^2 - k^2b^2 f(x)}\, dx\right] d\nu \ , \qquad (7.1.27)$$

where $\mu = \sqrt{k^2b^2 + \nu^2}$, $f(\theta)$ is a smooth, monotonously decreasing function of angle θ speaking at $f(0) = 1$ and having a prime zero at θ_p, $\mathbf{r} = \{r, \theta, \phi\}$ is the observation point, and

$$B(\mu, kr) = \sqrt{2/\pi kr} \cdot \cos(kr - \pi\mu/2 - \pi/4)$$

is the asymptotic of the Bessel function of the first kind, $J_\mu(kr)$, for $\mu \gg kr$. Numerical analyses with this formula have been outlined in [7.16, 17].

7.2 Caustic Integrals

7.2.1 Airy Function Based Integrals

If virtual rays form simple caustics, it would be natural to replace the WKB solutions having singularities at turning points with the Airy asymptotics which are finite everywhere. This procedure has been widely used in various problems. Suffice it to mention the results of *Brekhovskikh* [7.18] gathered in an analysis of reflection from a layered medium, and the approach of *Ludwig* [7.19] in which a smooth body is treated as a caustic of diffraction rays and the field is described as a superposition of Airy functions. However, a general treatment of the problem is due to *Orlov* [7.8].

For the reflecting slab considered in part (**a**) of Sect. 7.1.4a, in place of the ray integral (7.1.25) we obtain the caustic interference integral

$$G(x, z) = \frac{1}{2\pi} \int_{-\infty}^{\infty} W(\alpha, z_<) V(\alpha, z_>) e^{-ik\alpha x} d\alpha \ , \qquad (7.2.1)$$

where V and W are asymptotic solutions of the one-dimensional wave equation related to the Airy functions $\mathrm{Ai}(t)$ and $G(t) = \mathrm{Ai}(t) + i\,\mathrm{Bi}(t)$:

$$V(x, z) = C\left(\frac{t}{\varepsilon - \alpha^2}\right)^{1/4} \mathrm{Ai}(-t), \qquad W(t) = C\left(\frac{t}{\varepsilon - \alpha^2}\right)^{1/4} G(-t) \ ,$$

and

$$t = \begin{cases} \left(\frac{3}{2}\int_z^{\bar{z}} \sqrt{\varepsilon - \alpha^2}\, dz\right)^{2/3} , & \text{for } z < \bar{z} \ , \\ -\left(\frac{3}{2}\int_{\bar{z}}^{z} \sqrt{\alpha^2 - \varepsilon}\, dz\right)^{2/3} , & \text{for } z > \bar{z} \ . \end{cases}$$

Similar alterations may be introduced in the interference integral (7.1.26) for a radially inhomogeneous medium and in the integral (7.1.27) for a 2-D inhomogeneous medium. In the latter case, one may use as $B(\mu, kr)$ not only the Airy asymptotic $B(\mu, kr) \approx (2/\mu)^{1/3}\,\mathrm{Ai}[(\mu - kr)(2/\mu)^{1/3}]$ both for $\mu < kr$ and for $\mu > kr$, but also the rigorous solution of the radial part of the Helmholtz equation $B(\mu, kr) = J_\mu(kr)$, where J_μ is the Bessel function of the first kind [7.16].

7.2.2 Use of Miscellaneous Special Functions

For a layer whose dielectric permittivity has a minimum, the WKB solution in the integrand of (7.1.24, 25) may be replaced with asymptotic expressions relying on Weber functions which describe the tunnel effect. This approach has been implemented by *Tinin* [7.20] to obtain the bounds for the applicability domain of the geometrical optics approximation and some numerical data.

In principle one may develop a field representation in terms of more complex special functions associated with caustics. Similar to the functions of Airy, Bessel, and Weber such functions may themselves be represented with the aid of parametric integrals, see Chap. 4, so that multidimensional interference integrals would be an appropriate term. The main advantage of such integrals is that they allow specifying initial conditions in the form of caustic special functions, i.e., in a more complex form than the ray type of interference integrals admit.

7.2.3 Specific Problems

Orlov and *Demin* [7.2, 3] have compared the evidence gathered with the method of interference integrals with the results of the exact solution and geometric-optical method. Figure 7.3 shows the radiation pattern $|F(\phi)|$ of a filament with coordinates $r = r^0$, $\phi^0 = \pi$ calculated in the presence of a radially inhomogeneous cylinder of permittivity

$$\varepsilon(r) = \begin{cases} 1 - (R_0^2/r^2)(R^2 - r^2)/(R^2 - R_0^2), & \text{for } r < R\ , \\ 1, & \text{for } r > R\ , \end{cases}$$

where R_0 is the radius of $\varepsilon = 0$. At $kr^0 = 11$, $kR = 10$ and $kR_0 = 2, 5$ and 6 the ray and caustic integrals are seen to coincide with the exact solution accurate to 5% and 1%, respectively, that is, the efficiency of the method of interference integrals is pretty high.

In some problems of small electrical size, $ka \ll 1$, where the asymptotic formulas of ray and caustic integrals would seem to be invalid, the method yields quite satisfactory results. By way of example, Fig. 7.4 shows relative values of the full field $|U(r)/U_0|$ along the axis $\phi = 0$ in the region of shadow that occurs in irradiating an unlimited cylinder of $\varepsilon(r) = 1 - R_0^2/r^2$ by a plane wave incident from infinity to the right. At $kR_0 = 1$, i.e., at a relatively small value of the

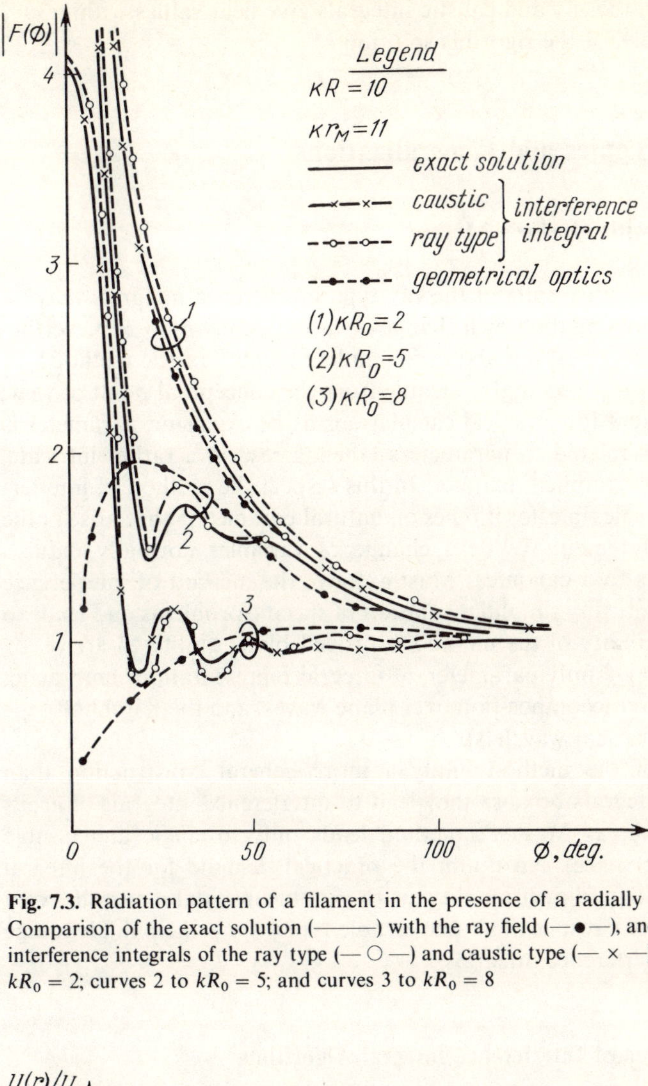

Fig. 7.3. Radiation pattern of a filament in the presence of a radially inhomogeneous cylinder. Comparison of the exact solution (———) with the ray field (—●—), and solutions in the form of interference integrals of the ray type (— ○ —) and caustic type (— × —). Curves 1 correspond to $kR_0 = 2$; curves 2 to $kR_0 = 5$; and curves 3 to $kR_0 = 8$

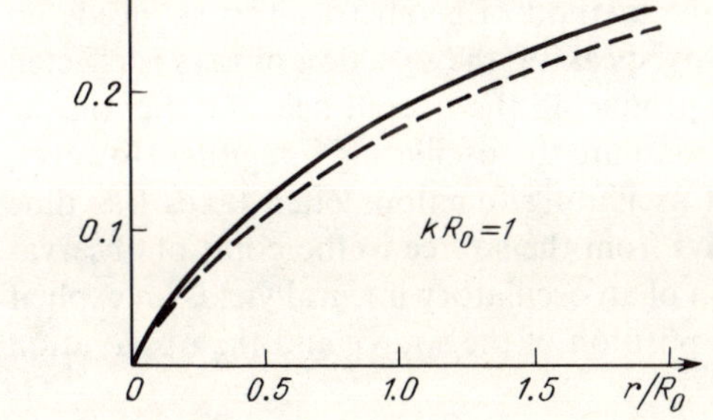

Fig. 7.4. Penetration of a plane wave field beyond a plasma cylinder of permittivity $\varepsilon(r) = 1 - (R_0/r)^2$; (———) exact solution, (– – –) caustic interference integral

electrical dimension, the ray and caustic integrals give field values within 30% and 20%, respectively, of the rigorous solution.

7.3 Additional Topics and Generalizations

7.3.1 Comparison with Maslov's Method

In a formal approach all results of the ray-type interference integrals may be obtained from Maslov's method by a change of variables, say $p_x \to \alpha$. Nevertheless, there is no way of speaking about a complete identity of the methods.

Maslov's method is exceedingly versatile from the conceptual point of view, but it is not convenient for practical calculations as its extension parameter is momentum which is related to parameters labeling rays in a rather intricate, more often than not – implicit, manner. In this respect the method of interference integrals is more flexible, for it relies on natural parameters that arise at the stage of problem statement. While a change of variables normally reduces interference integrals to a canonical Maslov form, the method of interference integrals often dramatically simplifies solution of specific problems and leads to the goal faster. Flexibility of the method is proved by its ability to arrive, by choice of a virtual-ray-family parameter, at integral representations embracing both the Fourier type (decomposition over plane waves) and the Kirchhoff type (superposition of spherical wavelets).

Caustic forms of the method imply a more general construction than Maslov's canonic integrals because they lead to interference integrals of arbitrary multiplicity whereas Maslov's method leads only to single and double integrals (in a three space). No doubt the practical demand for the integral representations of high dimensionality is by no means high, but the very possibility to embrace caustics of all perceivable types is a principal advantage of the method of interference integrals.

7.3.2 Implementation of Interference-Integral Algorithms

The key feature of the method is that it, like Maslov's method, yields a uniform asymptotic of the field and unlike the method of geometrical optics needs no preliminary "aiming" of rays. Roughly speaking the selection of rays is effected by the interference of partial waves producing the overall field. To pay off the freedom of such aiming one has to compute the oscillating integrals. However, the computation of the integrals of oscillating functions often takes less time than multiple runs and aiming of rays from the source to the point of observation. Moreover, the direct calculation of an oscillatory integral yields an explicit dependence of field intensity on the position of the source and the observation point.

To conclude this topic a few words will be in order on the comparison of the method with the summation of Gaussian beams see, e.g., the review paper [7.21]. Clearly the said method is at an advantage when the field at a given point is formed by a comparatively few beams, i.e., for not too complex caustics. On the other hand, the method of Gaussian beams involves a preliminary analysis of ray paths in much the same way as does the method of geometrical optics. Therefore, in adopting an algorithm of field computations one should seek a trade-off between the speed and accuracy of the computation.

7.3.3 Applicability Limits

As with Maslov's method, applicability of the method of interference integrals is controlled by two factors: (i) applicability of the method of geometrical optics for the major part of partial waves describing the resultant field, and (ii) the condition on virtual constancy of the amplitude $U(\mathbf{r}, \alpha)$ over the Fresnel interval $\Delta\alpha_f$ making the largest contribution in the resulting field. Note that while the general principles are quite clear, a detailed analysis of applicability of this method is yet to be done.

7.3.4 Some Generalizations

A generalization of the method of interference integral to the electromagnetic field can be effected by comparatively simple means: the partial wavs are now the vector, rather than sclar, fields satisfying the equations of geometrical optics for EM waves, e.g., see Sect. 2.1.

The generalization of the method on random inhomogeneous media is not that simple. If the eikonal of a partial wave is expanded in a series in small perturbations of medium parameters $\tilde{\varepsilon} = \varepsilon - \langle \varepsilon \rangle$ about the mean value $\langle \varepsilon \rangle$, then the resultant field (7.1.5) is capable of describing strong focusing of the field and thus offers a new way to handle strong fluctuations of fields in random inhomogeneous media. Specific results on strong fluctuations may be found in [7.5, 6] and in the review [7.7].

8 Penumbra Caustics

Inserting an opaque screen to obscure the passage of rays gives rise to penumbra caustics. In this case the standard integrals are incomplete Airy functions and their extensions. A close type of caustics is formed by the diffracted rays in the region of a diffraction penumbra. Both types of caustic, treated first by Yu.I. Orlov, correspond to the so-called edge catastrophes.

8.1 Broken Penumbra Caustics

8.1.1 Broken Caustics in Diffraction at Screens

Suppose that a system of rays forms a simple A_2 caustic. If the passage of rays is obstructed by a curvilinear wedge, as shown in Fig. 8.1a, then the caustic will be broken at the boundary between light and shadow. In the light region, the field is described by the sum of three ray fields one of which is associated with the edge diffraction rays leaving the edge of the wedge ($u_{\mathrm{d}} = U_{\mathrm{d}} \exp(ik\psi_{\mathrm{d}})$, and the two others, u_1 and u_2, are due to the primary wave that forms the caustic,

$$u(\mathbf{r}) = u_{\mathrm{d}}(\mathbf{r}) + u_1(\mathbf{r}) + u_2(\mathbf{r}) = U_{\mathrm{d}} \exp(ik\psi_{\mathrm{d}}) + \sum_{j=1}^{2} U_j \exp(ik\psi_j) \ . \qquad (8.1.1)$$

This formula is no longer valid in the vicinity of the light/shadow interface (here, $U_{\mathrm{d}} \to \infty$) and near the caustic where $U_{1,2} \to \infty$. In the place of a break of the caustic, both inapplicability domains of ray theory overlap.

In order to construct a uniform field asymptotic which would be equally valid near the lit-shadow boundary and near the caustic *Orlov* [8.1] has suggested to use an incomplete Airy function

$$I(\zeta, \eta) = \int_{\eta}^{\infty} \exp[i(t\zeta + t^3/3)]\, dt \qquad (8.1.2)$$

and its partial derivatives $\partial I/\partial\zeta$ and $\partial I/\partial\eta$. A motivation for this step has been that in the Kirchhoff approximation the screen acts as if it "cuts off" a part of the field associated with the simple caustic and the remaining part allows for a three-ray interpretation similar to (8.1.1). Two ray components are the result of application of the method of stationary phase to (8.1.2). These are waves

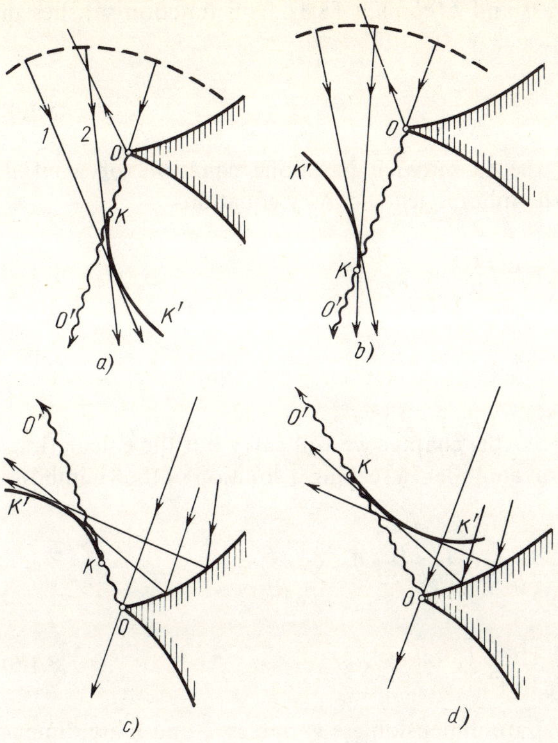

Fig. 8.1. Wave incident on a curved wedge forms a simple penumbra caustic (edge B_3 catastrophe): (**a, b**): incident rays; (**c, d**) for reflected rays

corresponding to the incident ray and the ray reflected from the caustic. The third component is due to the end effect and can be computed by the method of steepest descent. In the final analysis, the asymptotic of the integral (8.1.2) has the form

$$I(\zeta,\eta)=\mathrm{i}\sqrt{\pi}(-\zeta)^{-1/4}\{\exp[(2\mathrm{i}/3)(-\zeta)^{3/2}+\mathrm{i}\pi/4]$$
$$-\exp[(-2\mathrm{i}/3)(-\zeta)^{3/2}-\mathrm{i}\pi/4]\}$$
$$+\mathrm{i}(\eta^2+\zeta)^{-1}\exp[\mathrm{i}(\eta\zeta+\eta^3/3)]\ ,$$
$$-\zeta\gg 1,\qquad \sqrt{-\zeta}-|\eta|\gg 1\ , \tag{8.1.3}$$

adequate to the asymptotic (8.1.1).

In the regions where both equations (8.1.1, 3) are inapplicable, $I(\zeta,\eta)$ possesses the properties of two simpler standard functions, namely, the Fresnel integral which describes the field near the light-shadow boundary far from the caustic and the Airy function near the caustic but far from the shadow boundary. The properties of incomplete Airy functions have been analyzed by *Levey* and *Felsen* [8.2, 3] and *Agrest* and *Maksimov* [8.4]. This function satisfies in particular the equation

$$\frac{\partial^2 I}{\partial \zeta^2} - \zeta I - \mathrm{i}\frac{\partial I}{\partial \eta} = 0 \tag{8.1.4}$$

which is a modification of the Leontovich parabolic equation, or what is physically the same, satisfy the inhomogeneous Airy equation

$$\frac{\partial^2 I}{\partial \zeta^2} - \zeta I = -\mathrm{i}\exp[\mathrm{i}(\eta\zeta + \eta^3/3)] \ . \tag{8.1.5}$$

8.1.2 A Uniform Asymptotic

To diversify the presentation in this chapter we will carry out the calculations with rational powers of the wavenumber in seeking a solution of the Helmholtz equation in the form

$$\begin{aligned} u(\mathbf{r}) = \frac{k^{1/6}}{2\sqrt{\pi}}\Bigg[& I(\zeta,\eta)\sum_{m=0}\frac{A_m}{(\mathrm{i}k)^m} + \frac{\mathrm{i}}{k^{1/3}}\frac{\partial I(\zeta,\eta)}{\partial \zeta}\sum_{m=0}\frac{B_m}{(\mathrm{i}k)^m} \\ & + \frac{\mathrm{i}}{k^{2/3}}\frac{\partial I(\zeta,\eta)}{\partial \eta}\sum_{m=0}\frac{C_m}{(\mathrm{i}k)^m}\Bigg] \mathrm{e}^{\mathrm{i}k\chi - \mathrm{i}\pi/4} \ . \end{aligned} \tag{8.1.6}$$

Here, $\zeta = k^{2/3}\tilde{\zeta}$ and $\eta = k^{1/3}\tilde{\eta}$ are dimensionless variables, $\tilde{\zeta}$ and $\tilde{\eta}$ are dimensional variables, and the quantities A_m, B_m, C_m, ζ, η, and χ are to be determined.

Substituting (8.1.6) in the Helmholtz equation and eliminating the higher derivatives of $I(\zeta,\eta)$ with the aid of equations (8.1.4, 5) we set to zero the coefficients of the linearly independent functions I, $\partial I/\partial\zeta$ and $\partial I/\partial\eta$ and that of the like powers of ik. In the lowest approximation in k, we arrive at the following equations for functions $\tilde{\zeta}$, $\tilde{\eta}$, and χ

$$(\nabla\chi)^2 - \tilde{\zeta}(\nabla\tilde{\zeta})^2 = \varepsilon(\mathbf{r}), \qquad (\nabla\zeta\cdot\nabla\chi) = 0 \ , \tag{8.1.7}$$

$$(\tilde{\eta}^2 + \tilde{\zeta})(\nabla\tilde{\eta})^2 + 2\tilde{\eta}(\nabla\tilde{\zeta}\cdot\nabla\tilde{\eta}) + 2(\nabla\chi\cdot\nabla\tilde{\eta})(\nabla\tilde{\zeta})^2 = 0 \ . \tag{8.1.8}$$

In the next approximation, we obtain equations for the amplitude factors of zeroth approximation A_0, B_0, and C_0 whose zero subscripts will be dropped for brevity, viz.,

$$\mathscr{L}_1(A) - \mathscr{L}_2(B) = 0 \ , \qquad \mathscr{L}_2(A) + \tilde{\zeta}\mathscr{L}_1(B) + (\nabla\tilde{\zeta})^2 B = 0 \ , \tag{8.1.9}$$

$$\begin{aligned} -\mathscr{L}_3(A) + \mathscr{L}_1(B) + \tilde{\eta}\mathscr{L}_3(B) + (\nabla\tilde{\eta})^2 B + \mathscr{L}_2(C) + \tilde{\eta}\mathscr{L}_1(C) \\ + (\tilde{\eta}^2 + \tilde{\zeta})\mathscr{L}_3(C) + 2C[(\nabla\tilde{\zeta}\cdot\nabla\tilde{\eta}) + \tilde{\eta}(\nabla\tilde{\eta})^2] = 0 \ , \end{aligned} \tag{8.1.10}$$

where the operators $\mathscr{L}_{1,2,3}$ are defined by

$$\mathscr{L}_j(f) = 2(\nabla f\cdot\nabla\lambda_j) + f\Delta\lambda_j \ , \quad j = 1, 2, 3 \ ,$$

with $\lambda_1 = \tilde{\zeta}$, $\lambda_2 = \tilde{\eta}$, and $\lambda_3 = \chi$.

Structurally equations (8.1.7, 8) are in many aspects analogous to the equations of caustic asymptotics (Chap. 5) and may be reduced to the equation of eikonal, and equations (8.1.9, 10) to the transfer equation for the amplitudes of ray fields, $U_{1,2}$. Therefore, in full agreement with Sect. 5.5.1, from (5.1.7, 9) we have

$$\tilde{\zeta} = -\left[\tfrac{3}{4}(\psi_1 - \psi_2)\right]^{2/3}, \qquad \chi = \tfrac{1}{2}(\psi_1 + \psi_2)\ , \tag{8.1.11}$$

$$A = (-\tilde{\zeta})^{1/4}(\mathrm{i}U_2 + U_1), \qquad B = (-\tilde{\zeta})^{-1/4}(\mathrm{i}U_2 - U_1)\ , \tag{8.1.12}$$

where the subscript "1" relates to the ray which has not yet touched on the caustic.

The substitutions

$$\tilde{\eta}\tilde{\zeta} + \tilde{\eta}^3/3 = \psi_\mathrm{d} - \chi$$

and

$$C = (A - \tilde{\eta}B)/(\tilde{\eta}^2 + \tilde{\zeta}) - 2\sqrt{\pi}\mathrm{e}^{-\mathrm{i}\pi/4}k^{1/2}U_\mathrm{d} \tag{8.1.13}$$

reduce (8.1.8) for $\tilde{\eta}$ and (8.1.10) for C, respectively, to the equation of eikonal for the phase of the diffraction (edge) ray and to the equation of transport for the amplitude U_d. Thus, the sought function $\tilde{\eta}$ is expressed in terms of the eikonal $\psi_{1,2}$ of the ray fields and the eikonal of the diffraction field ψ_d by means of the caustic equation (8.1.13). An appropriate root of this equation is chosen subject to the condition $\tilde{\eta} = -\sqrt{-\tilde{\zeta}}$ at $\psi_\mathrm{d} = \psi_1$, and subject to $\tilde{\eta} = \sqrt{-\tilde{\zeta}}$ at $\psi_\mathrm{d} = \psi_2$.

In view of (8.1.11–13) the leading term of the penumbra caustic asymptotic will be written as

$$u(\mathbf{r}) = \frac{1}{2\sqrt{\pi}}\,\mathrm{e}^{\mathrm{i}k\chi - \mathrm{i}\pi/4}\left(AI + \mathrm{i}B\frac{\partial I}{\partial \zeta}\right) + \hat{C}\mathrm{e}^{\mathrm{i}k\psi_\mathrm{d}}\ , \tag{8.1.14}$$

where $\tilde{\zeta}$, χ, A and B are determined by (8.1.1, 12), $\tilde{\eta}$ is obtained from the first equation in (8.1.13), and

$$\hat{C} = U_\mathrm{d} + \frac{\exp(\mathrm{i}\pi/4)}{2\sqrt{\pi}}\,\frac{(B\tilde{\eta} - A)}{(\tilde{\eta}^2 + \tilde{\zeta})}\ .$$

8.1.3 Particular Cases

With reference to Fig. 8.1a, near the light-shadow boundary OO′, except for the neighborhood of caustic end $K(-\zeta \gg 1)$, the uniform asymptotic (8.1.6) recasts into the formulas of penumbral asymptotic [8.5–8]. On the segment OK

$$u(\mathbf{r}) = u_d(\mathbf{r}) + u_1(\mathbf{r})\,\Phi\left[\pm\sqrt{k(\psi_\mathrm{d} - \psi_\mathrm{inc})}\right]\ , \tag{8.1.15}$$

and on the segment KO′

$$u(\mathbf{r}) = u_d(\mathbf{r}) + u_1(\mathbf{r})\,\Phi\left[\pm\sqrt{k(\psi_\mathrm{d} + \psi_\mathrm{inc})}\right] + u_2(\mathbf{r})\ , \tag{8.1.16}$$

where

$$\Phi(x) = \pi^{-1/2}\exp(-\mathrm{i}\pi/4)\left[\int_x^{\infty} \exp(\mathrm{i}\tau^2)\mathrm{d}\tau + \exp(\mathrm{i}x^2)/2\mathrm{i}x\right],$$

and the plus (minus) sign is taken in the shadow (light) region of the respective wave.

Near the caustic, but outside of the penumbra region ($\eta \to -\infty$) the asymptotic (8.1.6) becomes the formulas of the uniform Airy asymptotic

$$u(\mathbf{r}) = u_{\mathrm{d}}(\mathbf{r}) + \sqrt{\pi}\exp(\mathrm{i}k\chi - \mathrm{i}\pi/4)[A\,\mathrm{Ai}(\zeta) + \mathrm{i}B\,\mathrm{Ai}'(\zeta)] \,. \tag{8.1.17}$$

With some modifications these formulas may be adapted to the description of broken caustics of other geometries shows in Fig. 8.1b–d. The modifications in particular imply the use of the complex conjugate $I^*(\zeta, \eta)$ in place of $I(\zeta, \eta)$ in the cases depicted in b and d [8.1]. Applicability bounds for various expressions are defined by the Fresnel criteria.

Combining the expressions for refracted and reflected waves, one may obtain manageable expressions also for the full field in arbitrary direction with respect to the edge of the wedge [8.1].

8.1.4 A Uniform Asymptotic for an EM Field

Such an asymptotic has been constructed by *Orlov* and *Vlasov* [8, 9]. It differs from (8.1.6) in having vector quantities in place of the scalar amplitude factors and the arguments $\tilde{\zeta}$, $\tilde{\eta}$, and χ obey the former relationships (8.1.11–13). Results of this work may be employed for analysis of polarization of EM fields in any sector of the wedge.

8.1.5 Broken Caustics of Higher Dimension

An obstacle inserted in the way of a wave that forms an A_m caustic of $m > 2$ brings about a broken caustic of higher dimension. As an example, Fig. 8.2 shows a broken trace of a swallowtail. For this type of caustics *Orlov* [8.10] has constructed a uniform asymptotic based on incomplete Airy functions

$$I(\zeta, \eta) = \int_{\eta}^{\infty} \exp\left[\mathrm{i}\left(\frac{t^{m+1}}{m+1} + \sum_{p=1}^{m-1} \frac{t^p}{p}\zeta_p\right)\right]\mathrm{d}t \,. \tag{8.1.18}$$

This uniformly asymptotic expression contains a standard integral (8.1.18), its derivatives with respect to parameters ζ_p, and the derivative with respect to the lower limit η

$$u(\mathbf{r}) = \left(Ak^{-\alpha}I + i\sum_{p=1}^{m-1} k^{-p\alpha}B_p\frac{\partial I}{\partial \zeta_p} + \mathrm{i}C\frac{\partial I}{\partial \eta}\right)\mathrm{e}^{\mathrm{i}k\chi}\,, \tag{8.1.19}$$

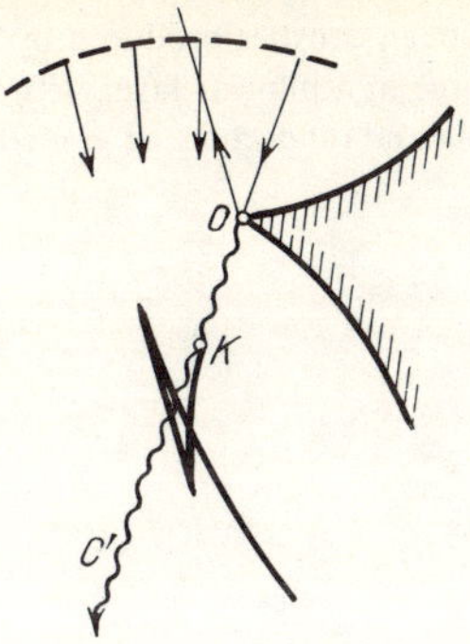

Fig. 8.2. Formation of a loop-like broken caustic (edge B_5 catastrophe) by rays diffracted at a wedge

where $\alpha = 1/(m+1)$. Letting $\zeta_p = \tilde{\zeta}_p k^{1-p\alpha}$ and $\eta = k^\alpha \tilde{\eta}$ we may derive for χ, $\tilde{\zeta}_p$, and $\tilde{\eta}$ and amplitude factors A, B_p, and C algebraic relations with the geometric-optical fields $u_j(\mathbf{r})$ and with the edge field $u_\mathrm{d}(\mathbf{r})$. Equation (8.1.19) becomes a caustic asymptotic far from the light-shadow boundary, and it becomes a penumbra asymptotic of the type (8.1.15) or (8.1.16) near line OO′ in Fig. 8.2 except for the neighborhood of the caustic.

8.1.6 Broken Caustics at Discontinuities of Phase-Front Curvature and Jumps of Refractive Index

If the radius of curvature of the phase front experiences a jump from R_1 to R_2 then this jump is reflected in the shape of the caustic as illustrated in Fig. 8.3. A similar phenomenon takes place in matching linear layers with different gradients of permittivity $\varepsilon(z) = n^2(z)$. A jump in the gradient of ε incurs a jumpwise variation of curvature radii of the rays and a break of the caustic. Two typical examples of rise and drop of $\partial\varepsilon/\partial z$ are presented in Fig. 8.4 borrowed from [8.11]. This kind of breaks of caustics often owes its origin not in jumps of $\partial\varepsilon/\partial z$ in the real medium, but rather in jumps caused by the approximation of

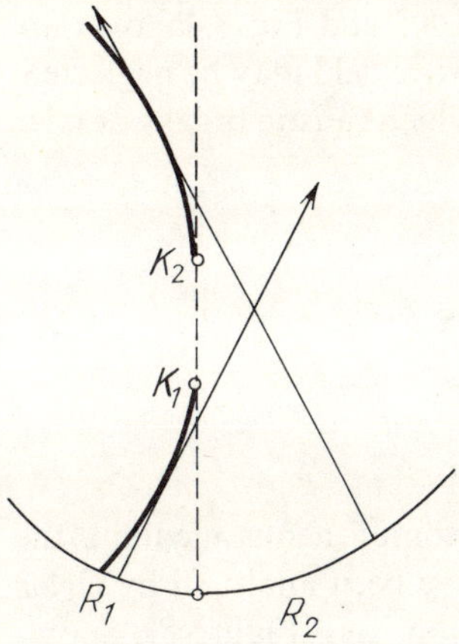

Fig. 8.3. Break of a caustic in a jumpwise variation of the phase-front curvature radius

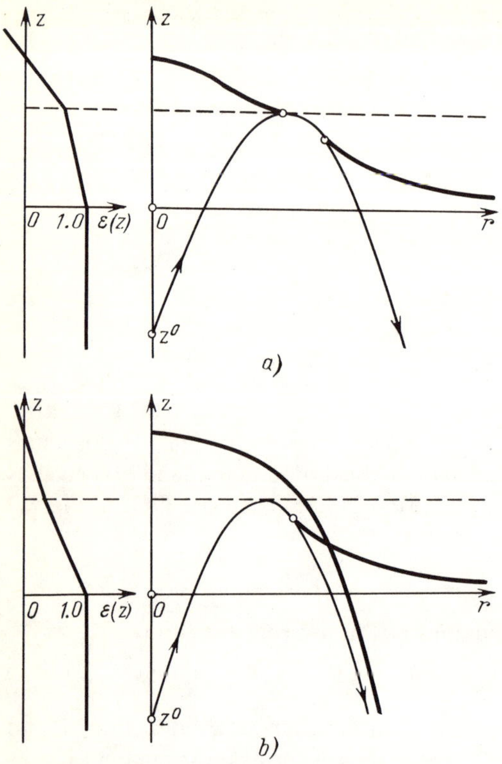

Fig. 8.4. Broken caustics produced by a point source in a bilinear layer with (**a**) increasing and (**b**) decreasing gradient

actual profiles of $\varepsilon(z)$ by layers with constant values of $\partial\varepsilon/\partial z$. In the last case one speaks of "false" caustics.

Standard integrals describing the field in the presence of such broken caustics may be derived by matching along OO′ the solutions induced by the diffraction at a half-plane, which with reference to Fig. 8.3, obstructs thc lcft half of the phase front in one case and then the right half in the other case.

Widths of caustic zones may be estimated by Fresnel considerations. If the caustic zones of two edge points, K_1 and K_2 in Fig. 8.3 and Fig. 8.4a, overlap, then the effect of the break on the magnitude of the wave field may be neglected. These considerations indicate the conditions under which caustic breaks or false caustics (Fig. 8.4b) should be taken into account.

8.2 Penumbra Caustics of Diffraction Rays

8.2.1 Generation of Caustics

If an obstruction, say a screen, is inserted in an inhomogeneous medium, the edge diffraction rays give birth to caustics which have been analyzed by *Orlov* [8.10, 12] and called penumbral caustics of diffraction rays. Figure 8.5 shows

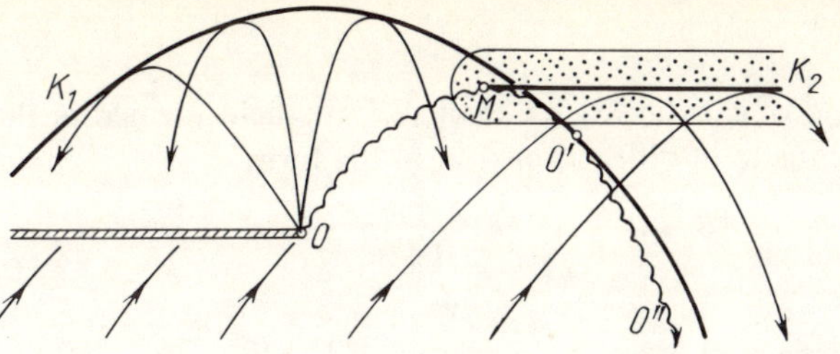

Fig. 8.5. Penumbral caustic of edge rays K_1 and broken caustic of incident rays K_2 (edge F_4 catastrophe). The boundary ray 00′0″ belongs to both configurations of rays

this kind of caustic for the edge of a screen inserted in a linear reflecting layer. The shadow part of the caustic is separated from the lit segment by point O′ belonging to the ray OO′O″ separating light from shadow. In addition, primary rays form a broken caustic, K_2, and in the vicinity of this caustic the field is described by an incomplete Airy function $I(\zeta, \eta)$.

If we exclude the neighborhood of the broken caustic (stippled in Fig. 8.5) from consideration, then the ray asymptotic of the field is represented by the contributions of two edge rays, u_{d1} and u_{d2}, and the contribution of the primary field u_{r1} and the field u_{r2} reflected from the caustic K_2, viz.,

$$u(\mathbf{r}) = \sum_{j=1}^{2} u_{d,j}(\mathbf{r}) + U_{r1}\exp[ik\psi_{r1}(\mathbf{r})] + U_{r2}\exp[ik\psi_{r2}(\mathbf{r})] \;. \tag{8.2.1}$$

A similar asymptotic behavior (least the field u_{r2}) is exhibited by the Airy–Fresnel function

$$V(\zeta, \eta) = \int_{-\infty + i\alpha^2}^{\infty + i\alpha^2} (t-\eta)^{-1}\exp[i(\zeta t + t^3/3)]\,dt \;. \tag{8.2.2}$$

It has been introduced by *Orlov* [8.10, 12] and is a generalization of the Airy function and like the incomplete Airy function (8.1.2), satisfies the Leontovich–Fock parabolic equation (8.1.4). For $-\zeta \gg 1$ (far from the caustic) and $\sqrt{-\zeta} - |\eta| \gg 1$ (far from the light-shadow boundary), we have approximately

$$V(\zeta, \eta) \cong -2\pi i\exp[i(\zeta\eta + \eta^3/3)] + \sqrt{\pi}(-\zeta)^{-1/4}$$
$$\times\left(\frac{\exp[i\tfrac{2}{3}(-\zeta)^{3/2} - i\pi/4]}{-\sqrt{-\zeta} - \eta} + \frac{\exp[-i\tfrac{2}{3}(-\zeta)^{3/2} + i\pi/4]}{\sqrt{-\zeta} - \eta}\right) . \tag{8.2.3}$$

This led *Orlov* [8.10, 12] to suggest $V(\zeta, \eta)$ as a standard integral for construction of a field asymptotic in the entire space least the neighborhood of caustic K_2.

8.2.2 Asymptotic Solution

Ignoring the neighborhood of caustic K_2 we shall seek the field formed in the presence of a penumbra caustic of diffraction rays in the form

$$u(\mathbf{r}) = \frac{k^{1/6}e^{ik\chi - i\pi/4}}{2\sqrt{\pi}}\left(AV + \frac{iB}{k^{1/3}}\frac{\partial V}{\partial \zeta} + \frac{iC}{k^{2/3}}\frac{\partial V}{\partial \eta}\right) + u_{r2}(\mathbf{r}) , \tag{8.2.4}$$

where A, B, and C are represented by series in powers of $1/ik$ like in (8.1.6). The field u_{r2} of the wave reflected from caustic K_2 will be forced to obey the equations of geometrical optics and will be kept in (8.2.4) in order that the general field structure should be represented.

Letting $\zeta = k^{2/3}\tilde{\zeta}$ and $\eta = k^{1/3}\tilde{\eta}$ we may obtain for the unknown functions ζ, $\tilde{\eta}$, and χ, and A_m, B_m, and C_m equations resembling (8.1.7–10). Like in the previous section these equations may be reduced to equations of geometrical optics for some combinations of $\tilde{\zeta}$, $\tilde{\eta}$, and χ and A, B, and C, so that the leading term of the asymptotic series will be

$$u(\mathbf{r}) = \frac{\exp(ik\chi - i\pi/4)}{2\sqrt{\pi}}\left[A_0V + iB_0\frac{\partial V}{\partial \zeta} + iC_0\frac{\partial V}{\partial \eta}\right] + u_{r2}(\mathbf{r}) , \tag{8.2.5}$$

where

$$\begin{aligned} A_0 &= (-\zeta)^{1/4}[iU_{d1}(-\eta - \sqrt{-\zeta}) + U_{d2}(-\eta + \sqrt{-\zeta})] , \\ B_0 &= (-\zeta)^{-1/4}[iU_{d1}(-\eta - \sqrt{-\zeta}) - U_{d2}(-\eta + \sqrt{-\zeta})] , \\ C_0 &= \frac{\exp(-i\pi/4)}{\sqrt{\pi}}\frac{1}{\eta^2 + \zeta}U_{r1} + (-\zeta)^{-1/4}(iU_{d1} - U_{d2}) . \end{aligned} \tag{8.2.6}$$

8.2.3 Properties of the Asymptotic Solution

Using power expansions in the neighborhood of the boundary ray OO′O″ consistent with $\eta = \pm\sqrt{-\zeta}$ and near the caustic of diffraction rays, $\zeta = 0$, one may verify that all amplitude factors A_0, B_0, and C_0 have no singularities there. In (8.2.6) the poles of the amplitudes $U_{d1,2}$ at the light-shadow boundary are compensated by the multipliers $-\eta \pm \sqrt{-\zeta}$, and the singularities of these amplitudes at the caustic $\zeta = 0$ by the factors $(-\zeta)^{\pm 1/4}$. Around point O′ the leading term in (8.2.5) is predominant.

In the vicinity of the shadow part of a diffraction-ray caustic, at a distance from the boundary point O′, i.e., for $|\eta| - \sqrt{-\zeta} \gg 1$, the formula (8.2.6) becomes the caustic asymptotic

$$\begin{aligned} u(\mathbf{r}) \approx{} & \sqrt{\pi}[(-\zeta)^{1/4}(iU_{d1} + U_{d2})(\zeta) + i(-\zeta)^{-1/4}(iU_{d1} - U_{d2})(\zeta)] \\ & \times \exp(ik\chi - i\pi/4) . \end{aligned} \tag{8.2.7}$$

In the neighborhood of the boundary ray, least the neighborhood of point O′, i.e., for $-\zeta \gg 1$, the penumbra asymptotics at the segments OO′ and O′O″ are respectively as follows:

$$u(\mathbf{r}) \cong u_{d1}(\mathbf{r}) + u_{d2}(\mathbf{r}) + u_{r} \cdot \Phi^{*}[\pm\sqrt{k(\psi_{r} - \psi_{d2})}] \ ,$$

$$u(\mathbf{r}) = u_{d2}(\mathbf{r}) + u_{d1}(\mathbf{r}) + u_{r}\Phi[\pm\sqrt{k(\psi_{d1} - \psi_{r})}] \ .$$

Thus, the solution (8.2.6) agrees with the formulas of the geometrical theory of diffraction [8.5–8].

Closer to caustic K_2 the component u_{r2} begins to grow without bound. If in the vicinity of a broken end of K_2, the components u_{r1} and u_{r2} may already be separated from the general solution (8.2.5), i.e., if the asymptotic (8.2.7) is already operable (to this end, in Fig. 8.5 point M on K_2 must leave the caustic zone of the edge rays of K_1) these functions may be used to construct with the aid of the incomplete Airy function $I(\zeta, \eta)$ a penumbra field consistent with the broken caustic K_2. If, on the other hand, point M finds itself inside the caustic zone of K_1, then in place of two simple special functions $I(\zeta, \eta)$ and $V(\zeta, \eta)$ one has to do with a more complex construction like an F_4 caustic to be considered below.

8.2.4 Some Generalizations

If a caustic of diffraction rays exhibits singularities, such as cusps, loops, etc., then their description should be effected with more complex Airy–Fresnel functions

$$V_m(\zeta, \eta) = \int_{-\infty}^{\infty} (t - \eta)^{-1} \exp\left(\mathrm{i} \sum_{p=1}^{m-1} \zeta_p \frac{t^p}{p} + \frac{t^{m+1}}{m+1}\right) \mathrm{d}t \qquad (8.2.8)$$

as suggested by *Orlov* [8.10]. Another construction, also initiated by this author, is an asymptotic for EM fields [8.13].

8.3 Penumbral Caustics and Edge Catastrophes

8.3.1 Simple Edge Catastrophes

As has been learned, the penumbral caustics are tied with the theory of edge catastrophes [8.14, 15]. The list of relevant singularities differs from that given in Chap. 3. Simple null-modal caustics are summarized in Table 8.1 as two infinite series B_{m+1} and C_{m+1} and an off-serial singularity F_4. Here, as before, the caustic indexes are due to the associated Lie groups. The table also shows the canonical form of generating function $\Phi(\zeta, \tau)$ which naturally occurs in the integral representations of the field.

Table 8.1. Classification of simple (null-modal) penumbra caustics

Type	m	$\Phi(\zeta, \mathbf{t})$	$\alpha_1, \ldots, \alpha_m$	σ_{foc}
B_{m+1}	≥ 2	$\pm \frac{1}{m+1} t_1^{m+1} \pm t_2^2 + \sum_{p=1}^{m} \frac{1}{p} \zeta_p t_1^p$	$\frac{m}{m+1}, \ldots, \frac{1}{m+1}$	$\frac{1}{2} - \frac{1}{m+1}$
C_{m+1}	≥ 2	$t_1 t_2 \pm \sum_{p=1}^{m} \frac{1}{p} \zeta_p t_2^p$	$\frac{m}{m+1}, \ldots, \frac{1}{m+1}$	0
F_4	3	$\pm t_1^2 + \frac{1}{3} t_2^3 + \zeta_1 t_1 + \zeta_2 t_2 + \zeta_3 t_1 t_2$	$\frac{1}{2}, \frac{2}{3}, \frac{1}{6}$	$\frac{1}{6}$

Simple penumbra caustics form at codimensions $m = 2$ and $m = 3$. Specifically, for $m \leq 3$, these are five types of caustics: B_3, C_3, B_4, C_4, and F_4. The B_2 type, equivalent to C_2, corresponds to a shadow without caustics. The series B_{m+1} and C_{m+1} are similar in the field phase structure to A_{m+1} and D_{m+1}, respectively.

For $m \geq 4$, unimodal, bimodal, and other penumbral caustics emerge beside their simple counterparts. For classification of these caustics the reader is referred to [8.14, 15] and to a group of publications [8.16–18] devoted to wave problems.

8.3.2 Typical Integrals of Edge Catastrophe Theory

Standard functions for the caustics of series B_{m+1} and C_{m+1} and the off-series caustic F_4 are the integrals

$$I(\zeta) = \int_0^\infty \mathrm{d}t_1 \int_0^\infty \mathrm{d}t_2 \exp[\mathrm{i}k\Phi(\zeta, t_1, t_2)] \, , \tag{8.3.1}$$

where $\Phi(\zeta, t_1, t_2)$ are the normal forms of generating functions of edge catastrophes displayed in Table 8.1.

For the B_{m+1} series, standard integrals of the type (8.3.1) are reducible to incomplete generalized Airy functions (8.1.18) which have been shown [8.1, 9, 10], prior to the development of edge catastrophe theory, to describe the field near broken caustics. For the C_{m+1} series, the integral (8.3.1) becomes a generalized Airy–Fresnel function (8.2.8). Such functions had been introduced in [8.10, 12, 13] to describe the field near penumbral caustics of diffraction rays also before catastrophe theory had emerged. The salient feature of integrals $V(\zeta, \eta)$ is that they describe weakly expressed focusing. Therefore, Table 8.1 shows for them a zero value of focusing index σ_{foc}.

The F_4 caustic combines, in certain sense, the properties of caustics B_3 and C_3. Specifically, the respective typical integral uniformly describes the field under the conditions when, with reference to Fig. 8.5, the end point M of caustic K_2 finds itself in the caustic zone of caustic K_1.

For edge catastrophes, the properties of typical integrals have been elucidated in great detail by *Kryukovskii* et al. [8.17, 18]. These papers give the plots of amplitude and phase reliefs of the typical integrals, describe evaluation techniques of global and local asymptotics for such integrals, and display dominant graphs for various caustics.

8.3.3 Angle Catastrophes

A screen with a cut-out sector inserted in the way of a wave brings about one more class of caustics – angular catastrophes. The respective class of typical integrals assumes integration over the quadrant $t_1 > 0, t_2 > 0$. Similar integrals occur, e.g., when a square aperture is inserted in an inhomogeneous medium. For respective properties, see [8.17, 18].

9 Modifications and Generalizations of Standard Integrals and Functions

The standard integrals derived in the catastrophe theory do not exhaust the list of standard functions suitable for a description of caustic fields. In this chapter we reflect on the immense possibilities which mathematics opens up to explore nature.

9.1 Nonpolynomial Phase Standard Integrals

9.1.1 Standard Integrals with Arbitrary Phase Functions

In addition to typical integrals of catastrophe theory discussed in Chaps. 4 and 5, descriptions of caustic fields invoke non-standard integrals and special functions. In this chapter we wish to delineate possibilities available in this respect.

Polynomials in the exponent of the standard integral (4.1.1) have been caused more by the convenience of classfication than by necessity. Suppose that the exponent contains a smooth function $\Psi(\zeta, \mathbf{t})$ having a desired number of stationary points coincident with the number of rays and being arbitrary in other respects. Clearly, a smooth change of variables can reduce the integral

$$I(\zeta) = (2\pi)^{-1/2} \int_{-\infty}^{\infty} \exp[\mathrm{i}\Psi(\zeta, \mathbf{t})]\, \mathrm{d}^l t \tag{9.1.1}$$

asymptotically to the integral (4.1.1) where $\Phi(\zeta, \mathbf{t})$ is an appropriate normal form of generating function having the same number of extrema as $\Psi(\zeta, \mathbf{t})$. Hence, the integrals (9.1.1) are suitable for the construction of asymptotics in much the same way as the standard integrals (4.1.1).

9.1.2 Uniform Asymptotics Based on Standard Integrals with Arbitrary Phase Functions

An asymptotic of this kind can be conveniently built by relying not upon the initial dimensionless integral (9.1.1), but rather upon the related integral

$$\tilde{I}(\tilde{\zeta}) = (k/2\pi)^{l/2} \int_{-\infty}^{\infty} \exp[ik\, \Psi(\tilde{\zeta}, \tilde{\mathbf{t}})]\, \mathrm{d}^l \tilde{t} \tag{9.1.2}$$

which in the general case operates with dimensional variables $\tilde{\zeta}$ and $\tilde{\mathbf{t}}$.

It would be reasonable to seek for a uniform asymptotic in a form similar to (5.2.1):

$$u(\mathbf{r}) = \left[A\tilde{I}(\tilde{\boldsymbol{\zeta}}) + \frac{1}{\mathrm{i}k} \sum_{p=1}^{m} B_p \frac{\partial \tilde{I}}{\partial \tilde{\zeta}_p} \right] \exp(\mathrm{i}k\chi) \ . \tag{9.1.3}$$

For $m + 1$ unknown phase functions $\chi(\mathbf{r}), \tilde{\zeta}_1(\mathbf{r}), \ldots, \tilde{\zeta}_m(\mathbf{r})$ and the same number of amplitude factors $A(\mathbf{r}), B_1(\mathbf{r}), \ldots, B_m(\mathbf{r})$ we may derive equations similar to that obtained in Chap. 5 and relate, likewise in a similar manner, the solution of these equations directly with the solutions of the ray equations ψ_j and U_j.

Let $\tilde{\mathbf{t}}_j(\tilde{\boldsymbol{\zeta}})$ be the stationary values of the integration variable $\tilde{\mathbf{t}}$ defined by $\partial \Psi(\tilde{\boldsymbol{\zeta}}, \tilde{\mathbf{t}})/\partial \tilde{t}_q = 0$. Then the unknown functions $\chi, \tilde{\zeta}_1, \ldots, \tilde{\zeta}_m$ are related to the phase ψ_j by nonlinear, and in the general case, transcendental equations:

$$\Psi[\tilde{\boldsymbol{\zeta}}, \tilde{\mathbf{t}}_j(\tilde{\boldsymbol{\zeta}})] = \psi_j(\mathbf{r}) \ , \tag{9.1.4}$$

whereas for $A(\mathbf{r})$ and $B(\mathbf{r})$ we obtain a system of linear equations

$$A + \sum_{p=1}^{m} B_p \frac{\partial \Psi[\tilde{\boldsymbol{\zeta}}, \tilde{\mathbf{t}}_j(\tilde{\boldsymbol{\zeta}})]}{\partial \tilde{\zeta}_p} = \det \left[\frac{\partial^2 \Psi}{\partial t_p \partial t_q} \right]_j U_j \ . \tag{9.1.5}$$

Now if a one-to-one correspondence is established between an arbitrary function $\Psi(\tilde{\boldsymbol{\zeta}}, \tilde{\mathbf{t}})$ and the polynomial function of $\Phi(\tilde{\boldsymbol{\zeta}}, \tilde{\mathbf{t}})$ of catastrophe theory, then the amplitudes A and B_p will be finite on caustics despite the singularity of ray amplitudes U_j. Thus, the use of non-polynomial phase functions does not change essentially the scheme of standard integrals. Changes are experienced only by the specific form of the equations for χ and ζ_p.

In his doctoral dissertion in 1972, G.V. Permitin suggested to use as standard integral the one with a phase function having a necessary number of stationary points equal to the number of rays. Similar considerations were put forth by *Connor* and *Southall* [9.1]. This idea may be generalized on multidimensional integrals. Use of non-polynomial phase functions may be of certain practical advantage if the respective integrals (9.1.1) are known better than the standard integrals of catastrophe theory, or if the integrals (9.1.1) yield a better approximation under specific circumstances.

9.1.3 Bessel Function Based Uniform Asymptotics near Simple Caustics

An example of a standard integral with a non-polynomial phase function may be the Bessel function

$$\begin{aligned} J_m(\zeta) &= \frac{(-\mathrm{i})^m}{\pi} \int_0^{\pi} \exp(\mathrm{i}\zeta \cos t_1) \cos(m t_1) \, \mathrm{d}t_1 \\ &= \frac{(-\mathrm{i})^m}{2\pi} \int_{-\pi/2}^{\pi/2} \mathrm{d}t \exp(\mathrm{i}\zeta \sin t) \\ &\quad \times [\exp(-\mathrm{i}mt + \mathrm{i}m\pi/2) + \exp(\mathrm{i}mt - \mathrm{i}m\pi/2)] \ . \end{aligned} \tag{9.1.6}$$

In the interval $(-\pi/2, \pi/2)$, for $\zeta > m$, the first term in the integrand can have only two stationary points $t_{1,2} = \pm \arccos(m/\zeta)$ which coincide at $\zeta = m$ and become complex-valued for $\zeta < m$. In other words, the Bessel function $J_m(\zeta)$ has properties similar to that of the Airy function, and may be used to advantage as a standard function in the construction of a uniform asymptotics in the case of a simple caustic.

Seeking the field as a linear combination of the Bessel function $J_m(k\tilde{\zeta}_{\mathrm{B}})$ and its derivative $J'_m(k\tilde{\zeta}_{\mathrm{B}})$ with thus far indeterminate amplitude factors A_{B} and B_{B} and phase functions $\tilde{\zeta}_{\mathrm{B}}$ and $\tilde{\chi}_{\mathrm{B}}$, viz.,

$$u(\mathbf{r}) = [A_{\mathrm{B}} J_m(k\tilde{\zeta}_{\mathrm{B}}) + \mathrm{i} B_{\mathrm{B}} J'(k\tilde{\zeta}_{\mathrm{B}})] \exp(\mathrm{i}k\chi_{\mathrm{B}}) \,, \tag{9.1.7}$$

one could carry out the familiar procedure and express $\tilde{\zeta}_{\mathrm{B}}$, χ_{B}, A_{B}, and B_{B} in terms of $\psi_{1,2}$ and $U_{1,2}$. However, in this case the evaluation of these quantities can be drastically simplified by recognizing that the Bessel function $J_m(\zeta)$ is expressed asymptotically through the Airy function, see Sect. 5.3.1. If we observe that the Bessel asymptotics (9.1.7) and the Airy asymptotics (5.1.1) are asymptotically equivalent (for $k \to \infty$), then the "Bessel" unknowns $\tilde{\zeta}_{\mathrm{B}}$, χ_{B}, A_{B} and B_{B} may be directly expressed through the respective parameters of the Airy asymptotic (subscript A), viz.,

$$\begin{aligned}
&\chi_{\mathrm{B}} = \chi_{\mathrm{A}} = (\psi_1 + \psi_2)/2 \,, \\
&\sqrt{\tilde{\zeta}_{\mathrm{B}}^2 - (m/k)^2} - (m/k)\arccos[m/(k\tilde{\zeta}_{\mathrm{B}})] = \tfrac{2}{3}(-\tilde{\zeta}_{\mathrm{A}})^{3/2} = (\psi_1 - \psi_2)/2 \,, \\
&A_{\mathrm{B}} = \frac{1}{2}\left(\frac{m}{k}\right)^{1/3}\left[\tilde{\zeta}_{\mathrm{B}}^2 - \left(\frac{m}{k}\right)^2\right]^{1/4} (\mathrm{i}U_1 + U_2)\mathrm{e}^{\mathrm{i}\pi/4} \,, \\
&B_{\mathrm{B}} = \frac{1}{2}\left(\frac{m}{k}\right)^{1/3}\left[\tilde{\zeta}_{\mathrm{B}}^2 - \left(\frac{m}{k}\right)^2\right]^{-1/4} (\mathrm{i}U_2 - U_2)\mathrm{e}^{\mathrm{i}\pi/4} \,.
\end{aligned} \tag{9.1.8}$$

Near a caustic ($\tilde{\zeta}_{\mathrm{B}} = m/k$ in Bessel variables and $\tilde{\zeta}_{\mathrm{A}} = 0$ in Airy variables) the quantities $\tilde{\zeta}_{\mathrm{B}}$ and $\tilde{\zeta}_{\mathrm{A}}$ are linearly related to each other, $\tilde{\zeta}_{\mathrm{A}} \approx -(2m/k)^{-1/3}(\tilde{\zeta}_{\mathrm{B}} - m/k)(3/2)^{4/3}$ so that the divergence of ray amplitudes $U_{1,2}$ at the caustic is now compensated for by the factors

$$[\tilde{\zeta}_{\mathrm{B}}^2 - (m/k)^2]^{\pm 1/4} \approx [(2m/k)(\tilde{\zeta}_{\mathrm{B}} - m/k)]^{\pm 1/4} \,.$$

Despite the asymptotic equivalence of (5.1.1) and (9.1.7), under certain conditions the solution (9.1.7) may have certain advantages. The point is that the Bessel function $J_m(kr)$ taken with the factor $\mathrm{e}^{\mathrm{i}m\alpha}$ is an exact solution of the wave problem for a circular caustic of radius $a = m/k$, see Sect. 5.3.1, whereas, the Airy function lends only an approximate (asymptotic) solution. If the radius of the caustic decreases, the Airy asymptotic (5.1.1) breaks down sooner or later; whereas, the Bessel asymptotic (9.1.7) remains valid even at $m = ka \to 0$.

This example delineates the class of problems for which the use of non-polynomial integrals would be reasonable: these are problems where a non-polynomial phase integral is closer to the exact solution than standard integrals of wave catastrophes with non-polynomial phase.

9.1.4 Contour Standard Integrals

Analysis indicates that the scheme of construction of uniform asymptotics does not change if the standard integral is a contour integral

$$\tilde{I}(\tilde{\zeta}) = \left(\frac{k}{2\pi}\right)^{1/2} \int_C \exp[ik\Psi(\tilde{\zeta},\tilde{t})]\,d\tilde{t} \ . \tag{9.1.9}$$

The phase function Ψ in this expression must possess the required number of stationary points, and at the ends of contour C, i.e., at $\tilde{t} = \tilde{t}_i$ and $\tilde{t} = \tilde{t}_f$, the periodicity condition must be met. Then the field asymptotic is given by (9.1.3) like in the general method of standard functions, and the phase function need not necessarily be a polynomial in $\tilde{t}$.

In a particular case of $\Psi(\tilde{\zeta},\tilde{t}) = \tilde{\zeta}\sin\tilde{t} - (m/k)\tilde{t}$, $\tilde{\zeta} > 0$, $-\pi < \tilde{t} < \pi$, the standard integral (9.1.9) is expressed in terms of a Bessel function

$$\tilde{I}(\tilde{\zeta}) = \sqrt{2\pi k}\, J_m(k\tilde{\zeta}) \ . \tag{9.1.10}$$

Whether or not this function can be used to describe fields with axially symmetric caustics will be discussed in the ensuing section.

9.2 Structurally Unstable Caustics

9.2.1 Structurally Stable and Unstable Objects

Perturbations involved in physical problems do not always fit within "small perturbations" of catastrophe theory, see Sect. 3.1.2. Quite appropriately, we have to distinguish the mathematical concept of structural stability from that of physical stability, i.e., stability of a wave field with respect to small perturbations of various physical nature. In the light of this distinction, it would be natural to analyze the place of structurally stable singularities in physical theories. It would be reasonable to carry out similar analysis for structurally unstable objects as well. Essentially the role of both in physical theories is decided by how well they model actual objects and the performance criterion should follow from the physical statement of the problem.

In the case of caustics, a model may be deemed appropriate if the perturbations of initial conditions and medium parameters characteristic to the problem lead only to small perturbations of the wave field rather than caustics, because experimental measurements are made for the wave field. Perturbations will be small in value if the perturbations of the caustic surface are small compared with the width of the caustic zone, see Sect. 2.3.6. When a perturbed caustic leaves the initial caustic zone, the model – no matter whether it is structurally stable or not – should be sophisticated by including the perturbing factor in the number of intrinsic parameters.

Thus, while structurally unstable caustics are not typical from the standpoint of catastrophe theory, they can serve as suitable models for some actual objects. In physical problems, structurally unstable constructions occur quite seldom. To these constructions belong, for example, plane, spherical, and cylindrical waves. A point focus in the case of a spherical wave and a focal line in the case of a cylindrical wave break down under however small perturbations of the phase front; the plane wave may be viewed as a limiting case for a spherical or cylindrical wave with an indefinitely large radius corresponding to focusing at infinity. *Berry* [9.2] has assigned to such objects infinite codimension implying that a distortion of the shape of the said caustics occurs under exceedingly weak perturbations described by polynomials of however high degree.

Structurally unstable caustics include also axial and axially symmetric caustics associated with wave beams of axial symmetry. This type of caustics will be discussed below.

9.2.2 Uniform Asymptotics for Axially Symmetric Caustics

Let us consider a family of edge rays that emerge when a plane wave is incident on a circular opening or on a circular disc, as shown in Fig. 9.1. Every point on the axis of the aperture KK is hit by an infinite number of rays (by way of example the figure shows a cone of rays passing through point M) so that the line KK is an axial or *sagittal* caustic. On the other hand, any off-axial point A is traversed by only two rays, labeled 1 and 2, one enroute to the caustic, the other after crossing the caustic.

If the ray field is projected on a plane perpendicular to the caustic, then all projections of the rays will pass through the central point K. Recognizing that in the two-dimensional problem the field near a focal spot is described by the Bessel function of zero order $J_0(kr)$, it would be reasonable to use this function as a standard one after having augmented it by another phase factor $e^{ik\chi}$

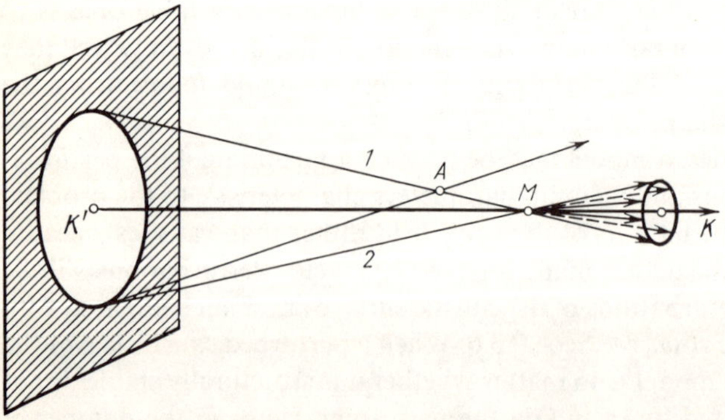

Fig. 9.1. Axial line is a sagittal caustic of edge rays that have touched the circular aperture

describing the wave motion along the axis KK. Along lines of this reasoning we will seek the field in the form

$$u(r) = \sqrt{\frac{\pi}{2}}[AJ_0(\zeta) + BJ'_0(\zeta)]\,\mathrm{e}^{\mathrm{i}k\chi - \mathrm{i}\pi/4}\ , \tag{9.2.1}$$

where ζ, χ, A, and B are indeterminate functions of coordinates. The factor $\sqrt{\pi/2}\cdot\exp(-\mathrm{i}\pi/4)$ is introduced for convenience of algebraic manipulations.

When the wave beam possesses an axial dependence $\mathrm{e}^{\mathrm{i}m\phi}$ (m is an integer, and ϕ is the azimuth angle in cylindrical coordinates z, ρ, ϕ), it may be also included in the scheme under consideration by taking as standard function the mth order Bessel function $J_m(\zeta)$ with factor $\mathrm{e}^{\mathrm{i}m\phi}$. In this case it would be reasonable to have a uniform asymptotic in the form

$$u(r) = \sqrt{\frac{\pi}{2}}[AJ_m(\zeta) + BJ'_m(\zeta)]\,\mathrm{e}^{\mathrm{i}k\chi + \mathrm{i}m\phi - \mathrm{i}\pi/4}\ . \tag{9.2.2}$$

Let us look first at a general case of $m \neq 0$, deferring the analysis of an axial caustic ($m = 0$) until the last stage. Here, we note that axially symmetric caustics corresponding to $m \neq 0$ occur in propagation of the mth azimuthal modes in axially symmetric waveguides, transmission lines, and open resonators. They emerge also in radiation of radio waves by axially symmetric antennas, open ends of circular waveguides, and axially symmetric radiating horns. Analysis of fields in such systems resulted in uniform asymptotics for the scalar field [9.3], for the electromagnetic field [9.4] including a possibly inhomogeneous medium, but keeping the requirement of axial symmetry, and finally, for the EM field radiated by a parabolic, axially symmetric antenna [9.5].

An easier way to find the unknown functions in (9.2.1) and (9.2.2) is by matching with the ray solution

$$u = U_1\exp(\mathrm{i}k\psi_1) + U_2\exp(\mathrm{i}k\psi_2)\ . \tag{9.2.3}$$

To do this we substitute in (9.2.2) the Debye asymptotics of Bessel function and its derivative

$$\begin{aligned} J_m(\zeta) &\cong (2/\pi)^{1/2}(\zeta^2 - m^2)^{-1/4}\cos(\eta - \pi/4)\ ,\\ J'_m(\zeta) &\cong -(2/\pi)^{1/2}(1/\zeta)(\zeta^2 - m^2)^{1/4}\sin(\eta - \pi/4)\ , \end{aligned} \tag{9.2.4}$$

where

$$\eta = \sqrt{\zeta^2 - m^2} - m\arccos(m/\zeta)\ .$$

Substituting (9.2.4) in (9.2.2) and comparing with (9.2.3) yields

$$\begin{aligned} &\chi = (\psi_1 + \psi_2)/2\ ,\\ &\sqrt{\tilde{\zeta}^2 - \tilde{\zeta}_m^2} - \tilde{\zeta}_m\arccos(\tilde{\zeta}_m/\tilde{\zeta}) = (\psi_2 - \psi_1)/2\ ,\\ &A = (\tilde{\zeta}^2 - \tilde{\zeta}_m^2)^{-1/4}(\mathrm{i}U_2 + U_1)\ ,\\ &B = \tilde{\zeta}(\tilde{\zeta}^2 - \tilde{\zeta}_m^2)^{-1/4}(\mathrm{i}U_1 - U_2)\ , \end{aligned} \tag{9.2.5}$$

where $\tilde{\zeta} = \zeta/k$, and $\tilde{\zeta}_m = m/k$. The quantity $\tilde{\zeta}_m = m/k$ corresponds to the radius of caustic in the scale of parameter $\tilde{\zeta}$: at $\tilde{\zeta} = \tilde{\zeta}_m$ the eikonals ψ_1 and ψ_2 coincide.

Having determined $\tilde{\zeta}(\rho)$ with the aid of the second line in (9.2.5), the actual radius of the circular section of caustic in the plane $z =$ constant can be determined from the condition $\tilde{\zeta}(\rho_m) = \tilde{\zeta}_m = m/k$. Thus, near an axially symmetric caustic the field is represented by the asymptotic formula

$$u(r) = \sqrt{\frac{\pi k}{2}}\exp(ik\chi + im\phi - i\pi/4)[(\tilde{\zeta}^2 - \tilde{\zeta}_m^2)^{1/4}(iU_2 + U_1)J_m(k) - i\tilde{\zeta}(\tilde{\zeta}^2 - \tilde{\zeta}_m^2)^{-1/4}(iU_1 + U_2)J'_m(k\tilde{\zeta})] \,. \tag{9.2.6}$$

If the azimuthal number m is sufficiently large, then by virtue of the uniform asymptotic formulas

$$\begin{aligned} J_m(\zeta) &= \sqrt{2}\left(\frac{t}{m^2 - \zeta^2}\right)^{1/4} \mathrm{Ai}(t), \\ J'_m(\zeta) &= -\frac{1}{\zeta}\sqrt{2}\left(\frac{t}{m^2 - \zeta^2}\right)^{-1/4} \mathrm{Ai}'(t) \,, \end{aligned} \tag{9.2.7}$$

(9.2.6) becomes an Airy asymptotic. Here,

$$t = \begin{cases} \left[\dfrac{3}{2}\left(\sqrt{\zeta^2 - m^2} - m\arccos(m/\zeta)\right)\right]^{2/3}, & \zeta > m \,, \\ \dfrac{3}{2}\left(m\ln\dfrac{m + \sqrt{m^2 - \zeta^2}}{\zeta} - \sqrt{m^2 - \zeta^2}\right), & \zeta < m \,. \end{cases} \tag{9.2.8}$$

9.2.3 A Uniform Asymptotic for an Axial Caustic

At $m = 0$ or, what is the same, at $\rho_0 = 0$, the axially symmetric caustic degenerates in a focal line – the axis of symmetry, z. Under these circumstances,

$$u(r) = \sqrt{\frac{\pi\zeta}{2}}\exp(ik\chi - i\pi/4)[(iU_2 + U_1)J_0(\zeta) + i(iU_2 - U_1)J_1(\zeta)] \,, \tag{9.2.9}$$

where it has been observed that $J'_0(\zeta) = -J_1(\zeta)$, For axial caustics,

$$\chi = \tfrac{1}{2}(\psi_1 + \psi_2) \,, \qquad \tilde{\zeta} = \tfrac{1}{2}(\psi_2 - \psi_1) \,. \tag{9.2.10}$$

Given the amplitudes U_1 and U_2 as a function of distance between the observation point and the axis of the caustic, one may compute the diffraction of an edge wave on a circular aperture or on a disc. Similar calculations have been made by *Gazazyan* et al. [9.6] but for an electromagnetic, rather than scalar, field. In the uniform asymptotic formulas, the vectorial nature of disturbance causes the amplitudes A and B to be replaced with vectors, and summation of vectorial amplitudes $\mathbf{U}_1^{E,H}$ and $\mathbf{U}_2^{E,H}$ results in an additional azimuthal dependence of $\mathbf{A}^{E,H}$ and $\mathbf{B}^{E,H}$.

9.2.4 Applicability of Axial Caustic Asymptotics in the Presence of Aberrations

Like the focal point, an axial caustic has an infinite codimension and from the formal point of view breaks down by any exceedingly small perturbation of the primary wavefront, i.e., by aberrations however small. From the physical point of view, smallness of a caustic perturbation is decided by whether or not it is small compared to the width of the caustic zone.

As an example consider the field of an edge wave at a distance z from the center of a circular aperture of radius a. If we resort to the Fresnel criterion, the width of the caustic zone embracing the z axis may be estimated from the condition $|\psi_1 - \psi_2| \sim \lambda/2$. In this particular case, the argument of the Bessel function $\zeta = k(\psi_2 - \psi_1)/2$ will be in the order of $k\lambda/4 \cong 1.57$. This figure is comparable with $\zeta \cong 2.3$ corresponding to the first zero of the Bessel function, which is another proof of validity of the Fresnel criteria.

For a circular aperture, $\tilde{\zeta} = (\psi_2 - \psi_1)/2 \cong \rho \sin \theta_0$, where $\theta_0 = \arctan(a/z)$ is the angle at which radius a is seen at a distance z. Therefore, the criterion $(\psi_2 - \psi_1) \approx \lambda/2$ is consistent with the distance $\rho_c \sim \lambda/(4 \sin \theta_0) \sim \lambda z/4a$ which should be viewed as the radius of the axial caustic zone. Any perturbation of the caustic smaller than ρ_c should be recognized insignificant as it virtually does not distort the field. When the caustic perturbation $\Delta\rho$ exceeds ρ_c, then the approximation of the axially symmetric caustic coinciding with the z axis should be reconsidered in favor of a more realistic pattern.

Let, for example, a circular caustic of radius a reshape into an ellipse of semiaxes $a_1 = a + \delta/2$ and $a_2 = a - \delta/2$. We wish to estimate the difference of semiaxes $\delta = a_1 - a_2$ beyond which the model of axial caustic will be no longer applicable. If we inscribe in the ellipse some circles near the vertices and near the equator, the centers of these circles mark off the extrema of the caustics of edge rays. These centers are shifted about the axis a distance $\sim \delta$; therefore, the condition allowing us to neglect a deformation of the caustic will be

$$\delta < \rho_c \sim \lambda z/4a \ . \tag{9.2.11}$$

Near the aperture where the distance z is comparable with radius a, this condition boils down to a requirement, common in optics and antenna engineering, that δ be small compared to $\lambda/4$.

At large distances z comparable with the diffraction length $z_d \sim a^2/\lambda$ it is admissible that the difference δ be comparable with radius a, $\delta \lesssim a/4$. Farther away where $z \gg a^2/\lambda$ the rays and caustics of the near field of edge waves give way to the rays and caustics of the Fraunhofer field of edge waves.

9.3 Standard Integrals with Amplitude Correction

9.3.1 Integrals of Weighted Rapidly Oscillating Functions

All standard integrals discussed thus far have been superpositions of rapidly oscillating exponential functions taken with identical weights. Some situations, however, bring about superpositions of waves with different weights. To describe such waves it would be reasonable to complicate the integrals (5.1.1) and (9.1.1) by introducing the amplitude factor $P(\mathbf{r}, t)$ in the integrand, namely,

$$\tilde{I}(\tilde{\zeta}) = \sqrt{\frac{ik}{2\pi}} \int P(\mathbf{r}, \tilde{t}) \exp[ik\,\Phi(\tilde{\zeta}, \tilde{t})]\, d\tilde{t} \ . \qquad (9.3.1)$$

We shall refer to such superpositions as standard integrals with amplitude correction.

Incorporation of a correcting factor alters the equations for amplitudes A and B_p in the uniform asymptotic representation (9.1.1). Now, they will be prefactored by $P(\mathbf{r}, t)$ representing different weights of the rays. In the ideal case, a choice of a factor $P(\mathbf{r}, t)$ may cause the representation (9.1.1) with integral (9.3.1) to describe the solution to the wave problem without additional terms containing derivatives $\partial\tilde{I}/\partial\tilde{\zeta}_p$, i.e., so that there will be A = constant and $B_p = 0$. In essence, with such a choice the correction of the solution by terms with $\partial\tilde{I}/\partial\tilde{\zeta}_p$ is replaced with the correction of a pre-exponential factor in the standard integral (9.3.1).

It is quite obvious that it would be infeasible and unreasonable to implement the correction of amplitude factor in full extent. In the final analysis, Maslov's method and the method of interference integrals lead to a representation of solution as a single integral (9.3.1) but then this integral has to be expressed in terms of simpler typical integrals at hand. Therefore, correction of amplitude factors is advantageous when it promises some new quality of the solution. Pertinent specific problems will be discussed below.

9.3.2 Uniform Penumbral Asymptotics near a Fuzzy Light–Shadow Boundary

This problem has no direct bearing on caustics. It is displayed here because this problem illustrates the main principles of amplitude correction and serves as a preparatory stage on the way to the construction of a field asymptotic for a caustic with diffused edge.

Suppose that an opaque (for $x < 0$) screen is inserted in the way of a fuzzy wave. In the Kirchhoff approximation, the field behind the screen is described by one of the forms of Fresnel integral

$$F(x) = \int_x^\infty \exp(\mathrm{i}t_1^2/2)\,\mathrm{d}t_1 = \exp(\mathrm{i}x^2/2)\int_0^\infty \exp(\mathrm{i}t^2/2 + \mathrm{i}tx)\,\mathrm{d}t \ . \quad (9.3.2)$$

If in the light region ($x < 0$) the amplitude of the incident wave increases by a power law, $A \sim x^\nu$, then in place of (9.3.2) we get an integral of the form

$$F_\nu(x) = \int_0^\infty t^\nu \exp(\mathrm{i}t^2/2 + \mathrm{i}tx)\,\mathrm{d}t \ . \quad (9.3.3)$$

This integral contains a pre-exponent correcting factor t^ν and describes the field for a diffused light–shadow boundary, since for $x > 0$ the illumination increases gradually, in a power law, rather than in a jump. At $\nu = 0$ this integral reduces to the Fresnel integral (9.3.2) corresponding to a field jump near the shadow boundary, and at $\nu \neq 0$ it is proportional to the parabolic cylinder function $D_{-\nu-1}(x\mathrm{e}^{\mathrm{i}\pi/4})$.

For a clear-cut light–shadow boundary, a uniform field asymptotic has been constructed in [9.7–9] on the base of Fresnel integral (9.3.2). By analogy with this development, *Orlov* and *Tropkin* [9.10] have constructed a field asymptotic for a diffused boundary. This construction is based on the function $D_{-\nu-1}(\zeta)$ and is as follows:

$$u(\mathbf{r}) = [AD_{-\nu-1}(\zeta) + BD'_{-\nu-1}(\zeta)]\,\mathrm{e}^{\mathrm{i}k\chi} \ . \quad (9.3.4)$$

Relevant functions $\zeta(r)$, $\chi(r)$, $A(r)$, and $B(r)$ may be expressed in terms of U_d and ψ_d of the diffraction (edge) field and parameters U_r and ψ_r of the ray field of primary wave as follows:

$$\begin{aligned} &\chi = (\psi_\mathrm{d} + \psi_\mathrm{r})/2, \qquad \zeta = \exp(\mathrm{i}\pi/4 \pm \mathrm{i}\pi/4)\sqrt{2k(\psi_\mathrm{d} - \psi_\mathrm{r})} \ , \\ &A = (Q_1 + Q_2)/2, \qquad B = (Q_1 - Q_2)/\zeta \ , \\ &Q_1 = (2\pi)^{-1/2}\,\Gamma(\nu + 1)\,\mathrm{e}^{\mathrm{i}\pi\nu}\zeta^{-\nu}\,U_\mathrm{r}, \qquad Q_2 = \zeta^{\nu+1}\,U_\mathrm{d} \ . \end{aligned} \quad (9.3.5)$$

It is assumed that in the lit region $\arg\zeta = 3\pi/4$ (upper sign in the expression for ζ), and in the shadow region $\arg\zeta = -\pi/4$.

Now the field (9.3.4) is a sum of the ray field and the field of edge wave in the lit region ($|\zeta| \gg 1$, $\arg\zeta = 3\pi/4$),

$$u = U_\mathrm{r}\exp(\mathrm{i}k\,\psi_\mathrm{r}) + U_\mathrm{d}\exp(\mathrm{i}k\,\psi_\mathrm{d}) \ ,$$

which goes over to the diffraction field $U_\mathrm{d}\exp(\mathrm{i}k\,\psi_\mathrm{d})$ in the shadow region ($|\zeta| \gg 1$, $\arg\zeta = -\pi/4$).

Modifications of the method of uniform asymptotics with the use of parabolic cylinder functions $D_\mu(\zeta)$ find their way in the description of propagation of wave beams, in investigations of near fields of aperture antennas with nonuniform amplitude excitation, and in the excitation of lateral waves at the interface, see the literature, e.g., [9.10].

9.3.3 Broken Caustics near Diffused Shadow

When the shadow region is diffused, the field near broken caustics, see Chap. 8, would be reasonable to describe with the aid of the standard function

$$F_\nu(\zeta,\eta) = \int_\eta^\infty (t-\eta)^\nu \exp[\mathrm{i}(t^3/3 + t\zeta)]\,\mathrm{d}t \;, \tag{9.3.6}$$

which emerges in the incorporation of a correcting amplitude factor $(t-\eta)^\nu$ into the integrand of the incomplete Airy function (8.1.2). This function satisfies the parabolic equation (8.1.4) and possesses both the properties of the Airy function (far from penumbra) and the properties of the function $D_{-\nu-1}$ (far from the caustic).

Tropkin [9.11] has suggested employing (9.3.6) as a standard function. Here, we quote only the final result of his development. If the wave field is sought in the form

$$u(\mathbf{r}) = \frac{\exp(\mathrm{i}k\chi - \mathrm{i}\pi/4)}{2\sqrt{\pi}}\left(AF_\nu + \mathrm{i}B\frac{\partial F_\nu}{\partial\zeta} + \mathrm{i}C\frac{\partial F_\nu}{\partial\eta}\right) , \tag{9.3.7}$$

the unknown functions χ, ζ, and η are related to the eikonal of the edge wave ψ_{d} and two eikonals of the primary ray field ψ_{r1} and ψ_{r2} by

$$\chi = (\psi_{\mathrm{r1}} + \psi_{\mathrm{r2}})/2 \;, \qquad \tfrac{2}{3}(-\zeta)^{3/2} = \tfrac{1}{2}(\psi_{\mathrm{r2}} - \psi_{\mathrm{r1}}) \;,$$

$$\eta^3/3 + \eta\zeta = k(\psi_{\mathrm{d}} - \chi) \;.$$

The amplitudes A, B, and C will be given in terms of the amplitudes $U_{\mathrm{r1,2}}$ and U_{d}, viz.,

$$A = (-\zeta)^{1/4}[\mathrm{i}(-\eta - \sqrt{-\zeta})^{-\nu}U_{\mathrm{r2}} + (-\eta + \sqrt{-\zeta})^{-\nu}U_{\mathrm{r1}}] \;,$$

$$B = (-\zeta)^{-1/4}[\mathrm{i}(-\eta - \sqrt{-\zeta})^{-\nu}U_{\mathrm{r2}} - (-\eta + \sqrt{-\zeta})^{-\nu}U_{\mathrm{r1}}] \;,$$

$$C = \frac{1}{\eta^2+\zeta}(A - B\eta) - \frac{2\sqrt{\pi}}{\Gamma(\nu+1)}\exp(-\mathrm{i}\pi/4 - \mathrm{i}\pi\nu/2)\cdot(\eta^2+\zeta)^\nu U_{\mathrm{d}} \;.$$

As before, we assume that ray 1 has not yet touched the caustic. In the inbreathe analysis of [9.11] it is emphasized that the integral (9.3.6) may be naturally generalized on caustics of higher dimension (by complicating the phase function) and on electromagnetic fields.

In general, diffraction rays have smaller amplitude for diffused shadow than for sharply defined shadow boundary. Therefore, the interference of the diffraction field with the ray field of the primary wave is defined weaker when the exponent ν is larger.

If the caustic is formed by diffracted rays, then in the case of diffused shadow, the standard integral (8.2.2) should be replaced with the corrected integral [9.12]

$$V_\nu(\zeta,\eta) = \int_{-\infty+\mathrm{i}\alpha^2}^{\infty+\mathrm{i}\alpha^2} (t-\eta)^\nu \exp[\mathrm{i}(t^3/3 + t\zeta)]\mathrm{d}t \;. \tag{9.3.8}$$

For $t < \eta$, $(t-\eta)^{\nu}$ should be thought of as the quantity $|\eta - t|\exp(\mathrm{i}\pi\nu)$. If caustics of diffracted rays and primary rays are considered simultaneously, then for diffused shadow $(t-\eta)^{\nu}$ should be incorporated in the integrand corresponding to an F_4 catastrophe, see Chap. 8. In this case, the diffusivity of the caustic will manifest itself as a weaker contrast of the interference pattern.

9.4 Reflection from a Barrier and Oscillations in a Potential Well

9.4.1 Weber Equation and Functions

As will be recalled, the standard integrals satisfy ordinary or partial differential equations. Specifically, the Airy function $\mathrm{Ai}(\zeta)$ satisfies the second order equation (4.2.14) which differs from the equation of harmonic oscillator $u'' + u = 0$ in having a linear function of $-\zeta$ in place of the constant coefficient of u.

The next complicated differential equation has a second order polynomial as a coefficient of $u(\zeta)$:

$$\mathrm{d}^2u/\mathrm{d}\zeta^2 + (\zeta^2 - b^2)u = 0 \ . \tag{9.4.1}$$

The substitution $\zeta = z\mathrm{e}^{\mathrm{i}\pi/4}/\sqrt{2}$ and $b = (2n+1)^{1/4}\mathrm{e}^{-\mathrm{i}\pi/4}$ reduces this equation to the Weber equation (Sect. 16.5 in [9.13])

$$\frac{\mathrm{d}^2w}{\mathrm{d}z^2} + \left(n + \frac{1}{2} - \frac{1}{4}z^2\right)w = 0 \ , \tag{9.4.2}$$

whose solutions play important role in the theory of potential wells and are called Weber functions or parabolic cylinder functions. Accordingly, we shall refer to the solutions of (9.4.1) also as Weber functions and denote them by $w(\zeta, b)$.

For positive values of parameter b, the coefficient of u in (9.4.1) vanishes at two turning points $\zeta = \pm b$ associated with the boundaries of the potential barrier. Beyond the barrier where $|\zeta| > b$, the WKB solutions to equation (9.4.1) exhibit oscillatory behavior, whereas inside the barrier, $-b < \zeta < b$, they increase or decay exponentially. For wave problem solving, it is convenient to take two functions $w_1(\zeta, b)$ and $w_2(\zeta, b)$ as two linearly independent solutions of (9.4.1). On the left from the barrier ($\zeta < -b$) these functions represent waves traveling to the barrier and backwards. The WKB approximations for these waves are

$$w_{1,2}(\zeta, b) \underset{\zeta\to\infty}{\cong} (\zeta^2 - b^2)^{-1/4}\exp\left[\pm\mathrm{i}\int_{\zeta}^{b}(\zeta^2 - b^2)^{1/2}\,\mathrm{d}\zeta\right] . \tag{9.4.3}$$

Conditionally we shall refer to w_1 as the incident wave and to w_2 as the reflected wave.

To the right from the barrier ($\zeta > b$) each of the functions $w_{1,2}$ is a superposition of waves traveling to and away form the barrier. In barrier problems the physical interest is only in a combination

$$w(\zeta, b) = w_1(\zeta, b) + R(b)w_2(\zeta, b) \ , \tag{9.4.4}$$

which for $\zeta \to \infty$ represents the outgoing wave

$$w(\zeta, b) \underset{\zeta\to\infty}{\cong} D(b)(\zeta^2 - b^2)^{-1/4} \exp\left[\mathrm{i} \int_b^\zeta (\zeta^2 - b^2)^{1/2} \,\mathrm{d}\zeta \right] \tag{9.4.5}$$

that is, satisfies the radiative condition at infinity. The quantity $R(b)$ plays the role of reflection factor and the quantity $D(b)$ the role of a sub-barrier penetration coefficient (tunneling).

From the properties of Weber functions, see [9.13] and also [9.14], it follows that

$$|R(b)| = [1 + \exp(-\pi b^2)]^{-1/2}, \qquad |D(b)| = [1 + \exp(\pi b^2)]^{-1/2} \ , \tag{9.4.6}$$

and $|R|^2 + |D|^2 = 1$ reflects conservation of energy in the tunnel effect. For a thick barrier when $b \to \infty$, by (9.4.6) we see that the penetration factor D vanishes and R tends to unity. Conversely, for $b \to 0$ when the barrier is no longer classically reflective, both R and D tend to 1/2, but such that R is always equal to or greater than D. We omit the expressions for the arguments of both $R(b)$ and $D(b)$ as they will not be needed any longer.

9.4.2 Asymptotic Solution to One-Dimensional Reflection from a Barrier

While the Airy function is suitable to approximate the solution of wave problems with one turning point (Langer's solution, Sect. 4.2.4), the Weber function allows an asymptotic description of the one-dimensional Helmholtz equation

$$\frac{\mathrm{d}^2u}{\mathrm{d}z^2} + k^2\varepsilon(z)u = 0 \tag{9.4.7}$$

in a more intricate situation with the effective permittivity turning to zero at two points, z_1 and z_2. In this case, the wave suffers partial reflection from the front turning point z_1 and partially penetrates behind the back turning point z_2 (tunneling). Detailed analysis of equation (9.4.7), under these circumstances, has been carried out by *Olver* [9.15] and *Pike* [9.16]. We will quote some results of this analysis which are worthwhile for the following:

Let $\varepsilon(z)$ be a monotonous function vanishing at z_1 and z_2. The leading term of the asymptotic solution is expressable in terms of a Weber function with variable amplitude $A(z)$, argument $\zeta(z) = k^{1/2}\tilde{\zeta}(z)$, and parameter $b = k^{1/2}a$, namely

$$u(z) \approx A(z)w[\zeta(z), b] = A(z)w[k^{1/2}\tilde{\zeta}(z), k^{1/2}a] \ . \tag{9.4.8}$$

The power of the wavenumber should be such that both terms in (9.4.7) upon substitution from (9.4.8) have the same order in k. Substituting (9.4.8) in (9.4.7) and equating to zero the coefficients of like powers in k, we obtain in the lowest order an equation for $\tilde{\zeta}$:

$$(\tilde{\zeta}^2 - a^2)\left(\frac{d\tilde{\zeta}}{dz}\right)^2 = \varepsilon(z) \ . \tag{9.4.9}$$

If we assume that the zeros z_1 and z_2 of the right side of (9.4.9) coincide with those of the left-hand side, i.e., if we let $\tilde{\zeta}(z_1) = -a$ and $\tilde{\zeta}(z_2) = a$, then the derivative $d\zeta/dz$ will be nonzero everywhere, and (9.4.9) may be treated as a smooth function that maps $\tilde{\zeta}$ onto z one-to-one. By (9.4.9) this mapping is defined by

$$\begin{aligned}
&\int_{\tilde{\zeta}}^{-a} \sqrt{\tilde{\zeta}^2 - a^2}\, d\tilde{\zeta} = \int_{z}^{z_1} \sqrt{\varepsilon(z)}\, dz, \quad \tilde{\zeta} < -a, \quad z < z_1 \ , \\
&\int_{-a}^{\tilde{\zeta}} \sqrt{a^2 - \tilde{\zeta}^2}\, d\tilde{\zeta} = \int_{z_1}^{z} \sqrt{|\varepsilon(z)|}\, dz, \quad -a \le \tilde{\zeta} < a, \quad z_1 < z < z_2 \ , \\
&\int_{a}^{\tilde{\zeta}} \sqrt{\tilde{\zeta}^2 - a^2}\, d\tilde{\zeta} = \int_{z_2}^{z} \sqrt{\varepsilon(z)}\, dz, \quad \tilde{\zeta} > -a, \quad z > z_2 \ ,
\end{aligned} \tag{9.4.10}$$

where parameter a is defined by the condition

$$\int_{-a}^{a} \sqrt{a^2 - \tilde{\zeta}^2}\, d\zeta = \pi a^2/2 = \int_{z_1}^{z_2} \sqrt{|\varepsilon(z)|}\, dz \ . \tag{9.4.11}$$

The integrals on the left-hand side of (9.4.10) are straightforward and we will not repeat them any more so that these relations define $\tilde{\zeta}(z)$ implicitly.

In the following order in k, the Olver–Pike procedure yields an equation for $A(z)$, viz.,

$$\frac{d}{dz}\left[A^2(z)\frac{d\zeta}{dz}\right] = 0 \ , \tag{9.4.12}$$

from this it follows:

$$A(z) = C\left(\frac{d\zeta}{dz}\right)^{-1/2} = C\left(\frac{\zeta^2 - a^2}{\varepsilon(z)}\right)^{1/4} \ . \tag{9.4.13}$$

It is worthwhile to note that $A(z)$ is nowhere singular including the turning points $z_{1,2}$ where $\varepsilon = 0$.

After a comparatively simple modification, the solution (9.4.8) may be extended to the case of over barrier reflection. Then the function $\varepsilon(z)$ of (9.4.7) will not vanish along the real z axis but exhibit a positive minimum, $\varepsilon_{\min} > 0$. The standard square function in Weber equation (9.4.1) may be $\zeta^2 + \beta^2$ having a single minimum and no real-valued zeros. In a formal way, a crossover from subbarrier penetration to over barrier reflection is described by a change of the positive parameter $b^2 = ka^2$ for the negative value $-\beta^2 = -k\alpha^2$, i.e., by a change of $b = k^{1/2}a \to i\beta = ik^{1/2}$ in all equations. Now the relation between $\tilde{\zeta}$ and z (again in the implicit form) is given by

$$\int^{\tilde{\zeta}} \sqrt{\tilde{\zeta}^2 + \alpha^2}\, d\tilde{\zeta} = \int^{z} \sqrt{\varepsilon(z)}\, dz \ , \tag{9.4.14}$$

while for the amplitude we have

$$A(z) = C\left(\frac{\zeta^2 + a^2}{\varepsilon(z)}\right)^{1/4} . \tag{9.4.15}$$

A sole essential change concerns the definition of parameters β and $\alpha = k^{-1/2}\beta$. Let z_1 and z_2 be the complex turning points closest to the real axis, i.e., complex zeros of $\varepsilon(z)$. We associate with these zeros of $\varepsilon(z)$ the imaginary values $a = i\alpha$ and $-a = -i\alpha$ of $\tilde{\zeta}(z)$. Relation between $a = i\alpha$ and $z_{1,2}$ can be established by extending (9.4.14) to complex values of $\tilde{\zeta}$ and z. Integrating (9.4.14) over both sides of the cut connecting z_1 and z_2 in the complex plane z (integration is denoted by a contour integral) yields

$$i\pi\alpha^2 = \oint \sqrt{\varepsilon(z)}\, dz \ . \tag{9.4.16}$$

The quantity α^2 is real-valued, for the contour integral assumes a purely imaginary value. Note that substituting $i\beta$ for b in (9.4.6) yields

$$|R(i\beta)| = |1 + \exp(\pi\beta^2)|^{-1/2}, \qquad |D(i\beta)| = |1 + \exp(-\pi\beta^2)|^{-1/2} ,$$

so that unlike subbarrier penetration, in over barrier reflection R is always less than D.

9.4.3 Penetration of a Plane Wave Through a Barrier

Suppose that the primary field u^0 is a plane wave given at the horizon $z = z^0$ as

$$u^0(x) \equiv u(x, z^0) = U^0 \exp(ikn^0 x \sin\theta^0) \ , \tag{9.4.17}$$

where $n^0 = \sqrt{\varepsilon(z^0)}$ is the refractive index, and θ^0 is the angle of incidence on the horizon z^0. The variables of this problem are separable: the dependence on x is given by the exponential factor in (9.4.17) and the dependence on z is determined from the equation (9.4.7) where $\varepsilon(z)$ should be treated as $\varepsilon(z) - (n^0 \sin\theta^0)^2$. Accordingly, the overall field has the form

$$u(x, z) = A(z) w[k^{1/2}\tilde{\zeta}(z), b] \exp(ikn^0 x \sin\theta^0) \ . \tag{9.4.18}$$

In the expression for amplitude, constant C is chosen such that at $z = z^0$ the amplitude of the incident wave will be U^0, viz.,

$$C = U^0[\varepsilon(z) - (n^0 \sin\theta^0)^2]^{-1/4} \exp\left[-ik \int_{z^0}^{z_1} \sqrt{\varepsilon(z) - (n^0 \sin\theta^0)^2}\, dz\right] .$$

Here, we have used the asymptotic of the incident wave w_1 from (9.4.3) and the first line in (9.4.10).

In the applicability domain of WKB asymptotics, the solution (9.4.18) allows for a ray interpretation. In the lit region lying below the level $z = z_1$ the field is a superposition of incident and reflected waves with relative intensities of 1 and $|R|^2$, as shown in Fig. 9.2a. Incomplete reflection ($|R|^2 < 1$) is caused by a loss of a portion of energy which penetrates through the barrier and behind the second turning point $z = z_2$ induces a traveling wave of intensity $|D|^2$. This is the region of secondary propagation.

In the subbarrier zone, $z_1 < z < z_2$, where the field decays exponentially, a ray interpretation can be achieved with the aid of complex rays [9.17]. Flows of energy are hard to localize in this region as the domain where complex rays are formed, is usually much larger than that of real rays [9.18, 19]. In caustic shadow, the size of the formation region, l in Fig. 9.2a, is defined by the imaginary part $\delta = \operatorname{Im} x^0$ on the shift of penetration point of the complex ray.

In the case of over barrier reflection, both branches, z_1 and z_2, of the penetrable caustic become complex valued; therefore, it is still harder to localize the domains of formation of transmitted and reflected waves than in subbarrier reflection. We are in a position to claim only that the reflected wave is formed in the neighborhood of the permittivity minimum, ε_{min}, as indicated in Fig. 9.2b by a wavy line.

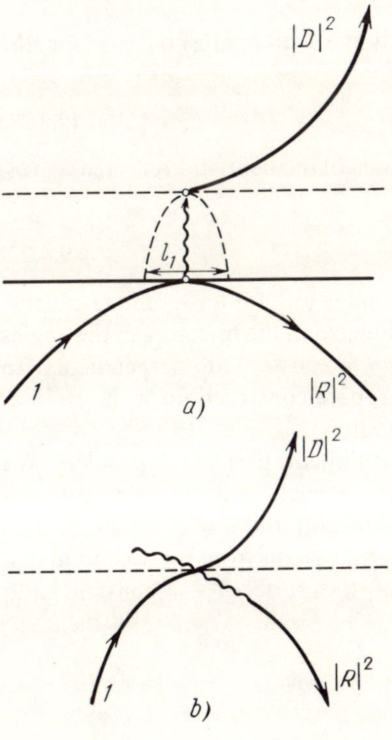

Fig. 9.2. (**a**) Caustic with penetration (sub-barrier penetration). (**b**) Ray interpretation of over-barrier reflection. In both situations an incident wave of unit intensity gives rise to reflected and transmitted waves of intensity $|R|^2$ and $|D|^2$, respectively

9.4.4 Asymptotic Representation of the Field for a Barrier with Variable Parameters

Suppose that the parameter $a = b/k^{1/2}$ characterizing the width of the potential barrier varies rather slowly along the axis of the barrier (slowness should in all probability be treated as the condition of smallness of variation over the length of field formation $l_{\|}$, i.e., $l_{\|}|\nabla_{\|}a| \ll a$, where $\nabla_{\|}$ implies differentiation along the axis). Then the solution to the Helmholtz equation may be represented in the form [9.20]

$$u = \left[Aw(k^{1/2}\tilde{\zeta}, b) + \frac{\mathrm{i}B}{[k(\tilde{\zeta}^2 - a^2)]^{1/2}} w'(k^{1/2}\tilde{\zeta}, b)\right] \mathrm{e}^{\mathrm{i}k\chi} , \tag{9.4.19}$$

where w' is the derivative of the Weber function (9.4.4), and A, B, ζ, and χ are indeterminate functions. In contrast to the solution (9.4.18) for a plane-stratified medium, (9.4.19) has a second term involving w' and the phase factor $\exp(\mathrm{i}k\chi)$ describing the propagation of the wave along the barrier.

In the lines of the standard procedure we develop the amplitudes A and B in series of negative powers of $\mathrm{i}k$ and set like powers in k equal to zero under an additional requirement that the terms involving a may be neglected. In the lowest order in k this leads to the system of equations for $\tilde{\zeta}$ and χ:

$$\begin{aligned} &(\nabla\chi)^2 + (\tilde{\zeta}^2 - a^2)(\nabla\tilde{\zeta})^2 = \varepsilon , \\ &\nabla\chi \cdot \nabla\tilde{\zeta} = 0 . \end{aligned} \tag{9.4.20}$$

These equations reduce to a single equation of eikonal $(\nabla\psi)^2 = \varepsilon$ for the functions

$$\psi_{1,2} = \chi \mp \int \sqrt{\tilde{\zeta}^2 - a^2}\, \mathrm{d}\tilde{\zeta} , \tag{9.4.21}$$

which would be natural to treat as the eikonals of incident and reflected waves. If ψ_1 and ψ_2 are known, then

$$\chi = (\psi_1 + \psi_2)/2 , \qquad \int \sqrt{\tilde{\zeta}^2 - a^2}\, \mathrm{d}\tilde{\zeta} = (\psi_2 - \psi_1)/2 . \tag{9.4.22}$$

These expressions are analogous to the formulas (5.1.15) for a simple caustic. Like in that case, the lines $\tilde{\zeta}$ = constant are directed in the bisectors of the angles formed by rays in the lit zone, and the lines χ = constant are perpendicular to these lines. These properties facilitate the graphic construction of the isolines ζ = constant and χ = constant in the light region.

Determination of parameter a is the most difficult part of the problem. We introduce orthogonal coordinates ξ, η, and σ such that $\tilde{\zeta}$ = constant and χ = constant on the coordinate planes ξ = constant and η = constant, respectively. The possibility of using coordinates ξ and η stems from the second line in (9.4.20) which establishes orthogonality of the surfaces $\tilde{\zeta}$ = constant and χ = constant. In the new coordinates

$$(\nabla\tilde{\zeta})^2 = h_1^{-2}\left(\frac{\partial\tilde{\zeta}}{\partial\xi}\right)^2, \qquad (\nabla\chi)^2 = h_2^{-2}\left(\frac{\partial\chi}{\partial\eta}\right)^2 ,$$

where $h_{1,2}$ are the respective Lame coefficients. Because we have assumed that parameter a is of slow variation, it should be viewed as a local quantity. This implies essentially that near a caustic we deal with a field of an almost plane wave reflected from an almost flat barrier.

Under the above assumptions the coordinates ξ, η, and σ will be close to Cartesian and we may let $h_1, h_2 \approx 1$. Then from (9.4.20) we obtain

$$(\tilde{\zeta}^2 - a^2)\left(\frac{\partial\tilde{\zeta}}{\partial\xi}\right)^2 = \varepsilon - \left(\frac{\partial\chi}{\partial\eta}\right)^2 . \tag{9.4.21}$$

Let at fixed values of η and σ the right side of (9.4.21) vanish at ξ_1 and ξ_2. From here on we may repeat all the reasoning of Sect. 9.4.1 to get by analogy with (9.4.11) the expression

$$\frac{\pi a^2}{2} = \int_{\xi_1}^{\xi_2} \sqrt{\varepsilon - (\partial\chi/\partial\eta)^2}\, d\xi . \tag{9.4.22}$$

For overbarrier reflection we obtain an analog of (9.4.16)

$$i\pi\alpha^2 = \oint \sqrt{\varepsilon - (\partial\chi/\partial\eta)^2}\, d\xi .$$

The next order in k yields two equations for the amplitudes A and B, which can be reduced to two ray optical transfer equations for the ray amplitudes $U_{1,2} = (A \pm B)(\tilde{\zeta}^2 - a^2)^{-1/4}$

$$\operatorname{div}(U_{1,2}^2 \nabla\psi_{1,2}) = 0 .$$

Thus, given the ray amplitudes U_1 and U_2,

$$A = \tfrac{1}{2}(U_1 + U_2)(\tilde{\zeta}^2 - a^2)^{1/4}, \qquad B = \tfrac{1}{2}(U_1 - U_2)(\tilde{\zeta}^2 - a^2)^{1/4} . \tag{9.4.23}$$

In one-dimensional case, when $U_1 = U_2$ and $\chi = 0$, we have $B = 0$ and the expression for A goes over to (9.4.12).

With the said values of $\tilde{\zeta}$, χ, A, and B, the field (9.4.19) becomes the sum of incident and reflected waves

$$u = U_1 \exp(ik\psi_1) + RU_2 \exp(ik\psi_2)$$

in the lit region, $\zeta < -a$, and becomes the transmitted wave

$$u = DU_1 \exp(ik\psi_1)$$

in the region of secondary propagation behind the barrier.

The quality of approximation achieved with (9.4.19) is higher the closer the actual problem is to the standard one of a plane wave incident on a flat barrier. Applicability conditions for this approximation may be written as

$$l \ll L_\varepsilon, \qquad l \ll \rho_c, \tag{9.4.24}$$

where L_ε is the typical length of variation of ε along the barrier, ρ_c is the curvature radius of the front branch of the caustic, and l is the size of the region responsible for the formation of the field at the back branch of the caustic

(Fig. 9.2a). The role of scale l may be played by the shift of the point of penetration of a complex ray into the complex space [9.18, 19]. This dimension is comparable with the width Δz of the barrier, hence, the second inequality in (9.4.24) is equivalent to a requirement that the radius of curvature be large compared with the width of the barrier.

These considerations agree qualitatively with the results of *Kaloshin* and *Orlov* [9.21] who indicated that a spherical wave incident on the barrier renders the back branch of the caustic complex valued so that in the region of secondary propagation one has to do with complex rather than real valued rays. However, the shift of the caustic into the complex space becomes insignificant if the separation of the point source from the barrier is large compared with the barrier width, that is, when the conditions (9.4.24) are met. These conditions were violated in the problem of "radio wave onset" of a satellite tackled in [9.20] as an example (relevant criticism may be found in [9.22]).

Another candidate for the typical dimension l emerges when the problem is approached from the light side, namely, this is the longitudinal Fresnel scale $l_{\|}$ that has been revealed in considering a simple caustic in Sects. 5.1.6, 13.

Recapitulating the statement that the quality of approximation (9.4.19) is better, the closer is the problem at hand to the reference conditions of a plane wave incident on a flat barrier. We point out specifically that a good approximation is observed when the orthogonal coordinates ξ, η, and σ tied to the barrier is close to Cartesian.

9.4.5 Waveguiding Caustics

The asymptotic field representation involving Weber functions can be modified so that it will be suited to describe waves in a potential well. To this end, as standard functions one should take functions satisfying the equation of a quantum harmonic oscillator

$$\frac{d^2w(\zeta,b)}{d\zeta^2}+(b^2-\zeta^2)w(\zeta,b)=0\ , \tag{9.4.25}$$

which differs from (9.4.1) in a permutation of ζ and b.

This equation is known to possess solutions fading out as $\zeta\to\pm\infty$ provided that parameter b assumes discrete values, $b_n=\sqrt{2n+1}$ with integer n [9.23]. The respective functions $w(\zeta,b_n)\equiv W_n(\zeta)$ are expressed through Hermite polynomials multiplied by a Gaussian function, viz.,

$$w(\zeta,b_n)=H_n(\zeta)\exp(-\zeta^2/2)\ . \tag{9.4.26}$$

A discrete set of functions (9.4.26) may be employed to approximately describe normal waves in refractive waveguides (Fig. 9.3) and normal oscillations of the bouncing-ball type in resonators (Fig. 9.4) [9.24]. In these circumstances, the asymptotic solution differs from (9.4.19) by a permutation of $\zeta=k^{-1/2}$ and

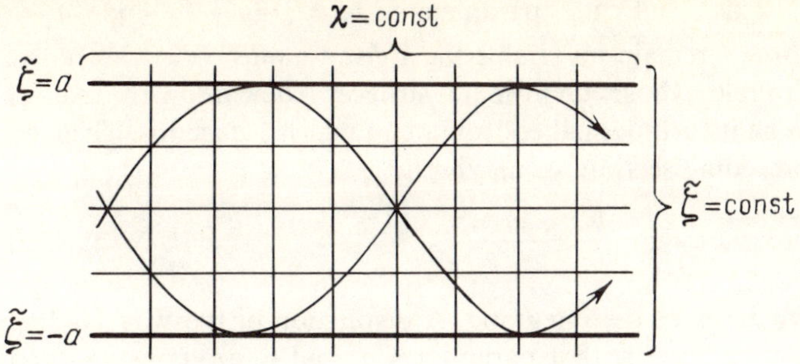

Fig. 9.3. Waveguiding caustic confining rays in a plane stratified wave duct. The grid represents isolines χ = constant and ζ = constant

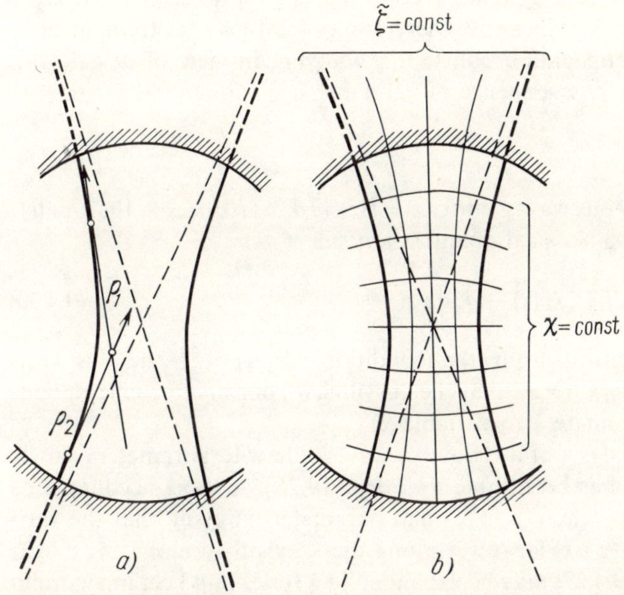

Fig. 9.4. Two-branched caustic confining "bouncing-ball" oscillations in an open resonator: (**a**) ray configuration, (**b**) configuration of isolines χ = constant and ζ = constant

$a_n = b_n k^{-1/2}$, namely,

$$u = \left\{ Aw(k^{1/2}\tilde{\zeta}, b_n) + \frac{\mathrm{i}B}{[k(a_n^2 - \tilde{\zeta}^2)]^{1/2}} w'(k^{1/2}\tilde{\zeta}, b_n) \right\} \mathrm{e}^{\mathrm{i}k\chi} . \tag{9.4.27}$$

Equations for $\tilde{\zeta}$, χ, A, and B are quite similar to the equations of preceding Sect. 9.4.4 and accurate to a substitution of $a_n^2 - \tilde{\zeta}^2$ for $\tilde{\zeta}^2 - a^2$.

An important feature in which the problem under consideration differs from subbarrier propagation is that both branches of the caustic are real and confine the lit region (Figs. 9.3, 4) whereas, behind the caustic there is a shadow region where the field decays exponentially. The actual width of the potential well varies along the axis, but its parameters $b_n = \sqrt{2n+1}$ and $a_n = k^{-1/2} b_n = k^{-1/2}\sqrt{2n+1}$ remain invariable for a given mode.

Let ξ be a variable whose coordinate surfaces coincide with that of $\tilde{\zeta}$ = constant, and η be its orthogonal coordinate of which χ depends. Then, by analogy with the preceding section we may write

$$a_n^2 = \frac{2}{\pi}\int_{\xi_1}^{\xi_2} \sqrt{\varepsilon - (\partial\chi/\partial\eta)^2}\, \mathrm{d}\xi = \frac{b_n^2}{k} = \frac{2n+1}{k} , \tag{9.4.28}$$

where ξ_1 and ξ_2 are zeros of the integrand. A distinction of the waveguiding problem from the barrier one is that parameters a_n and b_n are constant, and substituting (9.4.27) in the Helmholtz equation, we no longer have the derivatives of these quantities.

In the initial setting the system of coordinates ξ, η, and σ differs from Cartesian coordinates insignificantly; therefore, (9.4.28) may be thought of as an equation for the propagation constant $q = \partial\chi/\partial\eta$. In view of (9.4.28) this quantity assumes a discrete sequence of values $q_1, q_2, \ldots, q_n$, and for the nth mode,

$$\chi = \int q_n \,\mathrm{d}\eta . \tag{9.4.29}$$

For a strictly homogeneous waveguide of $\varepsilon = \varepsilon(z)$, (9.4.28) reduces to the familiar Bohr–Zommerfeld quasi-classical quantization rule

$$\pi b_n^2 = \pi(2n+1) = 2k\int_{z_1}^{z_2} \sqrt{\varepsilon(z) - q_n^2}\,\mathrm{d}z , \tag{9.4.30}$$

where $z_{1,2}$ are determined from the condition $\varepsilon(z_{1,2}) = q_n^2$. In this case, q_n = constant and $\chi = q_n x$ (recall that the curvilinear coordinate ξ corresponds to variable z, and coordinate η to variable x).

A more detailed analysis of a refractive waveguide with parameters slowly varying along the axis has been done by *Popov* [9.25]. A short account is as follows: Let $\varepsilon = \varepsilon(\mu x, z)$, where μ is a small parameter implying that the variation rate of permittivity ε is far lower along the x axis than along the z axis. Now, all quantities in (9.4.27) may be expanded in a series in μ keeping in mind that the amplitudes A and B can be simultaneously expanded in powers of $1/ik$. For the nth mode, the principal term in μ and k takes the form

$$u = \frac{\text{const.}\ (a_n^2 - \tilde{\zeta}^2)^{1/4}}{\sqrt{T}(\varepsilon - q_n^2)^{1/4}} \exp[(ik\int q_n \,\mathrm{d}x)\, W(k^{1/2}\zeta_0)] , \tag{9.4.31}$$

where $\zeta_0(z)$ is defined by

$$\int_{-a_n}^{\zeta_0} \sqrt{a^2 - \tilde{\zeta}^2}\,\mathrm{d}\tilde{\zeta} = \int_{z_1}^{z} \sqrt{\varepsilon - q_n^2}\,\mathrm{d}z ,$$

and the quantity

$$T_n = 2\int_{z_1}^{z_2} \frac{q_n\,\mathrm{d}z}{(\varepsilon - q_n^2)^{1/2}}$$

characterizes the period of longitudinal oscillation of the ray or the length of a ray cycle. The factor $T_n^{-1/2}$ in (9.4.31) describes the variation of a field amplitude in a waveguide of variable width and takes care of conservation of a flow of energy in such a waveguide.

In the following order in μ the function (9.4.31) acquires the phase factor $\exp(ik\mu\chi_1)$ caused by the reconstruction of the field in coming from a narrow to a wider cross section of the guide or from a wider into a narrower cross section [9.24]. An asymmetry of the waveguide adds to (9.4.31) another term involving W_n'.

9.4.6 Caustics Confining "Jumping Ball" Oscillations

Open and closed resonators support eigenmodes that *Keller* and *Rubinov* [9.26] called "bouncing-ball" oscillations (Fig. 9.4). A feature peculiar to such oscillations is families of rays confined by caustics on either side. The field oscillates between the caustic branches and falls off exponentially beyond the caustics. Such fields are adequately described by the product of a Weber function (9.4.26) of integer index by the phase factor $\exp(ik\chi)$ associated with the motion of the wave along the caustics. In other words, the construction (9.4.27) renders an asymptotic solution to the problem of jumping ball oscillations.

The quest of this problem has been the subject of many books, including those of *Vainstein* [9.27], and of *Babich* and *Buldyrev* [9.28]. Here we give a modification of the uniform asymptotic solution (9.4.27) adapted to the problem of eigenmodes in an open resonator filled with a homogeneous medium for which we let $\varepsilon = 1$ for definiteness [9.29]. This version is notorious for yielding specific amplitudes U_1 and U_2 for a two-dimensional symmetric caustic constituted by two branches which in a homogeneous medium have common X-shaped asymptotes. Points on the caustic may be labeled by a parameter χ which varies along the caustic.

Every point lying between the caustic branches is hit by two rays touching the caustic at points χ_1 and χ_2. Let ρ_1 and ρ_2 denote the distances from the point of observation to the points of tangency. By the common properties of two-dimensional fields the ray amplitudes are proportional to $1/\sqrt{\rho}$ (cylindrical divergence), namely,

$$U_{1,2} = F(\chi_{1,2})/\sqrt{\rho_{1,2}}\ , \tag{9.4.32}$$

where $F(\chi)$ characterizes the distribution of intensity along the caustic. Making use of (9.4.32, 23) we get

$$\begin{aligned} u = \Bigg[&(\tilde{\zeta}^2 - a^2)^{1/4} \left(\frac{F(\chi_1)}{\sqrt{\rho_1}} + \frac{F(\chi_2)}{\sqrt{\rho_2}} \right) W_n(k^{1/2}\tilde{\zeta}) \\ &+ ik^{-1/2}(\tilde{\zeta}^2 - a^2)^{-1/4} \left(\frac{F(\chi_1)}{\sqrt{\rho_1}} - \frac{F(\chi_2)}{\sqrt{\rho_2}} \right) W_n'(k^{1/2}\tilde{\zeta}) \Bigg] e^{ik\chi}\ . \end{aligned} \tag{9.4.33}$$

This expression represents a wave traveling along the caustic branches. This wave and its complex conjugate u^* may be combined to form a standing wave which may be viewed as the result of reflections from the curvilinear surfaces $\chi =$ constant, say $\chi = \chi^+$ and $\chi = \chi^-$. Eigenfunctions will correspond to surfaces which will meet the condition of longitudinal quantization

$$k(\chi^+ - \chi^-) = \pi N_\parallel \ , \tag{9.4.34}$$

while the condition (9.4.28) may be thought of as the one of transverse quantization.

Jumping ball oscillations are a particular case of a more general phenomenon – oscillations clustered in the neighborhood of a geodesic, i.e., a spatial line possessing extremal properties. *Babich* and *Buldyrev* [9.28] describe eigenmodes concentrated in the extremal ray on the basis of a parabolic equation. The development used in this book, see (7.2.12), does not involve a derivative of the Weber function, but instead, the argument of the Weber function is a series in powers of $k^{-1/2}$. Upon re-expansion this series gives rise to a combination of W_n and W_n'.

9.4.7 Applicability of the Weber Asymptotic

One of applicability conditions, mentioned above, concerns the slow rate of variation of barrier properties along the axis [inequalities of the type (9.4.24)]. Similar constraints should be met in waveguiding caustics. Specifically, the typical variation length of the amplitude factor $F(\chi)$ in (9.4.33) must be large compared with the longitudinal Fresnel scale $l_\parallel$ (Sect. 5.1.6).

Conditions imposed on the dependencies of medium and wave parameters transverse to the caustic are harder to formulate. Indeed one has to combine in a tractable form the essentially different conditions, namely, asymptotics of the Olver Pike type are applicable both for wide barriers and wells of arbitrary profile (parameter b is large) and for narrow barriers and wells with a narrow class of $\varepsilon(z)$ profiles close to parabolic. These conditions were perceived a long time ago, but unfortunately have not been formalized in an appropriate form.

In the absence of such a synthetic formulation we resort to a rather old example [9.20] related to the quantum mechanical problem of a particle oscillating in a potential well $V(z) = -\, V_0 \cosh^{-2}(\alpha z)$. Straightforward calculations with the formulas (9.4.30) yield for eigenvalues of particle energy E_n the approximate expression

$$\frac{E_n}{V_0} = -\frac{1}{\kappa}[\sqrt{\kappa} - (2n+1)]^2 \ , \tag{9.4.35}$$

where $\kappa = 8mV_0/\hbar^2\alpha^2$, m is the mass of the particle, and $\hbar$ is the Planck constant divided by 2π. Exact values of E_n/V_0 are given [9.23] by the relation

$$\frac{E_n}{V_0} = -\frac{1}{\kappa}[\sqrt{\kappa+1} - (2n+1)]^2 \ . \tag{9.4.36}$$

Figure 9.5 shows the plots of E_n/V_0 versus $\sqrt{\kappa} = \sqrt{8mV_0}/\hbar\alpha$. The plots of

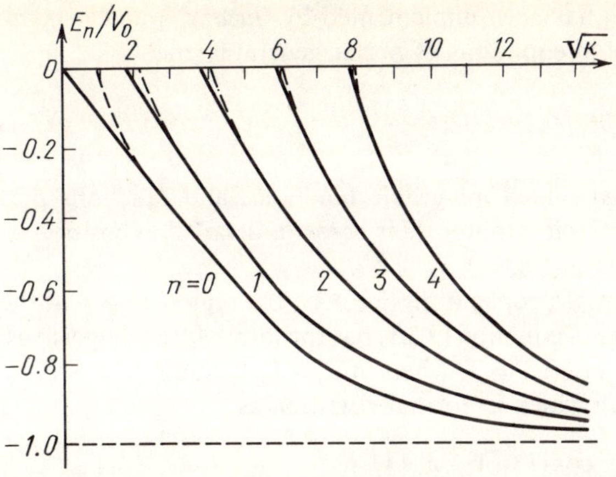

Fig. 9.5. Energy eigenvalues in the potential well $V = V_0/\cosh^2(\alpha x)$. Continuous lines represent the exact solution (9.4.36), and dashed lines represent approximate values calculated by (9.4.35)

approximate values of E_n (dashed lines) run close to the exact curves (continuous lines) as $\kappa \to \infty$ for both large and small quantum numbers including $n = 0$. Large values of κ are achievable in the case of a wide well ($1/\alpha \to \infty$) and also in the case of a narrow well, provided that the depth of well V_0 is sufficiently large. For a deep well, the fundamental state $n = 0$ corresponds to oscillations near the well's bottom where the well profile is close to parabolic so that the quality of the approximation (9.4.35) is rather high. For a shallow well ($\kappa \to 0$), the deviation from a parabolic profile becomes appreciable and the dashed lines run far from the continuous counterparts in Fig. 9.5.

All the above comments relate in equal measure to the problem of penetration through the barrier $V(z) = V_0 \cosh^{-2}(\alpha z)$. The approximate values of transmission and reflection coefficients, D and R in (9.4.6), calculated with the aid of (9.4.11) turn out to be close to the exact values of D and R given in [9.23] provided that (a) the barrier width $1/\alpha$ is large, or (b) the width of the barrier is small but the shape of the barrier between the turning points z_1 and z_2 is close to parabolic. These reservations should be kept in mind in face of the literature claims that the asymptotic formulas for transmittance and reflectance are valid for any profile of the barrier.

9.5 Standard Functions Induced by Ordinary Differential Equations

9.5.1 Using Second-Order Differential Equations as Standards

Thus far we have handled two simplest standard equations, Airy's equation $w'' - \zeta w = 0$ and Weber's equation $w'' \pm (\zeta^2 \pm b^2)w = 0$. An alternative point of departure may have been more complex second-order equations of a rather general form. This approach has been implemented by *Gazazyan* and *Ivanyan* [9.30]. They suggest the use of equations of rather general form

$$\frac{d^2f(\zeta)}{d\zeta^2}+\alpha(\zeta)\frac{df(\zeta)}{d\zeta}+k^2\beta(\zeta)f(\zeta)=0\ , \tag{9.5.1}$$

where $\alpha(\zeta)$ are arbitrary real-valued integrable functions, and functions $\beta(\zeta)$ must be indefinitely differentiable around their zeros (a detailed exposition of initial assumptions was given in [9.31]).

For $\beta > 0$, in the domain of propagation (let it be the region of $\zeta > a$) the WKB asymptotic of solutions of equation (9.5.1) has the form of traveling waves. To the right of the turning point $\zeta = a$, where $\beta(a) = 0$, the sum of traveling waves forms a standing wave which is normally written as

$$\begin{aligned} f(\zeta) &= C[\sqrt{\beta(\zeta)}\gamma(\zeta)]^{-1/2}\cos[ks(\zeta)-\pi/4] \\ &= (C/2)[\sqrt{\beta(\zeta)}\gamma(\zeta)]^{-1/2}\times[\exp(iks(\zeta)-i\pi/4) \\ &\quad + \exp(-iks(\zeta)+i\pi/4)]\ , \end{aligned} \tag{9.5.2}$$

where

$$s(\zeta)=\int_a^\zeta \sqrt{\beta(\zeta)}\cdot d\zeta \tag{9.5.3}$$

is the phase of the WKB asymptotic, and

$$\gamma(\zeta)=\exp[-\int\alpha(\zeta)d\zeta]\ .$$

On the left from the turning point, where $\beta(\zeta) < 0$, the wavefunction decays exponentially, viz.,

$$f(\zeta)=(C/2)[\sqrt{|\beta(\zeta)|}\gamma(\zeta)]^{-1/2}\exp[-k\int_\zeta^a\sqrt{|\beta(\zeta)|}d\zeta]\ . \tag{9.5.4}$$

Equations of the type (9.5.1) can be satisfied not only by harmonic function $\exp(i\zeta)$, Airy function $\mathrm{Ai}(\zeta)$ and Weber function $w(\zeta, b)$, but also by all known special functions of mathematical physics including cylinder functions, spherical functions (associated Legendre functions), Mathieu (elliptic cylinder) functions, spheroid functions, and ellipsoid (Lame) functions. Turning points, which is, zeros of $\beta_i(\zeta_i)$ with respect to ζ_i, correspond to various spatial caustics (cylinders, cones, ellipsoids, hyperboloids, etc.); therefore, combining functions f_i one can construct asymptotic solutions for a wide variety of problems related primarily to oscillations and eigenmodes in resonators.

9.5.2 Uniform Asymptotics of 3-D Wave Problems Developed with 1-D Standard Functions

Gazazyan and *Ivanyan* [9.30] have suggested looking for a uniform asymptotic solution to three-dimensional wave problems in the form of a product of three special functions $f_i(\zeta_i)$ with indeterminate amplitudes in combination with the derivatives of these functions $f_i'(\zeta_i) = f_i'$, namely,

$$u = A_{000} f_1 f_2 f_3 + A_{100} f_1' f_2 f_3 + A_{010} f_1 f_2' f_3 + A_{001} f_1 f_2 f_3'$$
$$+ A_{110} f_1' f_2' f_3 + A_{101} f_1' f_2 f_3' + A_{011} f_1 f_2' f_3' + A_{111} f_1' f_2' f_3' \ .$$

Introducing the indexes l, m, and n running from 0 to 1 and assuming that the zero-order derivative $\mathrm{d}^0 f_i / \mathrm{d}\zeta_i^0$ represents the function f_i we rewrite this sum in a compact form

$$u = \sum_{l,m,n=0}^{1} A_{lmn} \frac{\mathrm{d}^l f_1(\zeta_1)}{\mathrm{d}\zeta_1^l} \frac{\mathrm{d}^m f_2(\zeta_2)}{\mathrm{d}\zeta_2^m} \frac{\mathrm{d}^n f_3(\zeta_3)}{\mathrm{d}\zeta_3^n} \ . \tag{9.5.5}$$

The unknown amplitudes $A_{lmn}(\mathbf{r})$ are expanded in a series in powers of $1/k$ and sought, along with $\zeta_i(\mathbf{r})$, from the equations which are obtained by substituting (9.5.5) in the Helmholtz equation and equating to zero the coefficients of the like powers of k. The second and third derivatives of f_i that occur in differentiating (9.5.5) are eliminated by virtue of equation (9.5.1).

In the general case the equations for $\zeta_i(\mathbf{r})$ and $A_{lmn}(\mathbf{r})$ are rather involved. Fortunately, we can do without solving these equations. Like in the construction of uniform asymptotic involving standard integrals of catastrophe theory, the unknown functions ζ_i and A_{lmn} can be expressed in terms of ray amplitudes U_j and phases S_j.

Consider the WKB asymptotic of the wave field (9.5.5) in the region where all $\beta_i(\zeta_i) > 0$,

$$u(r) = \sum_{\lambda,\mu,\nu=0}^{1} \sum_{l,m,n=0}^{1} A_{lmn}(-1)^{\lambda l + \mu m + \nu n} \kappa_{lmn}$$
$$\times \exp\{ik[(-1)^\lambda s_1 + (-1)^\mu s_2 + (-1)^\nu s_3]\} \ . \tag{9.5.6}$$

The quantities κ_{lmn} involve the pre-exponential factors of the WKB representations of $f_i(\zeta_i)$ and their derivatives, viz.,

$$\kappa_{lmn} = \frac{\beta_1^{l/2} \beta_2^{m/2} \beta_3^{n/2}}{8(\sqrt{\beta_1 \beta_2 \beta_3 \gamma_1 \gamma_2 \gamma_3})^{1/2}} (ik)^{l+m+n} \ . \tag{9.5.7}$$

Equating all $2^3 = 8$ combinations of phases s_i in the exponent of (9.5.6) to the respective ray fields $U_{\lambda\mu\nu} \exp(ik\psi_{\lambda\mu\nu})$ we obtain equations for the three unknown phases $s_{1,2,3}$

$$(-1)^\lambda s_1(\zeta_1) + (-1)^\mu s_2(\zeta_2) + (-1)^\nu s_3(\zeta_3) = \psi_{\lambda\mu\nu}(\mathbf{r}) \ , \tag{9.5.8}$$

and equations for the eight amplitudes A_{lmn}

$$\sum_{l,m,n=0}^{1} A_{lmn}(-1)^{\lambda l + \mu m + \nu n} \kappa_{lmn}^{(l)} = U_{\lambda\mu\nu}(\mathbf{r}) \ . \tag{9.5.9}$$

From (9.5.8) it follows that of the eight quantities $\psi_{\lambda\mu\nu}$ only three are linearly independent, and the left-hand side of this equation satisfies the eikonal equation

$$[(-1)^\lambda \nabla s_1 + (-1)^\mu \nabla s_2 + (-1)^\nu \nabla s_3]^2$$
$$= [(-1)^\lambda \sqrt{\beta_1(\zeta_1)}\nabla\zeta_1 + (-1)^\mu \sqrt{\beta_2(\zeta_2)}\nabla\zeta_2 + (-1)^\nu \sqrt{\beta_3(\zeta_3)}\nabla\zeta_3]^2$$
$$= (\nabla\psi_{\lambda\mu\nu})^2 = \varepsilon \ , \tag{9.5.10}$$

which can be simplified by a suitable choice of orthogonal coordinates $\zeta_i(\mathbf{r})$ in which the cross-terms $\nabla\zeta_i \cdot \nabla\zeta_k$ vanish to take on the form

$$\sum_{i=1}^{3} \beta_i(\zeta_i)(\nabla\zeta_i)^2 = \varepsilon \ .$$

At the same time, the left-hand side of (9.5.9) satisfies the transfer equation

$$\operatorname{div}\left\{\left[\sum_{l,m,n=0}^{1} A_{lmn}(-1)^{\lambda l+\mu m+\nu n}\kappa_{lmn}\right]^2 \nabla\psi_{\lambda\mu\nu}\right\} = 0 \ . \tag{9.5.11}$$

It is important to emphasize that the amplitude factors A_{lmn} remain finite at caustics, while the ray amplitudes $U_{\lambda\mu\nu}$ are singular. Compensation of the singularities stems in effect from the behavior of the ray amplitudes near caustics where they increase as $1/\beta_i^{1/4}$, so that the product $U_j\beta_i^{1/4}$ remains finite.

9.5.3 Caustics for an Ellipsoid Cavity

Solutions of the type (9.5.5) are most valuable for description of eigenmodes in open resonators. However, despite the direct algebraic relations (9.5.8, 9) between certain combinations of sought functions and ray solutions, they are hard for practical implementation; therefore, the outlined scheme [9.30, 31] requires certain ingenuity and extreme attention because the eight ray congruence of this interplay go over to one another after they have touched caustics and are reflected from the walls of the resonators.

In order that ray congruences correspond to the eigenmodes of a cavity, certain relationships, or quantum conditions, between phase shifts must be met. For ellipsoidal resonators, such conditions have been derived by *Bykov* [9.32] from purely ray considerations developing the constructions of *Keller* and *Rubinov* [9.26] for two-dimensional problems. The findings of Bykov are in good agreement with Vainstein's analysis [9.33] of the asymptotic expressions for Lame's functions involved as eigenfunctions in ellipsoidal coordinates.

For a three-axial ellipsoid, four types of eigenmodes are feasible [9.32]. For the first type of oscillations, caustic surfaces are an ellipsoid and a two-sheet hyperboloid, shown schematically in Fig. 9.6a; for the second type, these are an ellipsoid and a one-sheet hyperboloid, the diagram is shown in Fig. 9.6b. These two types of oscillations have the form of whispering gallery waves. The third type of oscillations given in Fig. 9.6c is bounded by a one-sheet and a two-sheet hyperboloids, and the fourth type shown in Fig. 9.6d by two one-sheet hyperboloids. In this case we deal with the jumping ball type of oscillations.

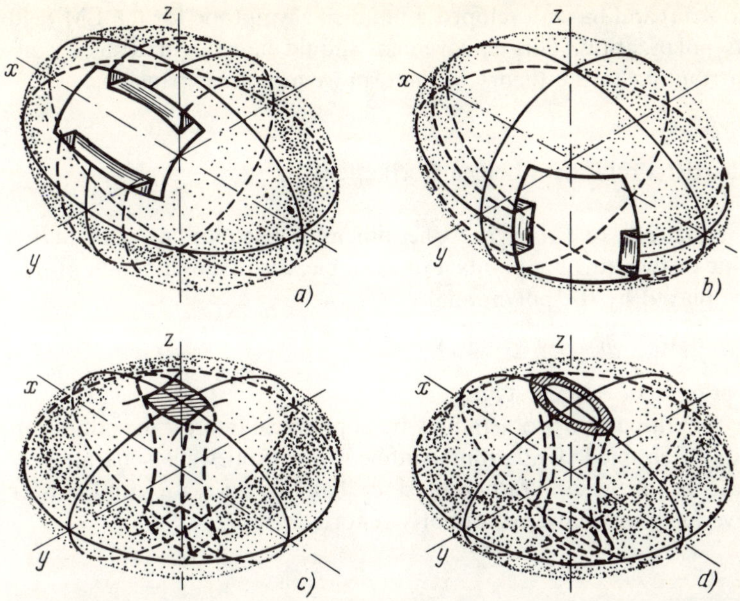

Fig. 9.6. Caustic surfaces corresponding to four types of oscillations in an ellipsoid cavity

In the case of an oblate ellipsoid of revolution, caustic surfaces are an ellipsoid and a one-sheet hyperboloid (Fig. 9.6b, second type) and two hyperboloids of one-sheet (Fig. 9.6d, fourth type). For an extended ellipsoid of revolution, only the first type of oscillations is realized; it is confined by an ellipsoid and a two-sheet hyperboloid (Fig. 9.6a). Finally, in the case of a spherical resonator, the ellipsoid surface degenerates into a spherical surface, and the two-sheet hyperboloid into a cone. If in the general case oscillations are described by Lame's functions, then for higher symmetries we are faced first with spheroid and Mathieu functions, then with spherical functions made up by Bessel functions of semi-integer index, associated Legendre functions, and a common harmonical function $\exp(\mathrm{i}\zeta)$.

In two-dimensional problems, the functions of Mathieu, Weber (for jumping-ball-type oscillations), and Airy (for whispering-gallery-type oscillations) come into play.

9.5.4 Extension of EM Oscillations

For electromagnetic waves, a solution is also sought in a form similar to that of (9.5.5) but with vectorial amplitudes. An analysis of EM fields for an ellipsoid cavity has been reported by *Gazazyan* and *Ivanyan* [9.34]. Although in ellipsoid polar coordinates the variables of Maxwell's equations are inseparable, they have exploited an approach based on Lame's functions (separating variables in the scalar problem) and have developed a uniform asymptotic for the EM field to analyze its polarization. This achievement should be recognized as a considerable contribution to the theory of uniform asymptotic expansions.

9.5.5 Multibarrier Problems: Coupled Oscillations

As standard equations one may use one-dimensional equations with many, rather than one or two, turning points. For example, the role of function $\beta(\zeta)$ in (9.5.1) may be played by the polynomial

$$\beta(\zeta) = (\zeta - b_1)(\zeta - b_2) \cdots (\zeta - b_m)$$

vanishing at m points $\zeta = b_1, \ldots, b_m$.

This type of multibarrier problem may be of practical interest in connection with the study of coupled oscillations confined in several wells separated by barriers. These problems are rather hard to handle, since for multibarrier problems wave functions have been poorly reported.

9.5.6 Caustics with Arbitrary Order of Ray Contact

Consider a one dimensional equation of the type (9.4.7) in the circumstances where the permittivity $\varepsilon(z)$ has a zero of order $\alpha > 0$:

$$\varepsilon(z) = -\left(\frac{|z - z_0|}{H}\right)^{\alpha} \operatorname{sgn}(z - z_0) \ . \tag{9.5.12}$$

This function, positive for $z < z_0$ and negative for $z > z_0$, for a particular case of $\alpha = 1$ becomes a usual linear layer of $\varepsilon(z) = -(z - z_0)/H$, whereas for $\alpha \neq 1$ the variation law is other than linear. In a problem of oblique incidence of a plane wave, $u = U^0 \exp(ikn^0 x \sin\theta^0)$ given at $z = z^0$, on a layer of permittivity ε when in (9.5.12) ε gives way to $\varepsilon(z) - \varepsilon^0 \sin^2\theta^0$, the order of ray contact with the caustic $z = z_0$ is variable; whereas, at $\alpha = 1$ at a point of tangency $z = z_0$ the ray has a shape of a common parabola, $z - z_0 \sim -(x - x_0)^2$, and in the general case $z - z_0 \sim -|x - x_0|^{2/(2-\alpha)}$. For example, at $\alpha = 1/2$ we have a parabola of order 4/3, and at $\alpha = 3/2$ a parabola of fourth order. When $\alpha \to 2$, the order of contact rises without bound, i.e., the ray seems as if it "sticks" to the caustic.

Orlov [9.35] has attempted to develop a uniform asymptotic of the field near such a caustic. He resorted to standard functions satisfying the equation

$$f''(\zeta) - (|\zeta|^{\alpha} \operatorname{sgn}\zeta) f(\zeta) = 0 \ .$$

Such functions are expressable in terms of Hankel's functions of order $\nu = 1/(2 + \alpha)$. In the particular case of $\alpha = 1$, these will be Airy functions which are represented by Hankel functions of order $\nu = 1/3$. In the general case, they are related to the Hardy integral and were quite appropriately called the Airy–Hardy functions.

It has been learned that Airy–Hardy functions change in form insignificantly when α is varied over a rather wide range, so that the order of ray contact to the caustic does not have an essential effect on the field structure near the caustic.

9.5.7 Standard Equations of Order Higher than Two

In one-dimensional inhomogeneous media, linear transformation of EM waves is known to be described by coupled second-order differential equations reducible to a single equation of fourth order (e.g., see [9.14, 36]). This problem is rather distant from the description of caustic fields, yet they reveal some points in common when it comes to asymptotic description of wave transformation. These have their origin in the complex turning points that occur in analysis of fourth order equations and contribute significantly in transformation of waves. Conditionally, to some degree, such turning points may be liken to complex caustics.

In addition, the method of standard functions is a valuable tool for both problems, for it allows solution of a wide array of problems with the aid of already known functions satisfying qualitatively similar conditions.

9.5.8 Interpolation Formulas for Oscillating Integrals

There is one more use of one-dimensional standard function, namely, approximation of oscillation integrals with a few stationary points. Relevant formulas involving a sum of four paired products of Airy functions (corresponding to traveling, rather than standing, waves) have been derived by *Markus* [9.37]. *Kreek* et al. [9.38] have suggested an alternative version based on a combination of Airy and Bessel functions. *Connor* [9.39] has reported an analysis of this approach where it claims that it gives excellent results when the saddle points lie close to one another, but is prone to error when these are scattered arbitrarily.

10 Caustics Revisited

The physical properties of a wave-propagating medium are often described by more complex wave-motion equations than Helmholtz's equation or dispersionless Maxwell's equations. Nevertheless manageable asymptotic solutions can be obtained for caustics in dispersive and anisotropic media, in nonlinear and random media and for related problems of quantum mechanics.

10.1 Caustics in Dispersive Media

10.1.1 Space–Time Caustics

Propagation of waves in dispersive media is described by a wave equation involving a non-local term characterizing a temporal or spatial dispersion. A wide class of problems for media with temporal (frequency dependent) dispersion obeys the scalar equation

$$\Delta u(t,\mathbf{r}) = \frac{1}{c^2}\frac{\partial^2}{\partial t^2}\int\limits_{-\infty}^{t} \varepsilon(t-t';t,\mathbf{r})u(t',\mathbf{r})\,\mathrm{d}t' \ . \tag{10.1.1}$$

Equations of geometrical optics closely related to (10.1.1) have been written down by *Felsen* and *Marcuvitz* [10.1] and *Kravtsov* and *Orlov* [10.2].

The field associated with a family of space–time rays $\mathbf{r} = \mathbf{R}(t, \xi)$ can be represented in parametric form as

$$\begin{aligned} u(t,\mathbf{r}) &= U(t,\boldsymbol{\xi})\,\mathrm{e}^{\mathrm{i}\phi(t,\boldsymbol{\xi})} \\ &= U^0(t^0,\boldsymbol{\xi})\left(\frac{n^0 g\mathscr{I}(t^0,\boldsymbol{\xi})}{ng^0\mathscr{I}(t,\boldsymbol{\xi})}\right)^{1/2}\exp\left[\mathrm{i}\phi^0(t,\boldsymbol{\xi}) + \int\limits_{t^0}^{t}(\mathbf{k}\cdot\mathbf{g}-\omega)\,\mathrm{d}t\right] , \end{aligned} \tag{10.1.2}$$

where $\boldsymbol{\xi} = (\xi_1, \xi_2, \xi_3)$ is a set of ray parameters, $\mathbf{k} = \nabla\phi$ is the local wave vector, $\omega = -\partial\phi/\partial t$ is the instantaneous frequency, and $\mathbf{g} = \mathrm{d}\mathbf{r}/\mathrm{d}t$ is the group velocity. An explicit dependence of the field on radius-vector $\mathbf{r}$ and time t may be obtained by eliminating $\boldsymbol{\xi}$ between this equation and $\mathbf{r} = \mathbf{R}(t, \boldsymbol{\xi})$.

In situations with multipath propagation, the overall field will be given by the sum of ray fields (10.1.2). Then there are regions where two or more rays merge and the Jacobians

$$\mathscr{J}(t, \boldsymbol{\xi}) = \det\left(\frac{\partial x_i(t, \boldsymbol{\xi})}{\partial \xi_m}\right) = \frac{\partial(x_1, x_2, x_3)}{(\xi_1, \xi_2, \xi_3)} \tag{10.1.3}$$

vanish. The equations $\mathscr{J}(t, \boldsymbol{\xi}) = 0$ and $\mathbf{r} = \mathbf{R}(t, \boldsymbol{\xi})$ together define a space–time caustic surface in the four-dimensional t, $\mathbf{r}$-space. At caustics the ray amplitude U grows without bound and the ray approximation (10.1.2) is not valid any longer.

Emergence of space–time caustics in dispersive media stems from the fact that signals of different frequency propagate at different group velocities. Consequently, if the initial signal $u^0 = U^0(t)\exp[i\phi^0(t^0)]$ is frequency-modulated, so that $\omega^0(t) = -\partial\phi^0/\partial t \neq$ constant, then on the z, t-plane (we take one dimension for simplicity) rays will have different slopes and can form a caustic.

Some possible situations are shown in Fig. 10.1. Suppose that the group velocity g increases with frequency ($\mathrm{d}g/\mathrm{d}\omega > 0$), and the frequency ω^0 increases within the initial pulse, $\mathrm{d}\omega^0/\mathrm{d}t > 0$. Then the faster rays of higher frequency overtake slower rays of lower frequency, thus, forming a caustic shown in the diagram of Fig. 10.1a. Behind the caustic the high frequency rays and low frequency rays change their places so that the final profile of frequency modulation is opposite to the initial one, i.e., $\mathrm{d}\omega/\mathrm{d}t < 0$.

If we choose an optimal variation law for initial frequency such that all portions of the signal meet simultaneously at one point, we can obtain

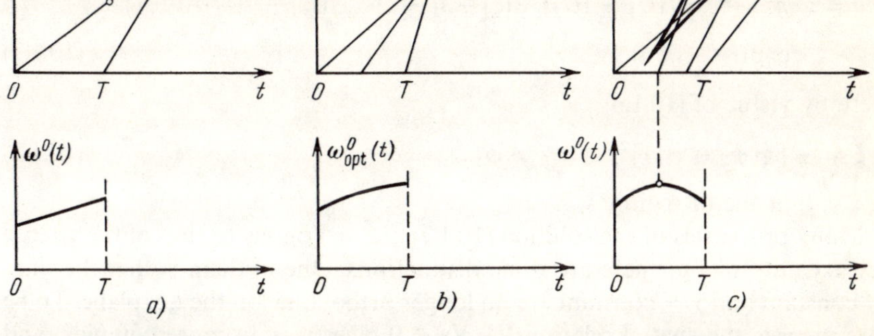

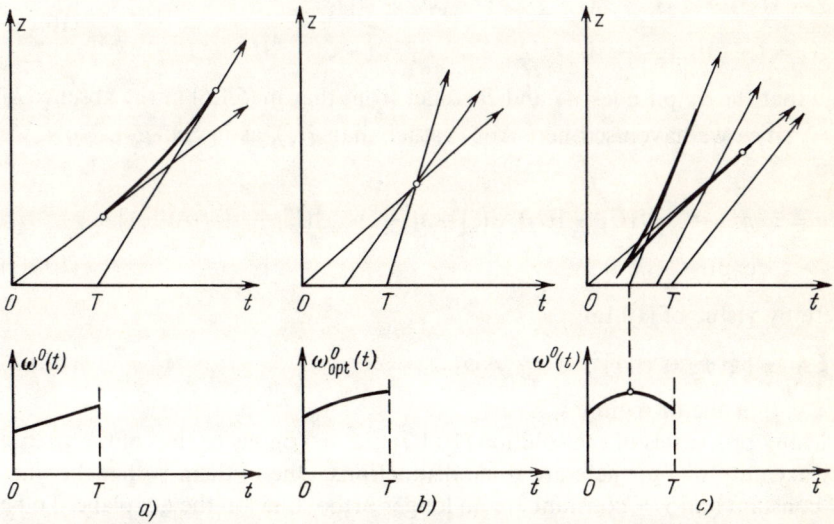

Fig. 10.1. Space-time caustics produced in the propagation of FM pulses of finite duration: **(a)** caustic broken at either side corresponds to a linear frequency modulation, **(b)** space-time focus corresponding to an optimal modulation, **(c)** broken caustic with a cusp, corresponds to a non-monotonous modulation

a space–time focus, as illustrated in Fig. 10.1b. Such a focus is realized in the optimal compression of a frequency-modulated pulse.

When the initial frequency increases in a non-optimal law, the situation may give birth to caustic cusps, as shown in Fig. 10.1c for the first stage at which the initial frequency decreases and no caustic emerges for $z > 0$, but instead space–time rays may form an imaginary caustic for $z < 0$ (not shown).

10.1.2 A Uniform Field Asymptotic for Space–Time Caustics

Let a given point of space–time $(t, \mathbf{r})$ be hit by two space–time rays with associated eikonals $\phi_{1,2}$ and amplitudes $U_{1,2}$. According to established custom we will assume that the wave $u_1 = U_1 \exp(i\phi_1)$ is yet to touch the caustic and the wave $u_2 = U_2 \exp(i\phi_2)$ has already touched it so that $\phi_1 < \phi_2$. As in the case of a space caustic, a uniform asymptotic of the field would be natural to represent as a combination of an Airy function and its derivative:

$$u = \sqrt{\pi}[A_1 \mathrm{Ai}(\zeta) + iB_1 \mathrm{Ai}'(\zeta)]\,e^{i\chi} \; . \tag{10.1.4}$$

The unknown functions ζ, χ, A_1, and B_1 may be found by comparing the asymptotic (10.1.4) with the ray field

$$u = U_1 \exp(i\phi_1) + U_2 \exp(i\phi_2) \; . \tag{10.1.5}$$

This comparison leads to relations similar to (5.1.15, 16), viz.,

$$\begin{aligned} &\chi = \tfrac{1}{2}(\phi_1 + \phi_2), \qquad \tfrac{2}{3}(-\zeta)^{3/2} = \tfrac{1}{2}(\phi_2 - \phi_1) \; , \\ &A_1 = (-\zeta)^{1/4}(U_1 + iU_2)e^{-i\pi/4}, \qquad B_1 = (-\zeta)^{-1/4}(U_1 - iU_2)e^{i\pi/4} \; , \end{aligned} \tag{10.1.6}$$

Note that the amplitudes A_1 and B_1 differ from that in (5.1.6) in the absence of $1/\sqrt{2}$, since we have used here $\mathrm{Ai}(\zeta)$ rather than $I(\zeta)$, as in Sect. 5.1. Now, we have

$$\begin{aligned} u = \sqrt{\pi}[&(-\zeta)^{1/4}(U_1 + iU_2)\mathrm{Ai}(\zeta) + i(-\zeta)^{-1/4}(U_1 - iU_2)\mathrm{Ai}'(\zeta)] \\ &\times \exp(i\chi - i\pi/4) \; , \end{aligned} \tag{10.1.7}$$

where by virtue of (10.1.6)

$$\zeta = -[\tfrac{3}{4}(\phi_2 - \phi_1)]^{2/3}, \quad \phi_2 > \phi_1 \; , \tag{10.1.8}$$

i.e., $\zeta < 0$ in the lit region.

Many properties of the solution (10.1.7) are analogous to that of the spatial Airy asymptotic, but there are some distinctions. One of them is that the lines ζ = constant and χ = constant are no longer orthogonal on the t, z-plane. To be more precise, the spatial relation $\nabla\zeta \cdot \nabla\chi = 0$ gives way to more complex, and practically unstudied expressions involving both spatial and temporal derivatives of ζ and χ. Another distinction, the possibility of emergence of caustics with anomalous phase shift deserves a separate consideration.

10.1.3 Caustics with Anomalous Phase Shift

Unlike the spatial case, it may so happen that wave u_1 has a larger phase ϕ_1 than wave u_2 notwithstanding ray 1 has not yet touched the caustic [10.3]. To ensure a correct structure of the field the argument ζ of the Airy function must be negative in the lit region. This requires that the phases ϕ_1 and ϕ_2 should be transposed in the above expressions for ζ, specifically, for $\phi_1 > \phi_2$ there must be

$$\tfrac{2}{3}(-\zeta)^{3/2} = \tfrac{1}{2}(\phi_1 - \phi_2), \quad \text{i.e.,} \quad \zeta = -[\tfrac{3}{4}(\phi_1 - \phi_2)]^{2/3} \ . \tag{10.1.9}$$

This transposition entails the substitution of U_2 for U_1 and U_1 for U_2

$$U = \sqrt{\pi}[(-\zeta)^{1/4}(U_2 + \mathrm{i}U_1)\mathrm{Ai}(\zeta) + \mathrm{i}(-\zeta)^{-1/4}(U_3 - \mathrm{i}U_1)\mathrm{Ai}'(\zeta)]\,\mathrm{e}^{\mathrm{i}\chi} \ . \tag{10.1.10}$$

However, in this case a crossover to the ray asymptotic (10.1.5) reveals that u_2 has a positive, rather than negative, phase shift $\pi/2$:

$$U_2 = |U_2|\,\mathrm{e}^{\mathrm{i}\pi/2} \ . \tag{10.1.11}$$

Based on an asymptotic analysis of a Fourier representation of the field, *Orlov* [10.4] has demonstrated that the anomalous phase shift can be observed in the conditions when the group velocity falls with frequency, $\mathrm{d}g/\mathrm{d}\omega < 0$. If under these circumstances the initial frequency decreases, $\mathrm{d}\omega^0/\mathrm{d}t < 0$, then the ray pattern will be as shown in Fig. 10.1a, but the caustic will have an anomalous phase shift of $\pi/2$.

To close this topic, we note that faced with space–time caustic cusps it would be natural to take the Pearsey integral as a standard function, while to handle loops one should take the A_4 integral (swallowtail) and so on.

10.1.4 Broken Space–Time Caustics

Pulses of finite duration form broken space–time caustics in the t, z-plane; as a matter of fact these types of caustic are depicted in Fig. 10.1. Allowing for edge space–time rays induced by the leading and trailing edges of a pulse, one may build a uniform asymptotic of the whole wavefield on the basis of incomplete Airy functions (Sect. 8.1). Edge space–time rays have been introduced by *Lewis* [10.5] in his pioneering effort on space–time diffraction theory. The status of this theory and available evidence may be found in a course of lectures due to *Orlov* [10.6] and in [10.7, 8].

10.1.5 Space–Time Lenses

In this context one more phenomenon is in order, namely, the focusing of waves with the aid of a parameter wave realized in the form of moving interfaces. In the

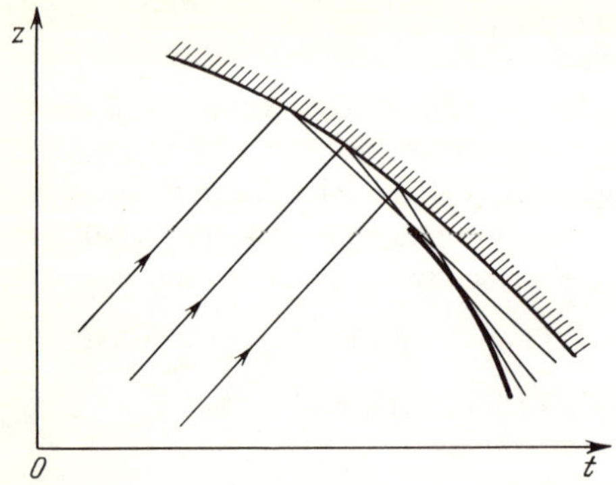

Fig. 10.2. Caustic formed in reflection of space–time rays from a moving interface

one-dimensional problem, the moving interface is represented by a curve which in Fig. 10.2 is emphasized by hachures.

A wave incident on a moving boundary is partially reflected from it and under certain conditions (a suitable law of dispersion, an optimal speed of the interface) can form a caustic or even a space–time focus. Space–time rays refracted at the interface may also undergo focusing. Effects of this kind have been reported in many articles, e.g., see [10.1–3] and references therein.

10.1.6 Uniform Asymptotics in Media with Spatial Dispersion

In the case of media with spatial dispersion, the Helmholtz equation contains a non-local term which, however, does not incur principal difficulties either for construction of a geometric-optical asymptotic or for development of a uniform caustic asymptotic. It is required only that the radius of spatial dispersion be small compared with the typical length of inhomogeneity of the medium.

Under the said circumstances a uniform Airy asymptotic of a scalar field has been constructed by *Kravtsov* [10.9]. The relation of the amplitude (A, B) and argument (ζ, χ) functions of this Airy asymptotic with the ray field parameters is the same as in the absence of spatial dispersion. These results may be extended to electromagnetic waves. An alternative approach to the description of waves in media with spatial dispersion is due to Maslov's method [10.10] which is applicable to both differential and integro–differential wave equations typical of waves in non-local media.

10.2 Caustics in Anisotropic Media

10.2.1 Description of Caustic Fields

In anisotropic media we are faced with a superposition of normal waves which are deemed to be noninteracting in most situations. The requirement of weakness of the interaction imposes certain constraints on the variation rates of wave and medium parameters. However, a quantitative formulation of these constraints is not an easy task. The easiest link in this chain is to specify a condition of weak interaction in the region free of caustics, namely, the typical length of spatial beats of normal waves, $\lambda/\Delta n$ (Δn is the difference of refractive indexes of normal waves), must be small compared to the scale L_ε of typical variation of the medium properties. Caustics also contribute caustic scales; however, the question on their role in the interaction of waves remains open.

In anisotropic media the refractive index depends on the direction in which the wave propagates (Sect. 2.1.3). This dependence per se does not interfere with the construction of caustic asymptotics, both local and uniform, but brings about certain peculiarities which should not be overlooked. One of them is concerned with the invalidated orthogonality of the lines ζ = constant (ζ is the argument of Airy functions) and χ = constant (χ is the longitudinal eikonal) in uniform asymptotic representations of the type (5.1.1). In the final analysis this stems from the misalignment of the direction of a ray, and simultaneously of the flow of energy, with the phase normal.

Note in passing that loop-like (swallowtail type) trajectories of the phase normal, that were revealed near the turning point of rays in a magnetically active plasma as far back as 40–50 years (for the status of the problem, see [10.1]) have not been analyzed from the standpoint of catastrophe theory.

10.2.2 Exceptional Directions of Radiative Transfer

If the wave surface (dependence of the refractive index on the direction of propagation) has points of inflection, then the anisotropic medium will have exceptional directions in which radiative transfer will be at a maximum intensity.

Let $u^0(x) = U^0(x)\exp[ik\psi^0(x)]$ be the initial field in the plane $z = 0$, and

$$F(q) = \frac{1}{2\pi}\int_{-\infty}^{\infty} U^0(x)\exp[ik\psi^0(x) - \mathrm{i}qx]\,\mathrm{d}x \tag{10.2.1}$$

be the Fourier transform of this field. For $z > 0$, each Fourier component $F(q)\exp(iqx)$ induces a wave $F(q)\exp[\mathrm{i}k_z(q)z + \mathrm{i}qx]$ so that the overall field is represented by the superposition

$$u(x, z) = \int_{-\infty}^{\infty} F(q)\exp[\mathrm{i}k_z(q)z + \mathrm{i}qx]\,\mathrm{d}q\ . \tag{10.2.2}$$

The function $k_z(q)$ is defined by the properties of the medium, which are described by the dependence of the refractive index on direction, $n(\mathbf{l})$. The ray field may be obtained from this expression with the aid of the method of stationary phase (for respective procedures, see [10.1, 4]).

If $k_z(q)$ has a point of inflection in (10.2.2), then the integrand allows three stationary values of q, rather than one. In the direction corresponding to the point of inflection, the second derivative $\partial^2 k_z(q)/\partial q^2$ vanishes, and the ray amplitude becomes infinite. Clearly, the real field in the exceptional directions is finite. It denies a representation in a ray form, but may be expressed in terms of an Airy function.

Here we deal with a caustic in a space of directions. While such objects have been under investigation for a comparatively long time, e.g., see [10.11] and relevant references quoted in [10.1, 2], in the space of wavenumbers caustics have been poorly documented.

10.2.3 Focusing of Waves at the Interface of Anisotropic and Isotropic Media

One more phenomenon is the result of anisotropic propagation, namely, the focusing of a point source field in an adjacent medium. For example, if a source is in a magnetically active plasma, then in a neighboring vacuum a caustic cusp can be formed, as shown in Fig. 10.3a. Conversely, when the source is the vacuum, then in the adjacent plasma there forms a cusp pointing toward the interface, as shown in the diagram of Fig. 10.3b. For asymptotic analyses of such caustics, the reader is referred to [10.1, 2]. Microwave engineering is a field where this type of caustics has been handled.

10.2.4 Caustics with Anomalous Phase Shift

An accurate stationary-phase calculation of the integral (10.2.2) yields for the terms associated with the rays a caustic phase shift S_c which depends on the sign of the derivative $\mathrm{d}^2 k_z(q)/\mathrm{d}q^2$ [10.4], namely,

$$S_c = \frac{\pi}{2}\operatorname{sgn}\frac{\mathrm{d}^2 k_z}{\mathrm{d}q^2} = \begin{cases} -\pi/2 & \text{for } \mathrm{d}^2 k_z/\mathrm{d}q^2 < 0 \ , \\ \pi/2 & \text{for } \mathrm{d}^2 k_z/\mathrm{d}q^2 > 0 \ . \end{cases} \tag{10.2.3}$$

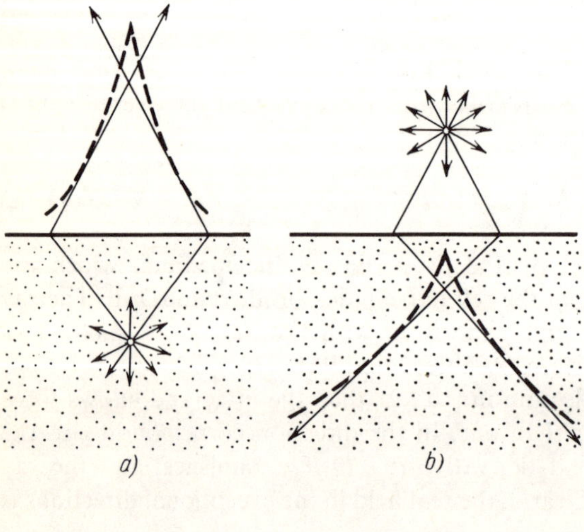

Fig. 10.3. Formation of a caustic cusp (**a**) in an isotropic half-space by a source embedded in an anisotropic half-space, and (**b**) in an anisotropic medium by a source embedded in an isotropic half-space

In isotropic media $k_z = (k^2 - q^2)^{1/2}$, the derivative $\mathrm{d}^2 k_z/\mathrm{d}q^2$ is always negative and the caustic shift is $-\pi/2$. In anisotropic media, $k_z(q)$ can have segments concave upwards. For such segments, an anomalous phase shift of $\pi/2$ is likely.

A uniform field asymptotic for such caustics should be constructed in the lines of the procedure outlined in Sect. 10.1.3, namely, the argument of the Airy function is calculated by formula (10.1.9) and the field by formula (10.1.10). From the physical point of view, the transposition $\phi_1 \leftrightarrow \phi_2$ has its origin in the fact that ray 2 which has touched the caustic acquires a smaller phase shift ϕ_2 than ray 1 which is yet to touch the caustic. In isotropic media this is impossible.

10.3 Complex Caustics

Similar to complex rays we may introduce the concept of complex caustics as the envelopes of complex rays in the six-dimensional complex space $(\mathrm{Re}\{R\}, \mathrm{Im}\{R\})$. An equation of a complex caustic is obtained by eliminating the parameter τ between the equation of the ray $\mathbf{r} = \mathbf{R}(\xi, \eta, \tau)$ and the vanishing condition of the Jacobian $D(\tau) = \partial(x, y, z)/\partial(\xi, \eta, \tau)$.

Near complex caustics the field may be obtained in a formal way by the methods outlined above with one distinction that the amplitudes and phases are shifted, i.e., analytically continued, in the complex space. These calculations are of a rather limited practical interest, as the main purpose of analysis is usually the fields in real space.

If a complex caustic is immersed in the complex space deeper than the width of the caustic zone, then a real space analysis may be limited with an ordinary and a complex ray approximation. When the caustic approaches the hyperplane $\mathrm{Re}\{R\}$, then one may expect a local rise of the field, i.e., emergence of a sort of halo in the respective places of real space. V.P. Maslov suggested to call them "light dots in the region of caustic shadow."

An example of such a situation may be the center of a circular caustic, which is a focus for complex rays. Strictly speaking, the field at the center of the caustic is zero, but enlightenment manifests itself another way, namely, in the profile of field drop near the center. While in the general case complex rays imply an exponential decay inward in the caustic shadow region but near the center the field falls off as r^ν with $\nu = ka = 2\pi a/\lambda$ being the dimensionless length of the caustic.

The Gaussian beam

$$u(x, y) = \left(1 + \frac{\mathrm{i}y}{ka^2}\right)^{-1/2} \exp\left[\mathrm{i}ky - \frac{x^2}{2a^2}\left(1 + \frac{\mathrm{i}y}{kz^2}\right)^{-1}\right] \qquad (10.3.1)$$

also exhibits a kind of enlightenment. In the beam waist section, i.e., in the plane $y = 0$, it has the form

$$u(x, 0) = \exp(-x^2/2a^2) , \quad (10.3.2)$$

whereas for $y \gg ka^2$ (far zone), the field (10.3.1) is a cylindrical wave with Gaussian profile:

$$u(x, y) = \left(\frac{ka^2}{ir}\right)^{1/2} \exp\left(ikr - \frac{x^2k^2a^2}{r^2}\right) ,$$

where $r = (x^2 + y^2)^{1/2}$, and ka^2 stands for the diffraction length.

The solution (10.3.1) which is derivable from the initial value (10.3.2) in the lines of geometrical optics [10.12], has a focus at the complex point $(0, ika^2)$ where the field (10.3.1) becomes infinite. In real space, the largest field is observed near the waist section $\mathrm{Re}\{y\} \approx 0$ closest to the complex focus. Therefore, the region of the beam waist may be treated as a light spot revealing the existence of a complex focus. A close treatment has been given by *Kogelnik* who noted that the cylindrical wave $\exp(ikr)/\sqrt{r}$ with $r = \sqrt{x^2 + (y - ika^2)^2}$ coincides with the Gaussian beam (10.3.1) for $y \gg ka^2$ and simultaneously serves as a solution to the wave problem for a point source embedded in the complex space.

A general insight in complex caustics is provided by the theory of singularities of differentiable mappings in complex spaces. Relevant elements of this theory may be gleaned from the works of *Vasilyev* [10.13].

10.4 Random Caustics

One may come across random caustics in handling various physical problems. Historically, the earlier problem was the one of passing a wave through a random phase screen which modulates the phase of the transmitted wave and retains its amplitude intact. Let $S(x, y)$ be a random phase of the wave leaving the screen in the plane $z = 0$. Farther from the screen one will observe amplitude fluctuations which become especially significant in the range of random focusing. If l_S is the correlation radius of the phase on the screen, i.e., the typical dimension of variation of the phase correlation function $K_S(\rho) = \langle S(\rho_1)S(\rho_1 + \rho)\rangle$, then the region of strong focusing will occur at a distance

$$z_{\mathrm{foc}} \sim kl_S^2/\sigma_S , \quad (10.4.1)$$

where $\sigma_S^2 = \langle S^2\rangle$ is the variance of phase fluctuations assumed to be $\gg 1$.

The scintillation index

$$\beta^2 = \frac{\langle (I - \bar{I})^2\rangle}{\bar{I}^2} \equiv \frac{\langle \Delta I^2\rangle}{\bar{I}^2} \quad (10.4.2)$$

characterizes the relative magnitude of fluctuations of the field intensity $I = |u|^2$. Immediately behind the screen ($z \ll z_{foc}$) these fluctuations are weak. In the region of strong focusing, $z \sim z_{foc}$, fluctuations of intensity are strongest, and for $z \gg z_{foc}$ where the number of rays coming to the observation point becomes large, fluctuations become Gaussian and the scintillation index asymptotically approaches unity; this is the region of saturated fluctuations. In going from weak fluctuations to strong and saturated, other statistics of the field (probability densities of intensity, amplitude and phase, their correlation functions and correlation radii) also undergo substantial variations (Ref. 10.14, Chap. 3] and the literature cited therein). Similar relationships are typical also of waves reflected from random surfaces [10.15].

In a random inhomogeneous medium, the wave has to interact virtually at every point of the medium. Therefore, the caustics have a more intricate form than those that result from passing through a screen or reflecting from a rough surface. An example of a longitudinal section of such a caustic is presented in Fig. 2.7, and a typical distribution of intensity in a cross section is shown in Fig. 1.3a (experiment) and Fig. 1.3b (computer simulation under similar conditions [10.16]). The figure shows also interference fringes representative of multiray caustics.

As in experiments with a wave passing through a screen in a random inhomogeneous medium we find three typical zones of weak, strong and saturated fluctuations (Fig. 2.7) but the evolution of caustics is as if extended in space. Thc immediate implication is that the scintillation index increases here slower than behind the screen, and the maximum value $\beta^2_{max} = 1.2$ to 1.4 is markedly lower than for the screen.

The region of strong fluctuations begins where early caustic cusps become noticeable [10.17], while the region of saturated fluctuations may be characterized by emergence of a host of caustics which is identical to a growth of the number of rays. Multiple new rays lead some authors to call this phenomenon "multipath propagation" (e.g., see [10.18]).

This concept is somewhat conditional in nature and is invoked to mark an exceedingly rough phase front of the wave. One may speak of the reality of microrays and about the possibility to distinguish them by physical means provided that the Fresnel volumes of rays coming to the observation point do not overlap [10.19]. Likewise, caustics will be clear-cut in the region of saturated fluctuations provided that individual branches of one and the same caustic be distant farther than the width of the caustic zone. Under the conditions of multipath propagation both conditions are practically equivalent.

The Fresnel volumes of rays increase and their spacing decreases as one moves deeper into a random inhomogeneous medium. Therefore, sooner or later one finds himself in a situation where microrays and the pattern of multiple caustics have no physical sense any longer and one has to completely renounce the ray interpretation of field fluctuations. The question on the limiting density of random caustics remains open. Of the recent works which are concerned with this topic we point out [10.20, 21].

10.5 Caustics in Quantum Mechanical Problems

Caustic theory quite unexpectedly has found its way to problems of atomic and molecular collisions. From the standpoint of wave theory, atomic collisions are described as a scattering of a wave function associated with the approaching atom at the potential characterizing the atom at rest. The classical paths of atoms perform here as analogs of rays, and caustics occur as the envelopes of the classical trajectories. In the vicinity of caustics the wave function increases, thus rising the probability of detection of a particle.

When one sort of atoms is scattered by another, in some directions caustics tend to infinity. Clearly, in these directions maxima may be expected to occur. Such maxima are called atomic rainbow. Figure 10.4a shows an angular distribution of scattered particles (experiments of Hundhausen and Pauli on scattering sodium atoms at mercury vapors). The main maximum and additional peaks are interpreted as being caused by caustics of the classical trajectories [10.22]. The diagram in Fig. 10.4b elucidates the trajectory pattern of colliding atoms. For comparatively large values of aiming parameter, atoms experience attraction which somewhat increases as the beam approaches the target. Therefore, the peripheral trajectories turn toward the scattering nucleus. However, near the target attraction gives way to repulsion because the screening action of the electronic shell is no longer present near the nucleus. This situation takes place in a head-on collision when the scattering is strictly backwards. If we move from the periphery toward the center zone, we come across a caustic. As in optical rainbow, its asymptote is the strongly deflected trajectory.

Analysis of angular distribution of scattered particles may yield valuable evidence on the scattering potential. The quasi-classical methods dramatically reduce the calculation of wave function over that involved in the rigorous procedures. Therefore, they prove to be of considerable value in handling

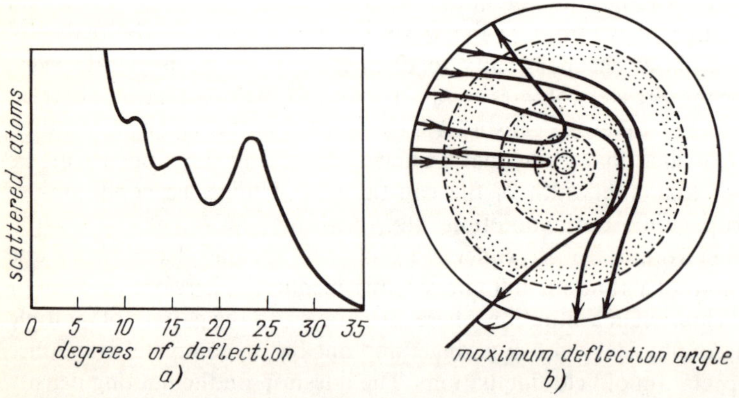

Fig. 10.4. (**a**) Typical angular distribution of Na atoms scattered by mercury vapor. (**b**) Trajectories of atoms forming a rainbow [10.22]

inverse problems, i.e., that of reconstruction of the scattering potential [10.23, 24].

The rainbow type of phenomena occur in inelastic collisions in chemically reacting systems. A rigorous quantum mechanical description of chemical transformations has to handle considerable difficulties, whereas the quasi-classic treatment of some simple reactions such as $H + F_2 = HF + F$ has proved to be quite successful [10.25].

Note also that if a collision carries the system into the bounded state (this transition is known as an inelastic collision), then in the classical context this will be treated as a capture of the approaching atom into a finite trajectory, while in the quasi-classical context it will be reduced to finding eigenfunctions "tied" to the caustic surface similar to oscillations of the whispering gallery type. The state of the art has been outlined in [10.26], and the general insight into the quasi-classical approximation applied to problems associated with caustics may be found in an excellent review of *Berry* and *Mount* [10.27].

10.6 Concluding Remarks

The material embodied in this book reflects only a small proportion of the physical phenomena and applications associated with caustics and caustic fields. In the ocean of caustic information we wished to highlight the key problem of "sewing wave flesh on a ray skeleton." Therefore, we put aside many other relevant topics.

We come acoss caustics wherever physical phenomena are wavelike in nature. Although the main emphasis in this text is on ordinary sound and electromagnetic waves, including light waves, one should not overlook the fact that caustic phenomena are intrinsic to both microspace and macrospace – recall gravitation lenses predicted by Einstein, and the salient features of the large-scale structure of the Universe [10.28, 29]. Caustics in moving media have already been reported [10.30]. The problem of estimation of fields at caustics, which has evolved in the context of linear wave theory, has become a subject of study in nonlinear wave theory [10.31–33], and this list of new recent fields and applications may be continued.

We are certain that nature will present researchers with many more beautiful caustic phenomena in the future, causing their hearts to beat faster; therefore, we are in no hurry to close this book with a final full stop. . .

References

Chapter 1

1.1 T.A. Croft: J. Geophys. Res. **72**, 2343 (1967)

1.2 F.D. Tapert: In *Wave Propagation and Underwater Acoustics*, ed. by J.B. Keller, J.S. Papadakis (Springer, Berlin, Heidelberg 1979) p. 224

1.3 A.S. Gurvich, M.A. Kalistratova, F.E. Martvel: Izv. VUZ Radiofiz. **20**, 1020 (1977) [English transl.: Radiophys. Quantum Electron **20**, 705 (1977)]

1.4 J.M. Martin, S.M. Flatté: Appl. Opt. **27**, 2111 (1988)

1.5 H. Whitney: Ann. Math. **62**, 374 (1955)

1.6 V.I. Arnold: Usp. Mat. Nauk **23**, 3 (1968); ibid. **29**, 11, 243 (1974); ibid. **30**, 3 (1975)

1.7 R. Thom: *Structural Stability and Morphogenesis* (Benjamin, New York 1972)

1.8 V.I. Arnold: *Catastrophe Theory*, 3rd edn. (Springer, Berlin, Heidelberg 1992)

1.9 V.I. Arnold: Sov. Phys. – Usp. **26**, 1024 (1983)

1.10 T. Poston, N. Stewart: *Catastrophe Theory and Its Applications* (Pitman, London 1978)

1.10 R. Gilmor: *Catastrophe Theory for Scientists and Engineers* (Wiley, New York 1981)

1.12 M. Golubitsky, V. Guillimin: *Stable Mappings and their Singularities*, Graduate Texts Math., Vol. 14 (Springer, Berlin, Heidelberg 1973)

1.13 V.I. Arnold, A.N. Varchenko, S.M. Gusein-Zade: *Singularities of Differentiable Mappings*, Vols. 1, 2 (Birkhäuser, Boston 1985, 1988)

1.14 J.J. Dustermaat: Commun. Pure Appl. Math. **27**, 207 (1974)

1.15 Yu.A. Kravtsov, Yu.I. Orlov: Sov. Phys. – Usp. **26**, 1038 (1983)

1.16 Yu.A. Kravtsov: Rays and caustics as physical objects, in *Progress in Optics* **26**, 227–348 (North-Holland, Amsterdam 1988)

1.17 Yu.L. Gazaryan: *Voprosy Dinamicheskoi Teorii Rasprostranenia Seismicheskikh Voln* (topics in dynamic theory of seismic wave propagation) (Izd. LGU, Leningrad 1961) Vol. 5, p. 73

1.18 R.N. Buchal, J.B. Keller: Commun. Pure Appl. Math. **13**, 85 (1960)

1.19 V.M. Babich, N.Y. Kirpicnikova: *The Boundary-Layer Method in Diffraction Problems*, Springer Ser. Electrophys., Vol. 3 (Springer, Berlin, Heidelberg 1979)

1.20 V.M. Babich, V.S. Buldyrev: *Short-Wavelength Diffraction Theory*, Springer Ser. Wave Phenom., Vol. 4 (Springer, Berlin, Heidelberg 1991)

1.21 L.B. Felsen, N. Marcuvitz: *Radiation and Scattering of Waves* (Prentice-Hall, Englewood Cliffs, NJ 1973) Vols. 1, 2

1.22 Yu.A. Kravtsov: Izv. VUZ Radiofiz. **7**, 664, 1049 (1964) [English transl.: Radiophys. (USSR) **7**, no. 4, 104; no. 6, 40 (1964)]

1.23 D. Ludwig: Commun. Pure Appl. Math. **19**, 215 (1966)

1.24 G. Dangelmayr, W. Veit: Ann. Phys. **118**, 108 (1979)

1.25 D.S. Ahluwalia, R.M. Lewis, J. Boersma: SIAM J. Appl. Math. **16**, 783 (1968)

1.26 V.P. Maslov: *Théorie des Perturbations et Méthode Asymptotique* (Dunod, Paris 1972)

1.27 V.P. Maslov, M.V. Fedoryuk: *Semiclassical Approximation in Quantum Mechanics* (Reidel, Hingham, MA 1981)

1.28 Yu.A. Kravtsov: Akust. Zh. **14**, 1 (1968) [English transl.: Sov. Phys. – Acoustics **14**, 1 (1968)]

1.29 J.M. Arnold: Radio Sci. **17**, 1181 (1982)

1.30 R. Ziolkowski, G.A. Deschamps: Radio Sci. **19**, 1001 (1984)
1.31 Yu.I. Orlov: Trudy MEI (Moscow Power Engineering Institute, Moscow 1977) No. 19, p. 82; Izv. VUZ Radiofiz. **17**, 1035 (1974) [English transl.: Radiophys. Quantum Electron. **17**, 788 (1974)]
1.32 L.A. Vainstein, E.S. Birger, N.V. Konyukhova, E.L. Kosarev: Sov. J. Plasma Phys. **26**, 362 (1976)
1.33 J.M. Arnold: J. Acoust. Soc. Am. **87**, 587 (1990)
1.34 V.B. Avdeyev, A.V. Demin, Yu.A. Kravtsov, M.V. Tinin, A.P. Yarygin: Izv. VUZ Radiofiz. **31**, 1279 (1988) [English transl.: Radiophys. Quantum Electron. **31**, 907 (1988)]
1.35 V.M. Babich, M.M. Popov: Izv. VUZ Radiofiz. **32**, 1447 (1987) [English transl.: Radiophys. Quantum Electron. **32**, 1063 (1989)]
1.36 M.V. Berry, K.E. Mount: Rept. Progr. Phys. **35**, 313 (1972)
1.37 M.V. Berry, C. Upstill: *Progress in Optics* **18**, 257–320 (North-Holland, Amsterdam 1980)
1.38 P.S. Theocaris, J.G. Michopoulos: Appl. Opt. **21**, 1080 (1982)
1.39 S. Cornbleet: Proc. IEEE **71**, 471 (1983)
1.40 J.J. Stamness: Opt. Acta **29**, 823 (1982)
1.41 P.L. Marston: Geometrical and catastrophe Boston optics method in scattering, in *Physical Acoustics* Vol 21, ed. by R.N. Thurston, A.D. Pierse (Academic, Boston 1992) pp. 1–234
1.42 J.J. Stamnes: *Waves in Focal Regions* (Adam Hilger, Bristol 1986)
1.43 Yu.A. Kravtsov, Yu.I. Orlov: *Geometrical Optics of Inhomogeneous Media*, Springer Ser. Wave Phenom., Vol. 6 (Springer, Berlin, Heidelberg 1990)

Chapter 2

2.1 M. Born, E. Wolf: *Principles of Optics*, 5th edn. (Pergamon, Oxford 1975)
2.2 Yu.A. Kravtsov, Yu.I. Orlov: *Geometrical Optics of Inhomogeneous Media*, Springer Ser. Wave Phenom., Vol. 6 (Springer, Berlin, Heidelberg 1990)
2.3 S.M. Rytov: Dokl. Akad. Nauk USSR **18**, 263 (1938)
2.4 Yu.A. Kravtsov: Dokl. Akad. Nauk USSR **183**, 74 (1968) [English transl.: Sov. Phys. – Dokl. **13**, 1125 (1968)]
2.5 V.V. Zheleznyakov, V.V. Kocharovskii, Vl.V. Kocharovskii: Usp. Fiz. Nauk **141**, 257 (1983) [English transl.: Sov. Phys. – Usp. **26**, 877 (1983)]
2.6 Yu.A. Kravtsov, Yu.I. Orlov: Sov. Phys. – Usp. **23**, 750 (1980); Radio Sci. **16**, 975 (1981)
2.7 Yu.A. Kravtsov: Rays and caustics as physical objects, in *Progress in Optics* **26**, 227–348 (North-Holland, Amsterdam 1988)
2.8 Yu.A. Kravtsov: Radiotekh. Elektron. **32**, 883 (1987) [English transl.: Sov. J. Commun. Technol. Electron. **32**, no. 4, 133 (1987)]
2.9 E.L. Feinberg: *Rasprostranenie Radiovoln Vdol Zemnoi Poverkhnosti* (propagation of radio waves over the Earth's surface) (Izd. Akad. Nauk USSR, Moscow 1961)
2.10 H.L. Bertoni, L.B. Felsen, A. Hessel: IEEE Trans. AP-**19**, 226 (1971)
2.11 F.B. Chernyi: *Rasprostranenie Radiovoln* (propagation of radio waves) (Sov. Radio, Moscow 1972)
2.12 A.A. Asatryan: Kratk. Soobsch. Fiz. FIAN No. 10, 1 (1985) [English transl.: Sov. Phys. – Lebedev Inst. Rept. No. 10, 1 (1985)]
2.13 A.A. Asatryan, Yu.A. Kravtsov: Wave Motion **10**, 45 (1988)
2.14 V.P. Maslov: *Théorie des Perturbations et Méthode Asymptotique* (Dunod, Paris 1972)
2.15 R.M. Lewis: Arch. Ration. Mech. Anal. **20**, 191 (1965)
2.16 Yu.I. Orlov: Izv. VUZ Radiofiz. **24**, 224 (1981) [English transl.: Radiophys. Quantum Electron. **24**, no. 2, 154 (1981)]
2.17 Yu.L. Gazaryan: *Voprosy Dinamicheskoi Teorii Rasprostranenia Seismicheskikh Voln* (topics in dynamic theory of seismic wave propagation) (Izd. LGU, Leningrad 1961) p. 73

2.18 V.Ya. Groshev, Yu.A. Kravtsov: Izv. VUZ Radiofiz. **11**, 1812 (1968) [English transl.: Radiophys. Quantum Electron. **11**, 1019 (1968)]
2.19 Yu.I. Orlov: Trudy MEI No. 194 (Power Institute, Moscow 1974) p. 103
2.20 A.A. Asatryan, Yu.A. Kravtsov, V.S. Filinov: Kratk. Soobschch. Fiz. FIAN No. 11, 30 (1968) [English transl.: Sov. Phys. – Lebedev Inst. Rept. No. 11, 20 (1988)]
2.21 O.N. Stavrodis: *The Optics of Rays, Wavefronts, and Caustics* (Academic, New York 1972)
2.22 S.M. Flatté, R. Dashen, W.H. Munk, K.M. Watson, F. Zakariasen: *Sound Transmission Through a Fluctuating Ocean* (Cambridge Univ. Press., Cambridge 1979)
2.23 Yu.A. Kravtsov: Zh. Eksp. Teor. Fiz. **55**, 1470 (1968) [English transl.: Sov. Phys. – JETP **28**, 769 (1969)]
2.24 M.V. Berry, C. Upstill: Catastrophe optics: Morphologies of caustics and their diffraction patterns, in *Progress in Optics* **18**, 257–320 (North-Holland, Amsterdam 1980)
2.25 J.B. Keller: A geometrical theory of diffraction, in *Calculus of Variations and its Applications*, Proc. Symp. Appl. Math., Vol. 1.8 (McGraw-Hill, New York 1958) pp. 27–52
2.26 B.D. Seckler, J.B. Keller: J. Acoust. Soc. Am. **31**, 192 (1959)
2.27 J.B. Keller, F.C. Karal: J. Appl. Phys. **31**, 1039 (1960)
2.28 V.P. Maslov: Sov. Phys. – Doklady **7**, 666 (1964)
2.29 V.M. Babich: *Voprosy Dinamicheskoi Teorii Rasprostraneniya Seismicheskikh Voln* (topics in dynamic theory of seismic waves) (Izd. LGU, Leningrad 1961) Vol. 5, p. 145
2.30 Yu.A. Kravtsov: Izv. VUZ Radiofiz. **10**, 1283 (1967) [English transl.: Radiophys. Quantum Electron. **10**, 719 (1967)]
2.31 G. Ghione, I. Montrosset, L.B. Felsen: IEEE Trans. AP-**32**, 68 (1984)
2.32 J.B. Keller, W. Streifer: J. Opt. Soc. Am. **61**, 40 (1971)
2.33 G.A. Deschamps: Electron. Lett. **7**, 684 (1971)
2.34 S. Choudhary, L.B. Felsen: Proc. IEEE **62**, 1530 (1974)
2.35 A.A. Asatryan, Yu.A. Kravtsov: Izv. VUZ Radiofiz. **31**, 1053 (1988) [English transl.: Radiophys. Quantum Electron. **31**, 746 (1988)]

Chapter 3

3.1 V.P. Maslov: *Théorie des Perturbations et Méthode Asymptotique* (Dunod, Paris 1972)
3.2 A.A. Andronov, A.A. Vitt, S.E. Haikin: *Teoriya Kolebanii* (theorie of oscillations) (Nauka, Moscow 1981)
3.3 V.I. Arnold: Usp. Mat. Nauk **23**, 3 (1968); **29**, 11, 243 (1974); ibid. **30**, 560 (1975)
3.4 R. Thom: *Structural Stability and Morphogenesis* (Benjamin, New York 1975)
3.5 T. Poston, N. Stewart: *Catastrophe Theory and its Applications* (Pitman, London 1978)
3.6 M. Golubitsky, V. Guillimin: *Stable Mappings and Their Singularities*, Graduate Texts Math., Vol. 14 (Springer, Berlin, Heidelberg 1973)
3.7 R. Gilmor: *Catastrophe Theory for Scientists and Engineers* (Wiley, New York 1981)
3.8 J.J. Dustermaat: Commun. Pure Appl. Math. **27**, 207 (1974)
3.9 V.I. Arnold, A.N. Varchenko, S.M. Gussein-Zade: *Singularities of Differentiable Mappings*, Vols. 1, 2 (Birkhäuser, Boston 1985, 1988)
3.10 V.I. Arnold: Sov. Phys. – Usp. **26**, 1024 (1983)
3.11 V.I Arnold: *Catastrophe Theory*, 3rd edn. (Spinger, Berlin, Heidelberg 1992)

Chapter 4

4.1 V.I. Arnold, A.N. Varchenko, S.M. Gussein-Zade: *Singularities of Differential Mappings*, Vols. 1, 2 (Birkhäuser, Boston 1985, 1988)
4.2 J.J Dustermaat: Commun. Pure Appl. Math. **27**, 207 (1974)

4.3 M.V. Berry: J. Phys. A **10**, 2061 (1977)
4.4 M.V. Berry, C. Upstill: *Progress in Optics* **18**, 257–320 (North-Holland, Amsterdam 1980)
4.5 Yu.A. Kravtsov, Yu.I. Orlov: *Geometrical Optics of Inhomogeneous Media*, Springer Ser. Wave Phenom., Vol. 6 (Springer, Berlin, Heidelberg 1990)
4.6 Yu.A. Kravtsov, Yu.I. Orlov: Sov. Phys. – Usp. **26**, 1038 (1983)
4.7 Yu.A. Kravtsov: Rays and caustics as physical objects. *Progress in Optics* **26**, 227–348 (North-Holland, Amsterdam 1988)
4.8 V.I. Arnold: Funkts. Anal. Prilozh. **6**, 61 (1972)
4.9 M.V. Fedoryuk: *Metod Perevala* (saddle-point method) (Nauka, Moscow 1977)
4.10 R.B. Dingle: *Asymptotic Expansions: Their Derivation and Interpretation* (Academic, London 1973)
4.11 A.A. Asatryan, Yu.A. Kravtsov: Wave Motion **10**, 45 (1988)
4.12 J.C.P. Miller: *The Airy Integral, Giving Tables of Solution of the Differential Equation* $y'' = xy$ (Cambridge Univ. Press, Cambridge 1971)
4.13 V.A. Fock: *Electromagnetic Diffraction and Propagation Problems* (Pergamon, New York 1965)
4.14 M. Abramovitz, J.A. Stegun: *Handbook of Mathematical Functions* (NBS, Washington, DC 1964)
4.15 R.E. Langer: Trans. Am. Math. Soc. **31**, 23 (1931)
4.16 T.M. Cherry: Trans. Am. Math. Soc. **68**, 224 (1950)
4.17 A.A. Dorodnitsyn: Usp. Mat. Nauk **7**, 3 (1952)
4.18 V.Ya. Groshev, Yu.A. Kravtsov: Izv. VUZ Radiofiz. **11**, 1812 (1968) [English transl.: Radiophys. Quantum Electron. **11**, 1019 (1968)]
4.19 T. Pearsey: Philos. Mag. **37**, 311 (1946)
4.20 L.M. Brekhovskikh: *Acoustics of Layered Media I, II*, Springer Ser. Wave Phenom., Vols. 5, 10 (Springer, Berlin, Heidelberg 1990, 1992)
4.21 A.S. Kryukovskii, D.S. Lukin, E.A. Palkin: Preprint No. 43, 415 (1984) Inst. Radiotech. Elecronics, Acad. Sci. USSR
4.22 J.N.L. Connor, D. Farrelly: J. Chem. Phys. **75**, 2831 (1981)
4.23 J.N.L. Connor, P.R. Curtis: J. Phys. A **15**, 1179 (1982)
4.24 Yu.A. Kravtsov, Yu.A. Orlov: Sov. Phys. – Usp. **23**, 750 (1980); Radio. Sci. **16**, 975 (1981)
4.25 D. Ludwig: Commun. Pure Appl. Math. **19**, 215 (1966)
4.26 H. Trinkaus, F. Drepper: J. Phys. A **10**, L11 (1977)
4.27 N.F. Dronov, E.B. Ipatyev, D.S. Lukin, E.A. Palkin: In *Rasprostraneniye Radiovoln v Ionosfere* (propagation of radio waves in the ionosphere) (IZMIRAN SSSR, Moscow 1978) p. 57
4.28 M.V. Berry, J.F. Nye, F.J. Wright: Philos. Trans. Roy. Soc. London **291**, 453 (1979)
4.29 D.S. Lukin, E.A. Palkin: *Chislennyi Kanonicheskii Metod v Zadachakh Difactsii i Rasprostraneniya Electromagnitnikh Voln v Neodnorodnykh Sredakh* (canonic numerical method in problems of diffration and propagation of EM waves in inhomogeneous media) (MFTI Press, Moscow 1982)
4.30 J.N.L. Connor, P.R. Curtis, D. Farrelly: J. Phys. A **17**, 283 (1984)
4.31 V.I. Arnold: Usp. Mat. Nauk **28**, 17 (1973)
4.32 J.N.L. Connor: Mol. Phys. **26**, 1371 (1973)
4.33 J.N.L. Connor: Faraday Discuss. Chem. Soc. **55**, 51 (1973)
4.34 J.N.L. Connor, D. Farrelly: Chem. Phys. Lett. **81**, 306 (1981)
4.35 R. Gilmor: *Catastrophe Theory for Scientists and Engineers* (Wiley, New York 1981)
4.36 M.V. Berry: Adv. Phys. **25**, 1 (1976)
4.37 J.N.L. Connor, P.R. Curtis, D. Farrelly: Mol. Phys. **48**, 1305 (1983)

Chapter 5

5.1 Yu.A. Kravtsov: Izv. VUZ Radiofiz. **8**, 659 (1965) [English transl.: Sov. Radiophys. **8**, 467 (1965)]

5.2 Yu.A. Kravtsov: Akust. Zh. **14**, 1 (1968) [English transl.: Sov. Phys. – Acoust. **14**, 1 (1968)]
5.3 Yu.A. Kravtsov: In *Analiticheskie Methody v Teorii Difraktsii i Rasprostraneniya Von* (analytic methods in diffraction and wave propagation theory) (Nauchnyi Soviet po Akustike, Akad. Nauk SSSR, Moscow 1970) pp. 258–362
5.4 G.B. Airy: Trans. Canbridge Philos. Soc. **6**, 379 (1838)
5.5 R.N. Buchal, J.B. Keller: Commun. Pure Appl. Math. **13**, 85 (1960)
5.6 Yu.L. Gazaryan: In *Voprosy Dinamicheskoi Teorii Rasprostranenia Seismicheskikh Voln* (topics in the dynamic theory of seismic waves), Vol. 5 (IZd. LGU, Leningrad 1961) p. 73
5.7 V.M. Babich, N.Y. Kirpicnikova: *The Boundary-Layer Method in Diffraction Problems*, Springer Ser. Electrophys., Vol. 3 (Springer, Berlin, Heidelberg 1979)
5.8 Yu.A. Kravtsov: Izv. VUZ Radiofiz. **7**, 664 (1964) [English transl.: Radiophys. Quantum Electron. **7**, no. 4, 104 (1964)]
5.9 C. Chester, B. Friedman, F. Ursell: Proc. Cambridge Philos. Soc. **53**, 599 (1957)
5.10 D. Ludwig: Commun. Pure Appl. Math. **19**, 215 (1966)
5.11 M.V. Berry: Proc. Philos. Soc. **89**, 479 (1966)
5.12 M.V. Berry: Sci. Prog. Oxf. **57**, 43 (1969)
5.13 M.V. Berry, K.E. Mount: Rep. Prog. Phys. **35**, 313 (1972)
5.14 Yu.A. Kravtsov, Yu.I. Orlov: Sov. Phys. – Usp. **26**, 1038 (1983)
5.15 Yu.A. Kravtsov: Izv. VUZ Radiofiz. **10**, 1283 (1967) [English transl.: Radiophys. Quantum Electron. **10**, 719 (1967)]
5.16 Yu.I. Orlov: Trudy MEI No. 194 (Moscow Power Engineering Institute, Moscow 1974) p. 103
5.17 V.M. Babic, V.S. Buldyrev: *Short-Wavelength Diffraction Theory*, Springer Ser. Wave Phenom., Vol. 4 (Springer, Berlin, Heidelberg 1991)
5.18 V.M. Babich, S.A. Egorov: In *Voprosy Dinamicheskoi Teorii Rasprostraneniya Seismicheskikh Voln* (topics in the dynamic theory of seismic waves) (Izd. LGU, Leningrad 1983) Vol. 3, p. 4
5.19 Yu.A. Kravtsov: Izv. VUZ Radiofiz. **7**, 1049 (1964) [English transl.: Radiophys. Quantum Electron. **7**, no. 6, 40 (1964)]
5.20 Yu.A. Kravtsov, Yu.I. Orlov: *Geometrical Optics of Inhomogenous Media*, Springer Ser. Wave Phenom., Vol. 6 (Springer, Berlin, Heidelberg 1990)
5.21 Yu.A. Kravtsov, L.A. Ostrovskii, N.S. Stepanov: Proc IEEE **62**, 1492 (1974)
5.22 V.Ya. Groshev, Yu.A. Kravtsov: Izv. VUZ Radiofiz. **11**, 1812 (1968) [English transl.: Radiophys. Quantum Electron. **11**, 1019 (1968)]
5.23 J.J. Dustermaat: Commun. Pure Appl. Math. **27**, 207 (1974)
5.24 M.V. Berry: Adv. Phys. **25**, 1 (1976)
5.25 G. Dangelmayr, W. Veit: Ann. Phys. **118**, 108 (1979)
5.26 J.B. Keller: A geometrical theory of diffraction, in *Calculus of Variations and its Applications*, Proc. Symp. Appl. Math., Vol. 8 (McGraw-Hill, New York 1958) pp. 27–52
5.27 D. Ludwig: SIAM Rev. **12**, 325 (1970)
5.28 F.W.J. Olver: Philos. Trans. Roy. Soc. London A **247**, 328 (1954)
5.29 B.T. Kormilitsyn: Radiotech. Elektron. **11**, 1130 (1966) [English transl.: Radio Eng. Electron. Phys. **11**, no. 6, 988–992 (1966)]
5.30 Yu.I. Orlov: Izv. VUZ Radiofiz. **9**, 497, 657 (1966) [English transl.: Sov. Radiophys. **9**, 307, 394 (1966)]
5.31 Yu.I. Orlov: Izv. VUZ Radiofiz. **11**, 317 (1968) [English transl.: Radiophys. Quantum Electron. **11**, 180 (1968)]
5.32 Yu.I. Orlov: Izv. VUZ Radiofiz. **10**, 30 (1967) [English transl.: Radiophys. Quantum Electron. **10**, 21 (1967)]
5.33 Yu.I. Orlov: Izv. VUZ Radiofiz. **20**, 1669 (1977) [English transl.: Radiophys. Quantum Electron. **20**, 1148 (1977)]
5.34 Yu.I. Orlov: Izv. VUZ Radiofiz. **13**, 412 (1970) [English transl.: Radiophys. Quantum Electron **13**, 320 (1970)]
5.35 Yu.I. Orlov, S.K. Tropkin: Izv. VUZ Radiofiz. **23**, 1473 (1980) [English transl.: Radiophys. Quantum Electron. **23**, 979 (1980)]

5.36 V.E. Grikurov: Izv. VUZ Radiofiz. **23**, 1038 (1980) [English transl.: Radiophys. Quantum Electron. **23**, 690 (1980)]
5.37 A.S. Kryukovskii, D.S. Lukin, E.A. Palkin: Izv. VUZ Radiofiz. **25**, 1375 (1982) (in Russian)
5.38 J.M. Martin, S.M. Flatte: Appl. Opt. **27**, 2111 (1988)
5.39 S.M. Rytov, Yu.A. Kravtsov, V.I. Tatarskii: *Principles of Statistical Radiophysics*, Vol. 4: Wave Propagation Through Random Media (Springer, Berlin, Heidelberg 1989)

Chapter 6

6.1 V.P. Maslov: *Théorie des Perturbations et Méthode Asymptotique* (Dunod, Paris 1972)
6.2 Yu.A. Kravtsov: Sov. Phys. – Acoust. **14**, 1 (1968)
6.3 J.M. Arnold: Radio Sci. **17**, 1181 (1982)
6.4 R. Ziolkowski, G.A. Deschamps: Rad. Sci. **19**, 1001 (1984)
6.5 B.R. Vainberg: In *Teoriya Rasprostraneniya Voln v Neodnorodnykh i Nelineinykh Sredakh* (theory of wave propagation in inhomogenous and nonlinear media) (Izd. IRE Akad. Nauk SSSR, Moscow 1979) p. 144
6.6 V.P. Maslov, M.V. Fedoryuk: *Semiclassical Approximation in Quantum Mechanics* (Reidel, Higham, MA 1981)
6.8 L. Hörmander: Fourier Integral Operators **127**, 79 (1971)
6.9 A.D. Gorman, R. Wells, G.N. Gleming: J. Phys. A **14**, 1519 (1981)
6.10 K. Hogo, Yu. Yi: Radio Sci. **22**, 357 (1987)
6.11 D.S. Lukin, E.A. Palkin: In *Teoreticheskoye i Experimentalnoye Issledovanie Rasprostraneniya Dekametrovykh Radiovoln* (theoretical and experimental investigation of decameter radiowave propagation) (IZMIRAN SSSR, Moscow 1976) p. 149
6.12 A.A. Kryukovsky, D.S. Lukin, E.A. Palkin: *Volnovaya teoriya katastrof* (wave theory of catastrophes) (Nauka, Moscow) in press

Chapter 7

7.1 Yu.I. Orlov: Trudy MEI No. 119 (Moscow Power Engineering Institue, Moscow 1972) p. 82
7.2 Yu.I. Orlov, A.V. Demin: Trudy MEI No. 553 (Moscow Power Engineering Institute, Moscow 1981) p. 5
7.3 Yu.I. Orlov, A.V. Demin: Trudy MEI No. 497 (Moscow Power Engineering Institute, Moscow 1980) p. 10
7.4 V.B. Avdeyev, A.P. Yarygin: Theses 12th All-Union Conf. Radiowave Propagation (Nauka, Moscow 1978) Vol. 4, p. 294 (in Russian)
7.5 V.N. Mirolyubov, M.V. Tinin: Theses 8th All-Union Conf. Diffraction and Wave Propagation (Acad. Nauk USSR, Moscow 1982) Vol. 2, p. 156 (in Russian)
7.6 M.V. Tinin: Izv. VUZ Radiofiz. **26**, 36 (1983) [English transl.: Radiophys. Quantum Electron. **26**, 29 (1983)]
7.7 V.B. Avdeyev, A.V. Demin, Yu.A. Kravtsov, M.V. Tinin, A.P. Yarygin: Izv. VUZ Radiofiz. **31**, 1279 (1988) [English transl.: Radiophys. Quantum Electron. **31**, 907 (1988)]
7.8 Yu.I. Orlov: Izv. VUZ Radiofiz. **17**, 1036 (1974) [English transl.: Radiophys. Quantum Electron. **17**, 788 (1974)]
7.9 L.A. Vainstein, E.S. Birger, N.V. Konyukhova, E.L. Kosarev: Sov. J. Plasma Phys. **26**, 362 (1976)
7.10 L.A. Vainstein, E.A. Tishchenko: Zh. Tekh. Fiz. **46**, 2271 (1976) [English transl.: Sov. Phys. – JTP **21**, 1338 (1976)]
7.11 L.A. Vainstein, P.Ya Ufimtsev: Radiotekn. Elektron. **25**, 625 (1982) [English transl.: Radio Eng. Electron. Phys. **25**, 411 (1982)]

7.12 J.M. Arnold: Radio Sci. **17**, 1181 (1982)
7.13 J.M. Arnold: IEE Proc. **133**, 165 (1986)
7.14 V.V. Stepanov: *Kurs Differentsialnykh Uravnenii* (course of differential equations) (Fizmatgiz, Moscow 1958); G.A. Korn, T.M. Korn: *Mathematical Handbook for Scientists and Engineers* (McGraw-Hill, New York 1961)
7.15 Yu.A. Kravtsov, Yu.I. Orlov: *Geometrical Optics of Inhomogeneous Media*, Springer Ser. Wave Phenom., Vol. 6 (Springer, Berlin, Heidelberg 1990)
7.16 V.B. Avdeyev, N.V. Shilov, A.P. Yarygin: Theses 12th All-Union Conf. on Radiowave Propagation (Nauka, Moscow 1981) Vol. 1, p. 302 (in Russian)
7.17 V.B. Avdeyev, A.P. Yarygin: *Rasseyanie Elektromagnitnykh Voln* (scattering of EM waves) (Institute of Radio Engineering, Taganrog 1983) Vol. 4, p. 40
7.18 L.M. Brekhovskikh: *Acoustics of Layered Media I, II*, Springer Ser. Wave Phenom., Vols. 6, 10 (Springer, Berlin, Heidelberg 1990, 1992)
7.19 D. Ludwig: Commun. Pure Appl. Math. **20**, 103 (1967)
7.20 M.V. Tinin: *Issledovaniya po Geomagnetizmu, Aeronomii i Fizike Solntsa* (studies in geomagnetism, aeronomy and solar physics) (Nauka, Moscow 1973) Vol. 29, p. 157
7.21 V.M. Babich M.M. Popov: Izv. VUZ Radiofiz. **32**, 1447 (1989) [English transl.: Radiophys. Quantum Electron. **32**, 1063 (1989)]

Chapter 8

8.1 Yu.I. Orlov: Radiotekh. Elektron. **20**, 242 (1975) [English transl.: Radio Eng. Electron. Phys. **20**, 145 (1975)]
8.2 L. Levey, L.B. Felsen: Radio Sci. **4**, 959 (1969)
8.3 L. Levey, L.B. Felsen: J. Instt. Math. Appl. **3**, 76 (1976)
8.4 M.M. Agrest, M.Z. Maksimov: *Teoriya Nepolnykh Tsilindrichekikh Funktsii i ikh Prilozheniya* (theory of incomplete cylindrical functions and their applications) (Atomizdat, Moscow 1965)
8.5 D.S. Ahluwalia, R.M. Lewis, J. Boersma: SIAM J. Appl. Math. **16**, 783 (1968)
8.6 R.M. Lewis, J. Boersma: J. Appl. Math. **10**, 2291 (1969)
8.7 D.S. Ahluwalia: SIAM J. Appl. Math. **18**, 287 (1970)
8.8 R.S. Hansen: *Geometrical Theory of Diffraction* (IEEE Press, New York 1981)
8.9 Yu.I. Orlov, S.A. Vlasov: Izv. VUZ Radiofiz. **21**, 422 (1978) [English transl.: Radiophys. Quantum Electron. **21**, 290 (1978)
8.10 Yu.I. Orlov: Radiotekh. Electron. **21**, 730 (1976) [English transl.: Radio Eng. Electron. Phys. **21**, no. 4, 50 (1976)]
8.11 Yu.A. Kravtsov, Yu.I. Orlov: *Geometricl Optics of Inhomogeneous Media*, Springer Ser. Wave Phenom., Vol. 6 (Springer, Berlin, Heidelberg 1990)
8.12 Yu.I. Orlov: Radiotekh. Elektron. **21**, 62 (1976) [English transl.: Radio Eng. Electron. Phys. **21**, no. 1, 50 (1976)
8.13 Yu.I. Orlov, S.A. Vlasov: Radiotekh. Elektron. **23**, 17 (1978) [English transl.: Radio Eng. Electron. Phys. **23**, no. 1, 14 (1978)
8.14 V.I. Arnold: Usp. Mat. Nauk **34**, 91, 206 (1978)
8.15 V.I. Arnold, A.N. Varchenko, S.M. Gussein-Zade: *Singularities of Differential Mappings* (Brikhäuser, Boston, MA 1985, 1988) Vols. 1, 2
8.16 E.B. Ipatov, A.S. Kryukovskii, D.S. Lukin, E.A. Palkin: Dokl. Akad. Nauk USSR **291**, 823 (1986) [English transl.: Sov. Phys. – Doklady **31**, 962 (1986)]
8.17 A.S. Kryukoskii, D.S. Lukin, E.A. Palkin: Sov. J. Numer. Anal. Math. Model **2**, 279 (1987)
8.18 A.S. Kryukovskii, D.S. Lukin, E.A. Palkin: *Krayevye i Uglovye Katastrofy v Zadachakh Difraktsii i Rasprostraneniya Voln* (edge and angle catastrophes in problems of diffraction and wave propagation) (Aviation Institute, Kazan 1988)

Chapter 9

9.1 J.N.L. Connor, W.J.E. Southall: J. Chem. Phys. **90**, 355 (1986)
9.2 M.V. Berry: Adv. Phys. **25**, 1 (1976)
9.3 V.I. Tokatly, B.E. Kinber: Izv. VUZ Radiofiz. **14**, 761 (1971) [English transl.: Radiophys. Quantum Electron. **14**, 390 (1971)]
9.4 E.D. Gazazyan, B.Ye. Kinber: Izv. VUZ Radiofiz. **14**, 1219 (1971) [English transl.: Radiophys. Quantum Electron. **14**, 601 (1971)]
9.5 E.D. Gazazyan, A.D. Ter-Pogosyan: Izv. Akad. Nauk Armenii – Fizika **9**, 169 (1974)
9.6 E.D. Gazazyan, M.I. Ivanyan, B.Ye. Kinber: Izv. Akad. Nauk Armenii – Fizika **13**, 87 (1978)
9.7 D.S. Ahluwalia, R.M. Lewis, J. Boersma: SIAM J. Appl. Math. **16**, 783 (1968)
9.8 R.M. Lewis, J. Boersma: J. Math. Phys. **10**, 2291 (1969)
9.9 D.S. Ahluwalia: SIAM J. Appl. Math. **18**, 287 (1970)
9.10 Yu.I. Orlov, S.K. Tropkin: Izv. VUZ Radiofiz. **24**, 334 (1981) [English transl.: Radiophys. Quantum Electron. **24**, 233 (1981)]
9.11 S.K. Tropkin: *Razrabotka metodov rascheta blizhnikh polei ploskikh aperturnykh antenn* (calculation of near fields of plane aperture antennas). Dr. Sci. Thesis, Moscow Power Engineering Institute (1984)
9.12 Yu.I. Orlov, S.K. Tropkin: Izv. VUZ Radiofiz. **24**, 1383 (1981) [English transl.: Radiophys. Quantum Electron. **24**, 936 (1981)]
9.13 E.T. Whittaker, G.N. Watson: *A Course of Modern Analysis* (Cambridge Univ. Press, Cambridge 1927)
9.14 V.L. Ginzburg: *The Propagation of Electromagnetic Waves in Plasmas* (Pergamon, Oxford 1970)
9.15 F.W.J. Olver: Philos. Trans. Roy. Soc. London A **247**, 307 (1956); ibid. A **249**, 65 (1958); ibid. A **250**, 479 (1958)
9.16 E.R. Pike: Quart. J. Mech. Appl. Math. **17**, 369 (1964)
9.17 Yu.A. Kravtsov: Izv. VUZ Radiofiz. **10**, 1283 (1967) [English transl.: Radiophys. Quantum Electron. **10**, 719 (1967)]
9.18 Yu.A. Kravtsov: Rays and caustics as physical objects. *Progress in Optics* **26**, 227–348 (North-Holland, Amsterdam 1988)
9.19 A.A. Asatryan, Yu.A. Kravtsov: Izv. VUZ Radiofiz. **31**, 1053 (1988) [English transl.: Radiophys. Quantum Electron. **31**, 746 (1988)]
9.20 Yu.A. Kravtsov: Izv. VUZ Radiofiz. **8**, 659 (1965) [English transl.: Sov. Radiophys. **8**, 467 (1965)]
9.21 V.A. Kaloshin, V.I. Orlov: Radiotekh. Elektron. **18**, 2028 (1973) [English transl.: Radio Eng. Electron. Phys. **18**, 1485 (1973)]
9.22 Yu.I. Orlov: Izv. VUZ Radiofiz. **9**, 1036 (1966) [English transl.: Sov. Radiophys. **9**, 604 (1966)
9.23 L.D. Landau, E.M. Lifshitz: *Quantum Mechanics, Nonrelativistic Theory* (Pergamon, Oxford 1962)
9.24 H.K.V. Lotsch: Optik **30**, 1, 181, 217, 563 (1969/70)
9.25 A.V. Popov: Dokl Akad. Nauk USSR **230**, 1322 (1976) [English transl.: Sov. Phys. – Dokl. **21**, 564 (1976)]
9.26 J.B. Keller, S. Rubinov: Ann. Phys. **2**, 24 (1960)
9.27 L.A. Vainstein: *Open Resonators and open Waveguides* (Golden, Boulder, CO 1969)
9.28 V.M. Babich, V.S. Buldyrev: *Short-Wavelength Diffraction Theory*, Springer Ser. Wave Phenom., Vol. 4 (Springer, Berlin, Heidelberg 1991)
9.29 V.P. Bykov: Izv. VUZ Radiofiz. **14**, 880 (1971) [English transl.: Radiophys. Quantum Electron. **14**, 692 (1971)]
9.30 E.D. Gazazyan, M.I. Ivanyan: Radiotekh. Elektron. **29**, 830 (1984) [English transl.: Radio Eng. Electron. Phys. **29**, no. 5, 12 (1984)]
9.31 E.D. Gazazyan: *Uniform Short-Wave Asymptotics of Scalar and EM Fields Based on One-Dimensional Canonic Functions* (Physical Institute, Erevan 1988) Preprint 1092-(55)-88 (in Russian)

9.32 V.P. Bykov: *Elekronika Bolshikh Moshchnostei* (power electronics) (Nauka, Moscow 1965) No. 4, p. 66
9.33 L.A. Vainstein: *Elektronika Bolshikh Moshchnostei* (power electronics) (Nauka, Moscow 1965) No. 4, p. 93
9.34 E.D. Gazazyan, M.I. Ivanyan: Radiotekh. Elektron. **21**, 2052 (1976) [English transl.: Radio Eng. Electron. Phys. **21**, no. 10, 16 (1976)]
9.35 Yu.I. Orlov: Izv. VUZ Radiofiz. **24**, 213 (1981) [English transl.: Radiophys. Quantum Electron. **24**, 154 (1981)]
9.36 V.V. Zheleznyakov, V.V. Kocharovskii, Vl.V. Kocharovskii: Usp. Fiz. Nauk **141**, 257 (1983) [English transl.: Sov. Phys. – Uspekhi **26**, 877 (1983)]
9.37 R.A. Markus: J. Chem. Phys. **57**, 4903 (1972)
9.38 H. Kreek, R.L. Ellis, R.A. Markus: J. Chem. Phys. **62**, 913 (1975)
9.39 J.N.L. Connor: Mol. Phys. **31**, 33 (1976)

Chapter 10

10.1 L.B. Felsen, N. Marcuvitz: *Radiation and Scattering of Waves* (Prentice-Hall, Englewood Cliffs, NJ 1973) Vols. 1, 2
10.2 Yu.A. Kravtsov, Yu.I. Orlov: *Geometrical Optics on Inhomogeneous Media*, Springer Ser. Wave Phenom., Vol. 6 (Springer, Berlin, Heidelberg 1990)
10.3 Yu.A. Kravtsov, L.A. Ostrovskii, N.S. Stepanov: Proc. IEEE **62**, 1492 (1974)
10.4 Yu.I. Orlov: Izv. VUZ Radiofiz. **24**, 224 (1981) [English transl.: Radiophys. Quantum Electron. **24**, 154 (1981)]
10.5 R.M. Lewis: *Electromagnetic Wave Theory* (Pergamon, Oxford 1967) Pt. 2, p. 845
10.6 Yu.I. Orlov: *Pryamye i Obratnye Zadachi Teorii Difraktsii* (direct and inverse problems of diffraction theory) (Izd. IRE AN USSR, Moscow 1979) p. 5
10.7 Yu.I. Orlov, A.P. Anyutin: Izv. VUZ Radiofiz. **17**, 1369 (1974); ibid. **19**, 335, 495 (1976) [English transl.: Radiophys. Quantum Electron. **17**, 1047 (1974); ibid. **19**, 235 and 347 (1976)]
10.8 A.P. Anyutin, Yu.I. Orlov: Radiotekh. Elektron. **22**, 2082 (1977) [English transl.: Radio Eng. Electron. Phys. **22**, no. 10, 62 (1977)]
10.9 Yu.A. Kravtsov: Izv. VUZ Radiofiz. **11**, 1582 (1968) [English transl.: Radiophys. Quantum Electron. **11**, 888 (1968)]
10.10 V.P. Maslov: *Théorie des Perturbations et Méthode Asymptotique* (Dunod, Paris 1972)
10.11 E. Arbel, L.B. Felsen: *Electromagnetic Theory and Antennas* (Pergamon, London 1963) ps. 391–421
10.12 Yu.A. Kravtsov: Radiophys. Quantum Electron. **10**, 719 (1967)
10.13 V.A. Vasilyev: Funkts. Anal. **11**, 1 (1979) ibid. **13**, 1 (1979)
10.14 S.M. Rytov, Yu.A. Kravtsov, V.I. Tatarskii: *Principles of Statistical Radiophysics IV* (Springer, Berlin, Heidelberg 1989)
10.15 M.V. Berry: J. Phys. A **10**, 2061 (1977)
10.16 J.M. Martin, S.M. Flatté: Appl. Opt. **27**, 2111 (1988)
10.17 Yu.A. Kravtsov: Sov. Phys. – JETP **28**, 769 (1969)
10.18 S.M. Flatté, R. Dashen, W.H. Munk, K.M. Watson, F. Zachariasen: *Sound Transmission Through a Fluctuating Ocean* (Cambridge Univ. Press, Cambridge 1979)
10.19 Yu.A. Kravtsov: Rys and caustics as physical objects, in *Progress in Optics* **26**, 227–348 (North-Holland, Amsterdam 1988)
10.20 V.A. Kulkarny, B.S. White: Phys. Fluids **25**, 1770 (1982)
10.21 D.I. Zwillinger, B.S. White: Wave Motion **7**, 207 (1985)
10.22 H.M. Nussenzweig: Sci. Am. **236**, 116 (April 1977)
10.23 V. Buck: Rev. Mod. Phys. **46**, 369 (1974)
10.24 J.N.L. Connor: In *Semiclassical Methods in Molecular Scattering and Spectroscopy*, ed. by M.S. Child (Reidel, Dordrecht 1980) p. 45

10.25 J.N.L. Connor, P.R. Curtis, C.J. Edge, A. Laganda: J. Chem. Phys. **80**, 1362 (1980)
10.26 M.V. Berry, M. Tabor: Proc. Roy. Soc. London A **349**, 101 (1976)
10.27 M.V. Berry, K.E. Mount: Rept. Progr. Phys. **35**, 315 (1972)
10.28 S.W. Hawking, G.F.R. Ellis: *Large Scale Structure of Space–Time* (Cambridge Univ. Press, Cambridge 1973)
10.29 Ya.B. Zeldovich, A.V. Shamaev, S.F. Shandarin: Usp. Fiz. Nauk **139**, 153 (1983) [English transl.: Sov. Phys. – Usp. **26**, 46 (1983)]
10.30 A.D. Gorman, R. Wills: J. Acoust. Soc. Am. **73**, 363 (1983)
10.31 J.I. Bobbit, E. Cumberbatch: Stud. Appl. Math. **55**, 229 (1976)
10.32 R.J. Smith: Fluid Mech. **77**, 417 (1976)
10.33 Y. Pomeau: Europhys. Lett. **11**, 713 (1990)

List of Symbols

Ai(.)	Airy function
a_f, a_{fx}, a_{fy}	Fresnel radii
E, H	EM field
$\mathscr{H}$	Hamiltonian
h	Hessian
$I(\zeta, t)$	dimensionless standard integral
$\tilde{I}(\tilde{\zeta}, \tilde{t})$	dimensional standard integral
$l_{\parallel}$	longitudinal caustic scale
$\mathscr{J}$	divergence
K	curvature; see K_{rel}, K_c, K_R
N_c	unit normal to a caustic
$p = \nabla\psi$	momentum
Q	initial surface
r^0, ψ^0, u^0	illustrating superscript
r_0, n_0	zero subscript
$u(\mathbf{r}, t)$	wave field
$U(\mathbf{r})$	amplitude
Δ	increment, Laplacian
ε	permittivity
Λ	transverse caustic scale
$\Phi(\zeta, t)$	generating or phase function
$\Psi(\zeta, t)$	non-polynomial phase function

Subject Index

Springer-Verlag and the Environment

We at Springer-Verlag firmly believe that an international science publisher has a special obligation to the environment, and our corporate policies consistently reflect this conviction.

We also expect our business partners – paper mills, printers, packaging manufacturers, etc. – to commit themselves to using environmentally friendly materials and production processes.

The paper in this book is made from low- or no-chlorine pulp and is acid free, in conformance with international standards for paper permanency.

Printing: Mercedesdruck, Berlin
Binding: Buchbinderei Lüderitz & Bauer, Berlin